Springer Monographs in Mathematics

T0181860

For further volumes published in this series see
www.springer.com/series/3733

Kazuhiko Aomoto · Michitake Kita

Theory of Hypergeometric Functions

With an Appendix by Toshitake Kohno

 Springer

Kazuhiko Aomoto
Professor Emeritus
Nagoya University
Japan
kazuhiko@aba.ne.jp

Toshitake Kohno (Appendix D)
Professor
Graduate School of Mathematical
    Sciences
The University of Tokyo
Japan
kohno@ms.u-tokyo.ac.jp

Michitake Kita (deceased 1995)

Kenji Iohara (Translator)
Professor
Université Claude Bernard Lyon 1
Institut Camille Jordan
France
iohara@math.univ-lyon1.fr

ISSN 1439-7382
ISBN 978-4-431-54087-8   ISBN 978-4-431-53938-4 (eBook)
DOI 10.1007/978-4-431-53938-4
Springer Tokyo Dordrecht Heidelberg London New York

Mathematics Subject Classification (2010): 14F40, 30D05, 32S22, 32W50, 33C65, 33C70, 35N10, 39B32

Cover design: deblik, Berlin

Printed on acid-free paper

Springer is part of Springer Science+Business Media (www.springer.com)

Preface

One may say that the history of hypergeometric functions started practically with a paper by Gauss (cf. [Gau]). There, he presented most of the properties of hypergeometric functions that we see today, such as power series, a differential equation, contiguous relations, continued fractional expansion, special values and so on. The discovery of a hypergeometric function has since provided an intrinsic stimulation in the world of mathematics. It has also motivated the development of several domains such as complex functions, Riemann surfaces, differential equations, difference equations, arithmetic theory and so forth. The global structure of the Gauss hypergeometric function as a complex function, i.e., the properties of its monodromy and the analytic continuation, has been extensively studied by Riemann. His method is based on complex integrals. Moreover, when the parameters are rational numbers, its relation to the period integral of algebraic curves became clear, and a fascinating problem on the uniformization of a Riemann surface was proposed by Riemann and Schwarz. On the other hand, Kummer has contributed a lot to the research of arithmetic properties of hypergeometric functions. But there, the main object was the Gauss hypergeometric function of one variable.

In contrast, for more general hypergeometric functions, including the case of several variables, the question arises: *What in fact are hypergeometric functions ?* Since Gauss and Riemann, many researchers tried generalizing the Gauss hypergeometric function. Those which are known under the names of Goursat, Pochhammer, Barnes, Mellin, and Appell are such hypergeometric functions. Although these functions interested some researchers as special objects, they didn't attract many researchers and no significant result came about. If anything, those expressed with the aid of some properties of hypergeometric functions appeared interestingly in several situations, either in partial or another form. The orthogonal polynomials studied in Szegö's book, several formulas that we can find everywhere in Ramanujan's enormous notebooks, spherical functions on Lie groups, and applications to mathematical physics containing quantum mechanics, are such examples. Simply, they were not considered from a general viewpoint of hypergeometric functions.

In this book, hypergeometric functions of several variables will be treated. Our point of view is that the hypergeometric functions are complex integrals of complex powers of polynomials. Most of the properties of hypergeometric functions which have appeared in the literature up to now can be reconsidered from this point of view. In addition, it turns out that these functions establish interesting connections among several domains in mathematics.

One of the prominent properties of hypergeometric functions is the so-called contiguity relations. We understand them based on the classical paper by G. D. Birkhoff [Bir1] about difference equations and their generalization. This is an approach treating hypergeometric functions as solutions of differ-ence equations with respect to shifts of parameters, and characterizing by analysis of asymptotic behaviors when the parameters tend to infinity. One sees a relation between the Padé approximation and the continued fractional expansion. For this purpose, we use either analytic or algebraic de Rham cohomology (twisted de Rham cohomology) as a natural form of complex integrals. In Chapter 2, several relations satisfied by hypergeometric func-tions will be derived and explained in terms of twisted de Rham cohomology. There, the reader may notice that the excellent idea due to J. Hadamard about a *"finite part of a divergent integral"* developed in his book [Had] will be naturally integrated into the theory. In Chapter 4, we will construct cycles via the saddle point method and apply the Morse theory on affine varieties to describe the global structure of an asymptotic behavior of solutions to difference equations.

Another prominent feature is a holonomic system of partial differential equations satisfied by hypergeometric functions, in particular, an infinitesimal concept called integrable connection (the Gauss−Manin connection) that has a form of partial differential equations of the first order, and a topological concept called monodromy that is its global realization. The latter means to provide a linear representation of a fundamental group, in other words, a local system on the underlying topological space. But here, what is important is not only the topological concept but the mathematical substance that provides it. Hypergeometric functions provide such typical examples. As a consequence, they also help us understand the fundamental group itself.

We will treat complex integrals of complex powers of polynomials, but the main point is not only to state general theorems in an abstract form but also to provide a concrete form of the statements. In Chapters 3 and 4, for linear polynomials, concrete formulas of differential equations, difference equations, integral representations, etc. will be derived, applying the idea from the invariant theory of general linear groups.

In the world surrounding hypergeometric functions, there are several sub-jects studying power series, orthogonal functions, spherical functions, dif-ferential equations, difference equations, etc. in a broad scope such as real (complex) analysis, arithmetic analysis, geometry, algebraic topololgy and combinatorics, which are mutually related and attract researchers. This book explains one such idea. In particular, micro-local analysis and the theory of holonomic $\mathcal{D}$-modules developed in Japan provided considerable impacts. In Chapter 3 of this book, we will treat a holonomic system of Fuchsian partial differential equations over Grassmannians satisfied by the hypergeometric functions, introduced by Gelfand et al., defined as integrals of complex powers of functions as described above. But there, we will explain them only by concrete computations. For a general theory of $\mathcal{D}$-modules, we propose that the reader consult the book written by Hotta and Tanisaki[1] in this series. Here, we will not treat either arithmetic aspects or the problem of the uniformization of complex manifolds. There are also several applications to mathematical physics such as conformal field theory, and solvable models in statistical mechanics. For these topics, the reader may consult Appendix D and the references in this book.

If this book serves as the first step to understanding hypergeometric functions and motivate the reader's interest towards further topics, we should say that our aim has been accomplished.

We asked Toshitake Kohno to write Appendix D including his recent result. We express our gratitude to him.

Lastly, our friends Takeshi Sasaki, Keiji Matsumoto and Masaaki Yoshida gave us precious remarks and criticisms on this manuscript. We also express our gratitude to them.

June, 1994.   *Kazuhiko Aomoto*
                                                          Michitake Kita

---

[1] The translation is published as [H-T-T].

Preface to English Edtion

After the publication of the original Japanese edition, hypergeometric functions attracted researchers both domestic and abroad, and some aspects are now fairly developed, for example in relation to arrangement of hyperplanes, conformal field theory and random matrix theory. Some related books have also been published: those by M. Yoshida [Yos3], which treats the uniformaization via period matrix, by M. Saito, B. Sturmfels and N. Takayama [S-S-T], which treats algebraic $\mathcal{D}$-modules satisfied by hypergeometric functions, and by P. Orlik and H. Terao [Or-Te3], which sheds light on hypergeometric functions from viewpoint of arrangements of hyperplanes, are particularly related to the contents of this book.

In this English edition, the contents are almost the same as the original except for a minor revision. In particular, in spite of its importance, hypergeometric functions of confluent type are not treated in this book (they can be treated in the framework of twisted de Rham theory but the situation becomes much more complicated). As for the references, we just added several that are directly related to the contents of this book. For more detailed and up-to-date references, the reader may consult the book cited above, etc.

The co-author of this book, who had been going to produce outstanding results unfortunately passed away in 1995. May his soul rest in peace.

Finally, I am indebted to Dr. Kenji Iohara, who has taken the trouble to translate the original version into English.

August, 2010.   *Kazuhiko Aomoto*

Contents

Notation

$\Omega^p(\mathbb{C}^n)$: the space of p-forms with polynomial coefficients

$$U(u) = \prod_{j=1}^{m} P_j(u)^{\alpha_j}, P_j(u) \in \mathbb{C}[u_1, \ldots, u_n]$$

D: the divisor defined by $P := P_1 \cdots P_m$

$M := \mathbb{C}^n \setminus D$: a variety related to an integral representation of hypergeometric functions

d: the exterior derivative

$$\omega := dU/U = \sum_{j=1}^{m} \alpha_j \frac{dP_j}{P_j}:$$ a completely integrable holomorphic connection form on M

$\nabla_\omega := d + \omega_\wedge$: the covariant differential operator with respect to ω

$\mathcal{L}_\omega$: the complex local system of rank 1 generated by solutions of $\nabla_\omega h = 0$

$\mathcal{L}_\omega^\vee$: the dual local system of $\mathcal{L}_\omega$

$\Omega^p(*D)$: the space of rational p-forms having poles along D

$\Omega^p(\log D)$: the space of logarithmic p-forms having poles along D

$\mathcal{O}_M$: the sheaf of germs of holomorphic functions on M

Ω^p_M: the sheaf of germs of holomorphic p-forms on M

$\mathcal{A}^p_M$: the sheaf of germs of C^∞ p-forms on M

$\mathcal{K}^p_M$: the sheaf of germs of currents of degree p on M

$\mathcal{S}(M)$: the space of sections of $\mathcal{S}$ over M

$\mathcal{S}_c(M)$: the space of sections with compact support of $\mathcal{S}$ over M

$H^p(M, \mathcal{S})$: the cohomology with coefficients in the sheaf $\mathcal{S}$

$H_p(M, \mathcal{L}_\omega)$: the homology with coefficients in the local system $\mathcal{L}_\omega$

$H_p^{\ell f}(M, \mathcal{L}_\omega)$: the locally finite homology with coefficient $\mathcal{L}_\omega$

$$r := \dim H^n(M, \mathcal{L}_\omega) = \binom{m-1}{n}$$

a lattice $L = \mathbb{Z}^\ell$

$e_1, \ldots, e_\ell$: a standard basis of L $e_i = (0, \ldots, \overset{i}{1}, 0, \ldots, 0), 1 \le i \le \ell$.

$$\nu = (\nu_1, \ldots, \nu_\ell) \in L, \nu! = \prod_{i=1}^{\ell} \nu_i!, |\nu| = \sum_{i=1}^{\ell} \nu_i$$

$|J| = n$ for a set of indices $\{j_1, \ldots, j_n\}$

For $x = (x_1, \ldots, x_\ell) \in \mathbb{C}^\ell$, $x^\nu = x_1^{\nu_1} \cdots x_\ell^{\nu_\ell}$

$\displaystyle\sum_\nu$:   the sum is taken over $\nu \in \mathbb{Z}_{\geq 0}^\ell$

$(\gamma; c) := \Gamma(\gamma + c)/\Gamma(\gamma), \; \gamma + c \notin \mathbb{Z}_{\leq 0}$

$L^\vee := \mathrm{Hom}_{\mathbb{Z}}(L, \mathbb{Z})$:   the dual lattice of L

$G(n + 1, m + 1)$:   Grassmannian of $(n + 1)$-dimensional subspaces of $\mathbb{C}^{m+1}$

For $x = \begin{pmatrix} x_{00} & \cdots & x_{0m} \\ \vdots & & \vdots \\ x_{n0} & \cdots & x_{nm} \end{pmatrix}$, $x(j_0 \cdots j_n) := \det \begin{pmatrix} x_{0j_0} & \cdots & x_{0j_n} \\ \vdots & & \vdots \\ x_{nj_0} & \cdots & x_{nj_n} \end{pmatrix}$

$Y := \bigcup_{0 \leq j_0 < \cdots < j_n \leq m} \{x(j_0 \cdots j_n) = 0\}$

$E(n + 1, m + 1; \alpha_0, \ldots, \alpha_m) = E(n + 1, m + 1; \alpha)$:   the system of hypergeometric differential equations of type $(n + 1, m + 1; \alpha)$

$\mathbb{C}[z] = \mathbb{C}[z_1, \ldots, z_m]$: the ring of polynomials

$\mathbb{C}(z) = \mathbb{C}(z_1, \ldots, z_m)$: the rational function field of $z_1, \ldots, z_m$

$\mathbb{C}[[z]] = \mathbb{C}[[z_1, \ldots, z_m]]$: the ring of formal power series

$\mathbb{C}((z)) = \mathbb{C}((z_1, \ldots, z_m))$: the ring of formal Laurent series

$GL_m(\mathbb{C}(z))$:   the group of the regular matrices of order n with components in $\mathbb{C}(z)$

$GL_m(\mathbb{C}((z)))$:   the group of the regular matrices of order n with components in $\mathbb{C}((z))$

$M_m(\mathbb{C}(z))$:   the algebra of the matrices of order m with components in $\mathbb{C}(z)$

$\mathcal{B}_m = \left\{ A = \begin{pmatrix} a_{11} & & * \\ & \ddots & \\ 0 & & a_{mm} \end{pmatrix} \in GL_m(\mathbb{C}) \right\}$:   a Borel subgroup of $GL_m(\mathbb{C})$

$\mathcal{U}_m = \left\{ A = \begin{pmatrix} 1 & & * \\ & \ddots & \\ 0 & & 1 \end{pmatrix} \in GL_m(\mathbb{C}) \right\}$:   a maximal unipotent Lie subgroup of $GL_m(\mathbb{C})$

$\mathcal{F}_m := GL_m(\mathbb{C})/\mathcal{B}_m$:   a flag manifold

$\mathcal{A}_m := GL_m(\mathbb{C})/\mathcal{U}_m$:   a principal affine space

$\triangle(0; \varepsilon) = \{z \in \mathbb{C} | \; |z| < \varepsilon\}$:   the open disk of center at the origin with radius ε

$\Re z, \Im z$:   the real and imaginary part of a complex number z

$\arg z$:   an argument of z

$\mathfrak{S}_m$:   the mth symmetric group

$\chi(M)$:   the Euler characteristic of M

$\hat{\varphi}(x; \alpha) = \displaystyle\int_\gamma U \cdot \varphi$

T, T_j: shift operators

Chapter 1
Introduction: the Euler−Gauss Hypergeometric Function

The binomial series

$$(1+x)^\alpha = \sum_{n=0}^\infty \frac{\alpha(\alpha-1)\cdots(\alpha-n+1)}{n!} x^n, \quad |x| < 1$$

is the generating function of binomial coefficients $\binom{\alpha}{n} = \dfrac{\alpha(\alpha-1)\cdots(\alpha-n+1)}{n!}$.
A hypergeometric function can be regarded as a generating analytic function of more complicated combinatorial numbers which generalizes the binomial series. By studying its analytic structure, it provides us with information such as relations among combinatorial numbers and their growth. The aim of this book is to treat hypergeometric functions of several variables as complex analytic functions. Hence, we assume that the reader is familiar with basic facts about complex functions.

$\Gamma(n) = (n-1)! = 1 \cdot 2 \cdot \cdots \cdot (n-1), n = 1, 2, 3, \cdots$ satisfies the recurrence formula $\Gamma(n+1) = n\Gamma(n)$ and $\Gamma(1) = 1$. Conversely, these two properties determine $\Gamma(n)$ uniquely. A question arises "Can we extend the function $\Gamma(z)$ for all $z \in \mathbb{C}$?" The answer is "No," if we do not restrict ourselves to $z \in \mathbb{Z}$. But, if the behavior of $\Gamma(z+m)$ is given as $m \mapsto +\infty$, $\Gamma(z)$ itself can be determined by considering $\Gamma(z+1)$, $\Gamma(z+2), \cdots, \Gamma(z+m), \cdots$ ($m \in \mathbb{Z}_{>0}$). That is, $\Gamma(z)$ can be determined by its behavior at infinity. As a phenomenon in analysis, it sometimes happens that a function or a vector is determined by its behavior at infinity. Such a situation is called a limit point and this is our basic idea to treat hypergeometric functions in this book.

In this Introduction, we shall study basic properties of the Euler−Gauss hypergeometric functions from several viewpoints. For detailed subjects, we may refer to the well-known books like [AAR], [Ca], [Er1], [Mag], [Ol], [Sh], [W-W], [Wat] etc. See also [I-K-S-Y] for a historical overview of analytic differential equations. First, we start from an infinite-product representation of the Γ-function.

K. Aomoto et al., *Theory of Hypergeometric Functions*, Springer Monographs in Mathematics, DOI 10.1007/978-4-431-53938-4_1, © Springer 2011

1.1 Γ-Function

1.1.1 Infinite-Product Representation Due to Euler

Consider a meromorphic function $\varphi(z)$ over $\mathbb{C}$ satisfying the difference equation

$$\varphi(z+1) = z\varphi(z), \quad z \in \mathbb{C}. \tag{1.1}$$

From this, we obtain

$$\varphi(z) = z^{-1}\varphi(z+1) \tag{1.2}$$

$$= z^{-1}(z+1)^{-1} \cdots (z+N-1)^{-1}\varphi(z+N),$$

for any natural number N. Take the limit $N \mapsto +\infty$: if we assume that an asymptotic expansion of $\varphi(z)$ as $|z| \mapsto +\infty$ has the form

$$\varphi(z) = e^{-z}z^{z-\frac{1}{2}}(2\pi)^{\frac{1}{2}}\left\{1 + O\left(\frac{1}{|z|}\right)\right\} \tag{1.3}$$

in the sector $-\pi + \delta < \arg z < \pi - \delta$ $(0 < \delta < \frac{\pi}{2})$ (here $O\left(\frac{1}{|z|}\right)$, called the Landau symbol, is a function asymptotically at most equivalent to $\frac{1}{|z|}$), applying (1.3) to $\varphi(z+N)$, we obtain

$$\varphi(z) = (2\pi)^{\frac{1}{2}} \lim_{N \mapsto +\infty} z^{-1}(z+1)^{-1} \tag{1.4}$$

$$\cdots (z+N-1)^{-1}e^{-z-N}(z+N)^{z+N-\frac{1}{2}}.$$

Now, an asymptotic expansion of the Γ-function $\varphi(z) = \Gamma(z)$ as $\Re z \mapsto +\infty$ is given by the Stirling formula ([W-W] Chap12 or [Er2])

$$\Gamma(z) = e^{-z}z^{z-\frac{1}{2}}(2\pi)^{\frac{1}{2}}\left\{1 + \frac{1}{12z} + \cdots\right\}, \tag{1.5}$$

$$-\pi + \delta < \arg z < \pi - \delta.$$

In particular, we have

$$\Gamma(N) = (N-1)! = e^{-N}N^{N-\frac{1}{2}}(2\pi)^{\frac{1}{2}}\left\{1 + O\left(\frac{1}{N}\right)\right\}, \quad (1.6)$$

$$(z+N)^{z+N-\frac{1}{2}} = N^{z+N-\frac{1}{2}}\left(1 + \frac{z}{N}\right)^{z+N-\frac{1}{2}} \quad (1.7)$$

$$= N^{z+N-\frac{1}{2}}e^z\left(1 + O\left(\frac{1}{N}\right)\right).$$

By the formula

$$\lim_{N\mapsto+\infty}\left(1 + \frac{1}{2} + \cdots + \frac{1}{N-1} - \log N\right) = \gamma \quad (1.8)$$

(γ is Euler's constant), the right-hand side of (1.4) becomes

$$(2\pi)^{\frac{1}{2}}\lim_{N\mapsto+\infty}z^{-1}(z+1)^{-1}\cdots(z+N-1)^{-1}e^{-z-N}(z+N)^{z+N-\frac{1}{2}} \quad (1.9)$$

$$= \lim_{N\mapsto+\infty}\frac{\Gamma(N)}{\prod_{j=0}^{N-1}(z+j)}N^z = e^{-\gamma z}\left\{z\prod_{j=1}^{\infty}\left(1 + \frac{z}{j}\right)e^{-\frac{z}{j}}\right\}^{-1} = \Gamma(z),$$

which coincides with an infinite-product representation of $\Gamma(z)$.

1.1.2 Γ-Function as Meromorphic Function

Similarly, if a meromorphic solution $\psi(z)$ of the difference equation (1.1) has an asymptotic expansion as $|z| \mapsto +\infty$

$$\psi(z) = e^{-z}(-z)^{z-\frac{1}{2}}e^{\pi\sqrt{-1}z}(2\pi)^{-\frac{1}{2}}\left\{1 + O\left(\frac{1}{|z|}\right)\right\} \quad (1.10)$$

in the sector $\delta < \arg z < 2\pi - \delta$, by the formulas

$$\psi(z) = (z-1)\cdots(z-N)\psi(z-N), \quad (1.11)$$

$$(N-z)^{\frac{1}{2}+z-N} = N^{\frac{1}{2}+z-N}\left(1 - \frac{z}{N}\right)^{\frac{1}{2}+z-N}$$

$$= N^{\frac{1}{2}+z-N}e^z\left(1 + O\left(\frac{1}{N}\right)\right),$$

we obtain

$$\psi(z) = e^{-\gamma z} e^{\pi \sqrt{-1} z} \prod_{j=1}^{\infty} \left(1 - \frac{z}{j}\right) e^{\frac{z}{j}} \qquad (1.12)$$

$$= \frac{e^{\pi \sqrt{-1} z}}{\Gamma(1 - z)}.$$

In this way, the function $e^{\pi \sqrt{-1} z}/\Gamma(1 - z)$ can be characterized as a meromorphic solution of (1.1) having the asymptotic behavior (1.10).

1.1.3 Connection Formula

Now, the ratio $P(z) = \psi(z)/\varphi(z)$ is a periodic function satisfying $P(z+1) = P(z)$ that can be expressed by the Gauss formula

$$P(z) = \frac{e^{\pi \sqrt{-1} z}}{\Gamma(z)\Gamma(1 - z)} = e^{\pi \sqrt{-1} z} \frac{\sin \pi z}{\pi} \qquad (1.13)$$

$$= e^{2\pi \sqrt{-1} z} z \prod_{j=1}^{\infty} \left(1 - \frac{z^2}{j^2}\right).$$

The relation

$$\psi(z) = P(z)\varphi(z) \qquad (1.14)$$

provides a linear relation between two solutions of (1.1) that each of them has an asymptotic expansion as $\Re z \mapsto +\infty$ or $-\infty$, called a connection relation, and the problem to find this relation is called a connection problem, and $P(z)$ is called a connection coefficient or a connection function. Any connection function is periodic.

1.2 Power Series and Higher Logarithmic Expansion

1.2.1 Hypergeometric Series

Consider the convergent series

$$_2F_1(\alpha, \beta, \gamma; x) = \sum_{n=0}^{\infty} \frac{(\alpha; n)(\beta; n)}{(\gamma; n)n!} x^n, \ \gamma \neq 0, -1, -2, \ldots \qquad (1.15)$$

on the unit disk $\Delta(0; 1) = \{x \in \mathbb{C}; |x| < 1\}$ $\big((\alpha; n) = \alpha(\alpha+1) \cdots (\alpha+n-1),$
the Pochhammer symbol[1] $\big)$. Here and after, we abbreviate ${}_2F_1(\alpha, \beta, \gamma; x)$ as
$F(\alpha, \beta, \gamma; x)$, which is called the Euler–Gauss hypergeometric function, or
simply Gauss' hypergeometric function. Setting $\beta = \gamma$, we get

$$F(\alpha, \beta, \beta; x) = \sum_{n=0}^{\infty} \frac{(\alpha; n)}{n!} x^n = (1-x)^{-\alpha},$$

and the specialization $\alpha = \beta = 1$, $\gamma = 2$ yields

$$xF(1, 1, 2; x) = \sum_{n=1}^{\infty} \frac{x^n}{n} = -\log(1-x).$$

Moreover, setting $\alpha = \gamma = 1$ and taking the limit $\beta \mapsto \infty$, we obtain an
elementary analytic function

$$e^x = \lim_{\beta \mapsto \infty} F\left(1, \beta, 1; \frac{x}{\beta}\right).$$

1.2.2 Gauss' Differential Equation

$y = F(\alpha, \beta, \gamma; x)$ satisfies the following second-order Fuchsian linear differential equation:

$$Ey = x(1-x)\frac{d^2y}{dx^2} + \{\gamma - (\alpha + \beta + 1)x\}\frac{dy}{dx} - \alpha\beta y = 0. \qquad (1.16)$$

In fact, it is sufficient to substitute y in (1.16) by (1.15). This is called Gauss'
differential equation. Here, $E = E(x, \frac{d}{dx})$ is a second-order Fuchsian differential operator. Conversely, assume that y can be expanded as a holomorphic
power series around the origin in the form

$$y = \sum_{n=0}^{\infty} a_n x^n, \qquad a_0 = 1. \qquad (1.17)$$

Comparing the coefficients of x^n, for y to be a solution of (1.16), we get the
recurrence relation

$$(n+1)(\gamma+n)a_{n+1} = (n+\alpha)(n+\beta)a_n,$$

i.e.,

---

[1] For $n < 0$, we set $(\alpha; n) = (\alpha + n)^{-1} \cdots (\alpha - 1)^{-1}$.

$$a_{n+1} = \frac{(n+\alpha)(n+\beta)}{(n+1)(n+\gamma)}a_n, \quad n \geq 0, \tag{1.18}$$

which implies that y coincides with $F(\alpha, \beta, \gamma; x)$. In this way, $F(\alpha, \beta, \gamma; x)$ can be characterized as a unique solution of (1.16) which is holomorphic around the origin and which gives 1 at $x = 0$.

1.2.3 First-Order Fuchsian Equation

Setting $y = y_1$, $x\frac{dy}{dx} = y_2\beta$, (1.16) can be transformed into the autonomous first-order Fuchsian equation

$$\frac{d}{dx}\begin{pmatrix} y_1 \\ y_2 \end{pmatrix} = \left(\frac{A_0}{x} + \frac{A_1}{x-1}\right)\begin{pmatrix} y_1 \\ y_2 \end{pmatrix}, \tag{1.19}$$

where we set

$$A_0 = \begin{pmatrix} 0 & \beta \\ 0 & 1-\gamma \end{pmatrix}, \quad A_1 = \begin{pmatrix} 0 & 0 \\ -\alpha & \gamma-\alpha-\beta-1 \end{pmatrix}.$$

This equation describes the horizontal direction of an integrable connection ([De]), called the Gauss–Manin connection, over the complex projective line $\mathbb{P}^1(\mathbb{C}) = \mathbb{C} \cup \{\infty\}$ which admits singularities at $x = 0, 1, \infty$.

1.2.4 Logarithmic Connection

Let a, b be two different points of $\mathbb{C} \setminus \{0, 1\}$. By a theory of linear ordinary differential equations, one can extend the solutions of equation (1.16) analytically along a path σ connecting a and b. Moreover, this extension depends only on the homotopy class of the path connecting a and b. This is the property known as the uniqueness of analytic continuation. From this, it follows that the function $F(\alpha, \beta, \gamma; x)$, a priori defined on $\Delta(0; 1)$, is extended analytically to a single-valued analytic function on the universal covering $\tilde{X}$ of the complex one-dimensional manifold $X = \mathbb{C} \setminus \{0, 1\}$. Let us think about expressing this extension more explicitly.

Now, we introduce parameters $\lambda_1 = \beta$, $\lambda_2 = \gamma - \alpha - 1$, $\lambda_3 = -\alpha$. Then, (1.19) can be rewritten in the form of a linear Pfaff system

$$d\begin{pmatrix} y_1 \\ y_2 \end{pmatrix} = (\lambda_1\theta_1 + \lambda_2\theta_2 + \lambda_3\theta_3)\begin{pmatrix} y_1 \\ y_2 \end{pmatrix}. \tag{1.20}$$

Here,

$$\theta_1 = \begin{pmatrix} 0 & 1 \\ 0 & 0 \end{pmatrix} d\log x + \begin{pmatrix} 0 & 0 \\ 0 & -1 \end{pmatrix} d\log(x-1), \tag{1.21}$$

$$\theta_2 = \begin{pmatrix} 0 & 0 \\ 0 & -1 \end{pmatrix} d\log x + \begin{pmatrix} 0 & 0 \\ 0 & 1 \end{pmatrix} d\log(x-1), \tag{1.22}$$

$$\theta_3 = \begin{pmatrix} 0 & 0 \\ 0 & 1 \end{pmatrix} d\log x + \begin{pmatrix} 0 & 0 \\ -1 & 0 \end{pmatrix} d\log(x-1), \tag{1.23}$$

are logarithmic differential 1-forms. We integrate (1.20) along a smooth curve σ connecting 0 and x. By the fact that $y_1 = 1$ and $y_2 = 0$ at $x = 0$, we obtain an integral form of (1.20):

$$\begin{pmatrix} y_1 \\ y_2 \end{pmatrix} = \begin{pmatrix} 1 \\ 0 \end{pmatrix} + \int_0^x (\lambda_1\theta_1 + \lambda_2\theta_2 + \lambda_3\theta_3) \begin{pmatrix} y_1 \\ y_2 \end{pmatrix}. \tag{1.24}$$

1.2.5 Higher Logarithmic Expansion

Solving (1.24) by Picard's iterative methods (cf. [In]), y_1 and y_2 can be expressed as convergent series of $\lambda_1, \lambda_2, \lambda_3$ around the origin of $\mathbb{C}^3$:

$$y_1 = L_\phi(x) + \sum_{r=1}^{\infty} \sum_{1 \leqslant i_1,\ldots,i_r \leqslant 3} \lambda_{i_1} \cdots \lambda_{i_r} L_{i_1\cdots i_r}(x), \tag{1.25}$$

$$y_2 = \sum_{r=1}^{\infty} \sum_{1 \leqslant i_1,\ldots,i_r \leqslant 3} \lambda_{i_1} \cdots \lambda_{i_r} L'_{i_1\cdots i_r}(x). \tag{1.26}$$

Here, $L_{i_1\cdots i_r}(x)$, $L'_{i_1\cdots i_r}(x)$ are the analytic functions on $\tilde{X}$ defined, along the path σ, by the recurrence relations:

$$L_\phi(x) = 1, \qquad L'_\phi(x) = 0 \tag{1.27}$$

$$L_{1i_2\cdots i_r}(x) = \int_0^x d\log x \cdot L'_{i_2\cdots i_r}(x), \quad r \geq 1 \tag{1.28}$$

$$L_{2i_2\cdots i_r}(x) = L_{3i_2\cdots i_r}(x) = 0, \quad r \geq 1 \tag{1.29}$$

$$L'_{1i_2\cdots i_r}(x) = -\int_0^x d\log(x-1)L_{i_2\cdots i_r}(x), \quad r \geq 1 \tag{1.30}$$

$$L'_{2i_2\cdots i_r}(x) = \int_0^x d\log\left(\frac{x-1}{x}\right) L'_{i_2\cdots i_r}(x), \quad r \geq 1 \tag{1.31}$$

$$L'_{3i_2\cdots i_r}(x) = -\int_0^x d\log(x-1)L_{i_2\cdots i_r}(x) \tag{1.32}$$

$$+ \int_0^x d\log x \cdot L'_{i_2\cdots i_r}(x), \quad r \geq 1.$$

The series of functions $L_{i_1\cdots i_r}(x)$, $L'_{i_1\cdots i_r}(x)$ appearing here has been studied by Lappo-Danilevsky and Smirnov ([La], [Sm]) in detail, and the functions are called hyper logarithms by them. Today, these functions, which are analytic functions on $\tilde{X}$, are also called polylogarithms or higher logarithms. By the way, as we have

$$L_1(x) = L_2(x) = L_3(x) = 0$$

$$L_1'(x) = L_2'(x) = 0, \quad L_3'(x) = -\log(1-x)$$

$$L_{11}(x) = L_{21}(x) = L_{31}(x) = L_{12}(x) = L_{22}(x)$$

$$= L_{32}(x) = L_{23}(x) = L_{33}(x) = 0,$$

$$L_{13}(x) = -\int_0^x \frac{\log(1-x)}{x} dx,$$

$$L_{11}'(x) = L_{21}'(x) = L_{31}'(x) = L_{12}'(x) = L_{22}'(x) = L_{32}'(x) = 0,$$

$$L_{13}'(x) = \frac{1}{2}\log^2(1-x), \quad L_{23}'(x) = -\frac{1}{2}\log^2(1-x) + \int_0^x \frac{\log(1-x)}{x} dx,$$

$$L_{33}'(x) = -\int_0^x \frac{\log(1-x)}{x} dx,$$

$$L_{113}(x) = \frac{1}{2}\int_0^x \frac{\log^2(1-x)}{x} dx,$$

$$L_{123}(x) = -\frac{1}{2}\int_0^x \frac{\log^2(1-x)}{x} dx + \int_0^x \frac{dx}{x}\left(\int_0^x \frac{\log(1-x)}{x} dx\right),$$

$$L_{133}(x) = -\int_0^x \frac{dx}{x}\left(\int_0^x \frac{\log(1-x)}{x} dx\right)$$

etc., from (1.25), we obtain the expansion

$$F(\alpha, \beta, \gamma; x) = 1 - \lambda_1\lambda_3 \int_0^x \frac{\log(1-x)}{x} dx \qquad (1.33)$$

$$+ \frac{1}{2}\lambda_1\lambda_3(\lambda_1 - \lambda_2) \int_0^x \frac{\log^2(1-x)}{x} dx$$

$$+ \lambda_1\lambda_3(\lambda_2 - \lambda_3) \int_0^x \frac{dx}{x}\left(\int_0^x \frac{\log(1-x)}{x} dx\right) + \cdots$$

In particular, the function

$$-\int_0^x \frac{\log(1-x)}{x} dx = \sum_{n=1}^{\infty} \frac{x^n}{n^2} \qquad (1.34)$$

is called a di-logarithm or the Abel–Rogers–Spence function ([Lew]) and some interesting arithmetic properties are known. In general, $L_{i_1\cdots i_r}(x)$ and $L_{i_1\cdots i_r}'(x)$ are expressed in terms of iterated integrals à la K.T.-Chen with logarithmic differential 1-forms $d\log x$ and $d\log(x-1)$, and are extended to higher-dimensional projective spaces ([Ch], [Ao5]). See also Remark 3.12 in § 3.8 of Chapter 3.

1.2.6 𝒟-Module

One can re-interpret (1.16) as follows. Denoting the sheaf of holomorphic functions on X by $\mathcal{O}$ and of the ring of holomorphic differential operators by $\mathcal{D}$ (for sheaves, see, e.g., [Ka]), y gives a local section of $\mathcal{O}$ around a neighborhood of each point x_0 of X. Now, one can apply to y a partial differential operator $P(x, \frac{d}{dx})$ which is a section of $\mathcal{D}_{x_0}$, a germ of $\mathcal{D}$ at x_0, and one obtains a morphism

$$
\begin{array}{ccc}
\mathcal{D}_{x_0} & \longmapsto & \mathcal{O}_{x_0} \\
\cup & & \cup \\
P\left(x, \dfrac{d}{dx}\right) & \longmapsto & P\left(x, \dfrac{d}{dx}\right) y.
\end{array}
\tag{1.35}
$$

A necessary and sufficient condition for the equality

$$
P\left(x, \frac{d}{dx}\right) y = 0
\tag{1.36}
$$

to be satisfied is that P can be rewritten in the form

$$
P\left(x, \frac{d}{dx}\right) = Q\left(x, \frac{d}{dx}\right) E\left(x, \frac{d}{dx}\right)
\tag{1.37}
$$

$\left(Q\left(x, \frac{d}{dx}\right) \in \mathcal{D}_{x_0}\right)$ in a neighborhood of x_0, and a morphism (1.35) induces a homomorphism of $\mathcal{D}_{x_0}$-modules

$$
\begin{array}{ccc}
\mathcal{D}_{x_0}/\mathcal{D}_{x_0} E & \longmapsto & \mathcal{O}_{x_0} \\
\cup & & \cup \\
P\left(x, \dfrac{d}{dx}\right) & \longmapsto & P\left(x, \dfrac{d}{dx}\right) y,
\end{array}
\tag{1.38}
$$

i.e., an element of $\mathrm{Hom}_{\mathcal{D}_{x_0}}(\mathcal{D}_{x_0}/\mathcal{D}_{x_0} E, \mathcal{O}_{x_0})$. From such a viewpoint, one obtains a structure of $\mathcal{D}$-modules satisfied by hypergeometric functions. But here, we do not discuss such a structure further. For a more systematic treatment, see, e.g., [Pha2], [Kas2], [Hot].

1.3 Integral Representation Due to Euler and Riemann ([AAR], [W-W])

1.3.1 Kummer's Method

Let us rewrite some of the factors which have appeared in coefficients of the power series (1.15) as the Euler integral representation. Assuming $\Re\alpha > 0$, $\Re(\gamma - \alpha) > 0$, we have

$$\frac{(\alpha; n)}{(\gamma; n)} = \frac{\Gamma(\gamma)\Gamma(\alpha + n)}{\Gamma(\alpha)\Gamma(\gamma + n)} \tag{1.39}$$

$$= \frac{\Gamma(\gamma)}{\Gamma(\alpha)\Gamma(\gamma - \alpha)} \int_0^1 u^{\alpha+n-1}(1 - u)^{\gamma-\alpha-1}du,$$

and by the binomial expansion

$$\sum_{n=0}^{\infty} \frac{(\lambda; n)}{n!}u^n = (1 - u)^{-\lambda}, \quad |u| < 1,$$

from (1.15), we obtain

$$F(\alpha, \beta, \gamma; x) = \frac{\Gamma(\gamma)}{\Gamma(\alpha)\Gamma(\gamma - \alpha)} \int_0^1 u^{\alpha-1}(1 - u)^{\gamma-\alpha-1} \tag{1.40}$$

$$\left\{ \sum_{n=0}^{\infty} u^n x^n \frac{(\beta; n)}{n!} \right\} du$$

$$= \frac{\Gamma(\gamma)}{\Gamma(\alpha)\Gamma(\gamma - \alpha)} \int_0^1 u^{\alpha-1}(1 - u)^{\gamma-\alpha-1}(1 - ux)^{-\beta}du$$

with the interchange of limit and integral. Here, for that the domain of the integral can avoid the singularity of $(1 - ux)^{-\beta}$, we assume $|x| < 1$. This method of finding an elementary integral representation is by Kummer. The integral (1.40) has been studied by Riemann in detail ([Ri1], [Ri2]), and one of our purposes is to extend this integral to higher-dimensional cases and to reveal systematically the structure of generalized hypergeometric functions.

1.4 Gauss' Contiguous Relations and Continued Fraction Expansion

1.4.1 Gauss' Contiguous Relation

The functions $F(\alpha, \beta, \gamma; x)$ and $F(\alpha + l_1, \beta + l_2, \gamma + l_3; x)$ obtained by shifting the parameters α, β, γ by integers $(l_1, l_2, l_3 \in \mathbb{Z})$ are linearly related. In particular, the following formulas are referred to as Gauss' contiguous relations [Pe]:

$$F(\alpha, \beta, \gamma; x) = F(\alpha, \beta + 1, \gamma + 1; x) \tag{1.41}$$

$$- \frac{\alpha(\gamma - \beta)}{\gamma(\gamma + 1)} x F(\alpha + 1, \beta + 1, \gamma + 2; x),$$

$$F(\alpha, \beta, \gamma; x) = F(\alpha + 1, \beta, \gamma + 1; x) \tag{1.42}$$

$$- \frac{\beta(\gamma - \alpha)}{\gamma(\gamma + 1)} x F(\alpha + 1, \beta + 1, \gamma + 2; x).$$

Indeed, one can check these formulas by expanding $F(\alpha, \beta, \gamma; x)$, $F(\alpha, \beta + 1, \gamma + 1; x)$, $F(\alpha + 1, \beta, \gamma + 1; x)$, $F(\alpha + 1, \beta + 1, \gamma + 2; x)$ as power series in x with the aid of (1.15) and comparing the coefficients of each term. For example, by (1.15), the coefficient of x^n $(n \geq 1)$ in the right-hand side of (1.41) is given by

$$\frac{(\alpha; n)(\beta + 1; n)}{(\gamma + 1; n)n!} - \frac{\alpha(\gamma - \beta)}{\gamma(\gamma + 1)} \frac{(\alpha + 1; n - 1)(\beta + 1; n - 1)}{(\gamma + 2; n - 1)(n - 1)!}$$

$$= \frac{(\alpha; n)(\beta + 1; n - 1)}{(\gamma + 1; n)n!} \left\{ \beta + n - \frac{n(\gamma - \beta)}{\gamma} \right\}$$

$$= \frac{(\alpha; n)(\beta; n)}{(\gamma; n)n!}$$

and as the constant term is 1, the right-hand side of (1.41) is equal to the left-hand side of (1.41). By (1.41), we have

$$\frac{F(\alpha, \beta, \gamma; x)}{F(\alpha, \beta + 1, \gamma + 1; x)} \tag{1.43}$$

$$= 1 - \frac{\alpha(\gamma - \beta)}{\gamma(\gamma + 1)} x \frac{F(\alpha + 1, \beta + 1, \gamma + 2; x)}{F(\alpha, \beta + 1, \gamma + 1; x)}.$$

On the other hand, by (1.42), we obtain

$$\frac{F(\alpha,\beta+1,\gamma+1;x)}{F(\alpha+1,\beta+1,\gamma+2;x)} \tag{1.44}$$

$$= 1 - \frac{(\beta+1)(\gamma-\alpha+1)}{(\gamma+1)(\gamma+2)}x\frac{F(\alpha+1,\beta+2,\gamma+3;x)}{F(\alpha+1,\beta+1,\gamma+2;x)}.$$

Using symbols expressing continued fractions such as $a\pm\frac{b}{c} = a\pm\frac{b|}{|c}$, $a\pm\frac{b}{c\pm\frac{d}{e}} =$

$a\pm\frac{b|}{|c}\pm\frac{d|}{|e}$, it follows from (1.43), (1.44) that

$$\frac{F(\alpha,\beta,\gamma;x)}{F(\alpha,\beta+1,\gamma+1;x)} = 1 - \frac{\frac{\alpha(\gamma-\beta)x}{\gamma(\gamma+1)}|}{|1} - \frac{\frac{(\beta+1)(\gamma-\alpha+1)}{(\gamma+1)(\gamma+2)}x|}{|\frac{F(\alpha+1,\beta+1,\gamma+2;x)}{F(\alpha+1,\beta+2,\gamma+3;x)}}. \tag{1.45}$$

By the shift $(\alpha,\beta,\gamma)\mapsto(\alpha+1,\beta+1,\gamma+2)$, we can repeat this transformation several times, namely, we obtain the finite continued fraction expansion

$$\frac{F(\alpha,\beta,\gamma;x)}{F(\alpha,\beta+1,\gamma+1;x)} = 1 + \frac{a_1 x|}{|1} \tag{1.46}$$

$$+\cdots+\frac{a_{2\nu}x|}{|\frac{F(\alpha+\nu,\beta+\nu,\gamma+2\nu;x)}{F(\alpha+\nu,\beta+\nu+1,\gamma+2\nu+1;x)}},$$

$$\begin{cases} a_{2\nu} = -\frac{(\beta+\nu)(\gamma-\alpha+\nu)}{(\gamma+2\nu-1)(\gamma+2\nu)} \\ a_{2\nu+1} = -\frac{(\alpha+\nu)(\gamma-\beta+\nu)}{(\gamma+2\nu)(\gamma+2\nu+1)} \end{cases}. \tag{1.47}$$

Since the left-hand side of (1.46) is holomorphic in a neighborhood of $x = 0$, it can be expanded as a power series in x. The infinite continued fraction expansion

$$1 + \frac{a_1 x|}{|1} + \frac{a_2 x|}{|1} + \cdots = 1 + a_1 x - a_1 a_2 x^2 + \cdots \tag{1.48}$$

makes sense as a formal power series at $x = 0$.

1.4.2 Continued Fraction Expansion

The identity

$$\frac{F(\alpha,\beta,\gamma;x)}{F(\alpha,\beta+1,\gamma+1;x)} = 1 + \frac{a_1 x|}{|1} + \frac{a_2 x|}{|1} + \cdots, \tag{1.49}$$

as a formal power series at $x = 0$, always makes sense, in fact, it also makes sense as a convergent power series, as we shall show below. For this purpose, we use the Sleshinskii–Pringsheim criterion.

Sleshinskii–Pringsheim Criterion ([Pe])

Let a_ν, b_ν ($b_\nu \neq 0$), $\nu \geq 1$, be a pair of sequences of complex numbers, and consider a continued fraction

$$\lambda = \frac{a_1|}{|b_1} + \frac{a_2|}{|b_2} + \frac{a_3|}{|b_3} + \cdots . \tag{1.50}$$

If $|b_\nu| \geq |a_\nu| + 1$ holds for any $\nu \geq 1$, then the right-hand side of (1.50) converges and the absolute value of λ does not exceed 1.

Proof. Define A_ν, B_ν by the recurrence relations

$$\begin{cases} A_\nu = b_\nu A_{\nu-1} + a_\nu A_{\nu-2} \\ B_\nu = b_\nu B_{\nu-1} + a_\nu B_{\nu-2} \end{cases} (\nu = 1, 2, 3, \cdots), \tag{1.51}$$

$A_{-1} = 1$, $A_0 = 0$, $A_1 = a_1$, $B_{-1} = 0$, $B_0 = 1$, and $B_1 = b_1$. The ν-th approximated continued fraction of λ

$$\lambda_\nu = \frac{a_1|}{|b_1} + \cdots + \frac{a_\nu|}{|b_\nu} \tag{1.52}$$

is equal to $\frac{A_\nu}{B_\nu}$. Now, by

$$|B_\nu| \geq |b_\nu| \, |B_{\nu-1}| - |a_\nu| \, |B_{\nu-2}|$$

$$\geq |b_\nu| \, |B_{\nu-1}| - (|b_\nu| - 1)|B_{\nu-2}|,$$

i.e.,

$$|B_\nu| - |B_{\nu-1}| \geq (|b_\nu| - 1)(|B_{\nu-1}| - |B_{\nu-2}|)$$

$$\geq (|b_\nu| - 1) \cdots (|b_1| - 1) \geq |a_1 a_2 \cdots a_\nu|,$$

in other words,

$$|B| \leq |B_2| \leq |B_3| \leq \cdots .$$

By $A_\nu B_{\nu-1} - A_{\nu-1} B_\nu = (-1)^{\nu-1} a_1 a_2 \cdots a_\nu$, we obtain

$$\frac{A_\nu}{B_\nu} = \frac{A_1}{B_1} + \left(\frac{A_2}{B_2} - \frac{A_1}{B_1}\right) + \cdots + \left(\frac{A_\nu}{B_\nu} - \frac{A_{\nu-1}}{B_{\nu-1}}\right) \tag{1.53}$$

$$= \frac{A_1}{B_0 B_1} - \frac{a_1 a_2}{B_1 B_2} + \cdots + (-1)^{\nu-1} \frac{a_1 a_2 \cdots a_\nu}{B_{\nu-1} B_\nu}.$$

But then

$$\left| \frac{a_1 a_2 \cdots a_\nu}{B_{\nu-1} B_\nu} \right| \leq \frac{|B_\nu| - |B_{\nu-1}|}{|B_{\nu-1} B_\nu|} = \frac{1}{|B_{\nu-1}|} - \frac{1}{|B_\nu|}$$

implies

$$\sum_{\nu=2}^{\infty} \left| \frac{A_\nu}{B_\nu} - \frac{A_{\nu-1}}{B_{\nu-1}} \right| + \left| \frac{A_1}{B_1} \right| \qquad (1.54)$$

$$\leq \sum_{\nu=2}^{\infty} \left(\frac{1}{|B_{\nu-1}|} - \frac{1}{|B_\nu|} \right) + \left| \frac{A_1}{B_1} \right| = \frac{1 + |a_1|}{|B_1|} \leq 1.$$

Thus, $\lambda = \lim_{\nu \to \infty} \frac{A_\nu}{B_\nu}$ converges and its absolute value is at most 1.

1.4.3 Convergence

We return to our situation. For $|x| < 1$, our continued fraction expansion can be formally written as

$$1 + \frac{a_1 x}{\lceil 1} + \frac{a_2 x}{\lceil 1} + \cdots = \frac{1}{2} \left(2 + \frac{4a_1 x}{\lceil 2} + \frac{4a_2 x}{\lceil 2} + \cdots \right). \qquad (1.55)$$

By $\lim_{\nu \to \infty} a_{2\nu} = \lim_{\nu \to \infty} a_{2\nu+1} = \frac{1}{4}$, for a sufficiently big N, we have $4|a_{2\nu} x| < 1$, $4|a_{2\nu+1} x| < 1$ for $\nu \geq N$ which implies that the continued fraction expansion

$$\frac{4a_{2N} x}{\lceil 2} + \frac{4a_{2N+1} x}{\lceil 2} + \cdots \qquad (1.56)$$

satisfies Sleshinskii–Pringsheim criterion. Hence, (1.56) converges uniformly on every compact subset of $|x| < 1$. Hence, (1.55) converges uniformly on every compact subset of $|x| < 1$. Expanding this continued fraction as a power series around $x = 0$, it coincides with the power series $\frac{F(\alpha, \beta, \gamma; x)}{F(\alpha, \beta+1, \gamma+1; x)}$, namely, the identity (1.49) holds for $|x| < 1$.

Actually, it is J. Worpitzky who first proved the convergence of (1.48) under the condition $4|a_{2\nu} x| \leq 1$, $4|a_{2\nu+1} x| \leq 1$, see [J-T-W].

The continued fraction expansion (1.49) will be rediscussed in § 4.2 from a general consideration of difference equations and an asymptotic expansion of their solutions.

1.5 The Mellin–Barnes Integral

1.5.1 Summation over a Lattice

Setting $\Phi(z) = x^z \frac{\Gamma(\alpha+z)\Gamma(\beta+z)}{\Gamma(\gamma+z)\Gamma(1+z)}$, $\Phi(z)$ is meromorphic on $\mathbb{C}$ as function of z. By the identity of Γ-function $\Gamma(z+1) = z\Gamma(z)$, we see that $\Phi(z)$ satisfies the difference equation

$$\Phi(z+1) = b(z)\Phi(z). \tag{1.57}$$

Here, $b(z) = x\frac{(\alpha+z)(\beta+z)}{(\gamma+z)(1+z)}$ is a rational function. As $\Phi(n) = 0$ $(n \in \mathbb{Z}_{<0})$, $F(\alpha, \beta, \gamma; x)$ in (1.15) can be described as the sum over a one-dimensional lattice as follows:

$$F(\alpha, \beta, \gamma; x) = \frac{\Gamma(\gamma)}{\Gamma(\alpha)\Gamma(\beta)} \sum_{n=-\infty}^{\infty} \Phi(n), \quad |x| < 1. \tag{1.58}$$

1.5.2 Barnes' Integral Representation

Let us rewrite (1.58) as a contour integral in $\mathbb{C}$ by applying a residue formula. For simplicity, we assume $\Re\alpha > 0$, $\Re\beta > 0$, $\Re\gamma > 0$ and $0 < \arg x < 2\pi$. For $-\pi+\delta < \arg z < \pi-\delta$ (δ a small positive number), it follows from (1.5) that an asymptotic expansion as $|z| \mapsto +\infty$ is given, for any constant $c \in \mathbb{C}$, by

$$\Gamma(c+z) \sim e^{-c-z} e^{(c+z-\frac{1}{2})\log(c+z)} \sqrt{2\pi} \left\{1 + O\left(\frac{1}{|z|}\right)\right\}, \tag{1.59}$$

which implies

$$\Gamma(-z) = -\frac{\pi}{\sin \pi z} \frac{1}{\Gamma(z+1)} \tag{1.60}$$

$$\sim -\frac{\pi}{\sin \pi z} e^{1+z-(\frac{1}{2}+z)\log(1+z)} \frac{1}{\sqrt{2\pi}} \left\{1 + O\left(\frac{1}{|z|}\right)\right\}.$$

Hence, we have

$$\frac{\Gamma(\alpha+z)\Gamma(\beta+z)\Gamma(-z)}{\Gamma(1+z)} \sim \tag{1.61}$$

$$-\frac{\pi}{\sin \pi z} e^{(\alpha+\beta-\gamma-1)\log z} \left\{1 + O\left(\frac{1}{|z|}\right)\right\}.$$

On the other hand, we have

$$(-x)^z = e^{z \log(-x)} = e^{z\{\log|x| + \sqrt{-1}\arg(-x)\}}$$

but since $-\pi + \delta < \arg(-x) < \pi - \delta$, $e^{\sqrt{-1}z \arg(-x)} \frac{\pi}{\sin \pi z}$ decays exponentially with respect to $|z|$ as $|z| \mapsto +\infty$. Here, $e^{z \log|x|}$ is bounded for $|\arg z| \leq \frac{\pi}{2}$ and, when $\arg z < \frac{\pi}{2}$, it decays exponentially with respect to $|z|$. Now, let σ_0 and σ_1 be contours in the complex plane of z defined by

$$\sigma_0 = \left\{ z \in \mathbb{C}; \arg z = -\frac{\pi}{2}, |z| \geq \frac{1}{2} \right\} \cup \left\{ z \in \mathbb{C}; |z| = \frac{1}{2}, \frac{\pi}{2} \leq \arg z \leq \frac{3\pi}{2} \right\}$$

$$\cup \left\{ z \in \mathbb{C}; \arg z = \frac{\pi}{2}, |z| \geq \frac{1}{2} \right\},$$

$$\sigma_1 = \left\{ z \in \mathbb{C}; \Re z \geq 0, \Im z = -\frac{1}{2} \right\} \cup \left\{ z \in \mathbb{C}; \Re z \geq 0, \Im z = \frac{1}{2} \right\}$$

$$\cup \left\{ z \in \mathbb{C}; |z| = \frac{1}{2}, \frac{\pi}{2} \leq \arg z \leq \frac{3}{2}\pi \right\}.$$

See Figure 1.1: σ_0 is drawn with a solid line, σ_1 with a dotted line (with orientation).

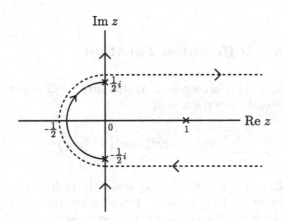

Fig. 1.1

Then by the above estimate and Cauchy's integral formula, we have

$$\frac{1}{2\pi i} \int_{\sigma_0} \frac{\Gamma(\alpha+z)\Gamma(\beta+z)\Gamma(-z)}{\Gamma(\gamma+z)} (-x)^z dz \qquad (1.62)$$

$$= \frac{1}{2\pi i} \int_{\sigma_1} \frac{\Gamma(\alpha+z)\Gamma(\beta+z)\Gamma(-z)}{\Gamma(\gamma+z)} (-x)^z dz.$$

For $|\arg z| \leq \frac{\pi}{2}$, we remark that the integrand admits simple poles only at $z = 0, 1, 2, 3, \ldots$. Hence, the right-hand side of (1.62) can be rewritten by the residue formula

$$\mathrm{Res}_{z=n}\Gamma(-z) = \frac{(-1)^{n+1}}{n!}, \quad n = 0, 1, 2, \cdots$$

as follows:

$$\sum_{n=0}^{\infty} \frac{\Gamma(\alpha + n)\Gamma(\beta + n)}{\Gamma(\gamma + n)n!} x^n = \frac{\Gamma(\alpha)\Gamma(\beta)}{\Gamma(\gamma)} F(\alpha, \beta, \gamma; x). \tag{1.63}$$

Namely, we obtain

$$F(\alpha, \beta, \gamma; x)$$
$$= \frac{\Gamma(\gamma)}{\Gamma(\alpha)\Gamma(\beta)} \frac{1}{2\pi\sqrt{-1}} \int_{\sigma_0} \frac{\Gamma(\alpha + z)\Gamma(\beta + z)\Gamma(-z)}{\Gamma(\gamma + z)} (-x)^z dz \tag{1.64}$$

$$(0 < \arg x < 2\pi, \ |x| < 1).$$

This is the Barnes formula. Although, here, we imposed some restrictions on α, β, γ, by deforming the cycle σ_0 appropriately, α, β, γ can be chosen arbitrarily except for the condition $\gamma \neq 0, -1, -2, \cdots$.

1.5.3 Mellin's Differential Equation

In the summation (1.58), more generally, choosing $\xi \in \mathbb{C}$ arbitrarily, one may consider the bilateral convergent sum

$$\sum_{n=-\infty}^{\infty} \Phi(\xi + n)\varphi(\xi + n) \left(=: \langle\varphi\rangle_\xi\right) \tag{1.65}$$

over the lattice $L_\xi = \{\xi + n; n \in \mathbb{Z}\} \cong \mathbb{Z}$ passing through ξ. Now, we introduce the shifting operator $T : T\varphi(z) = \varphi(z+1)$ and define the covariant difference $\nabla_{\mathrm{disc}}\varphi(z) = \varphi(z) - b(z)\varphi(z+1)$. As it can be seen easily from (1.65), we have $\langle\nabla_{\mathrm{disc}}\varphi\rangle_\xi = 0$. In particular, setting $\hat{\Phi} = \langle 1\rangle_0$, by

$$\left\langle \nabla_{\mathrm{disc}}((\gamma + z - 1)z) \right\rangle_0 = 0$$

and $x\frac{d}{dx}\langle\varphi(z)\rangle_0 = \langle z\varphi(z)\rangle_0$, we obtain immediately the differential equation:

$$0 = \left(\gamma - 1 + x\frac{d}{dx}\right) x\frac{d}{dx}\hat{\Phi} - x\left(\alpha + x\frac{d}{dx}\right)\left(\beta + x\frac{d}{dx}\right)\hat{\Phi}. \tag{1.66}$$

(1.66) defines the same differential equation as (1.16).

At the beginning of the 20th century, Mellin extended equation (1.66) to the case of several variables based on this idea. In modern language, this is a holonomic system of linear partial differential equations. For this and its further developments, see Appendix A.

1.6 Plan from Chapter 2

The integral representation (1.40) of Gauss' hypergeometric function $F(\alpha, \beta, \gamma; x)$ initiated by Riemann, is one of the most powerful tools to study its global behavior. In this book, we will generalize (1.40) to the case of several variables and reveal its topological and analytic properties. In Chapter 2, we will present some fundamental basis. First, we will recall the relation between algebraic de Rham complex and topological properties over affine varieties. Next, we will state some basic facts about (co)homology with coefficients in a local system (in this book, we call it twisted (co)homology). We will also show how the structure of the twisted de Rham cohomology, representing analytic properties of hypergeometric functions, reflects algebraic analytic properties of them. In Chapter 3, we will give more concrete results on the twisted de Rham (co)homology arising from arrangements of hyperplanes. We will present the so-called Gelfand system (denoted by $E(n+1, m+1; \alpha)$ in this book) over a Grassmannian satisfied by hypergeometric functions together with a logarithmic Gauss−Manin connection, and clarify relations among them. In Chapter 4, we will first show the existence theorem due to G.D. Birkhoff that is a fundamental theorem about difference equations, and will explain about its generalization to the case of several variables. Next, we will derive difference equations on the parameter α satisfied by hypergeometric functions, and through its integral representation, we will see its relation to topological properties of affine varieties. Finally, we will mention about the connection problem arising from solutions of a difference equation.

In the appendices, we will present Mellin's general hypergeometric functions by using the Bernstein−Sato b-function (Appendix A) and present hypergeometric equations associated to the Selberg integral, which was introduced by J. Kaneko (Appendix B). Moreover, we will give the simplest example of the monodromy representations of hypergeometric functions treated in Chapter 3 (Appendix C). Lastly, T. Kohno will explain a recent topic in other domains such as mathematical physics which is related to hypergeometric functions (Appendix D).

Chapter 2
Representation of Complex Integrals and Twisted de Rham Cohomologies

In integral representations of Euler type of classical hypergeometric functions of several variables or of hypergeometric functions which are studied these days, integrals of the product of powers of polynomials appear. We will establish a framework to treat such integrals, and after that, we will study hypergeometric functions of several variables as an application of the theory. Since ordinary theory of integrals of single-valued functions is formalized under the name of the de Rham theory, by modifying this theory, we will constuct a theory suitable for our purpose in this chapter. As the key to the de Rham theory is Stokes theorem, we will start by posing the question how to formulate Stokes theorem for integrals of multi-valued functions.

2.1 Formulation of the Problem and Intuitive Explanation of the Twisted de Rham Theory

2.1.1 Concept of Twist

To be as concrete as possible and sufficient for applications below, here and after, we assume that M is an n-dimensional affine variety obtained as the complement, in $\mathbb{C}^n$, of zeros $D_j := \{P_j(u) = 0\}$ of a finite number of polynomials $P_j(u) = P_j(u_1, \cdots, u_n)$, $1 \le j \le m$

$$M := \mathbb{C}^n \setminus D, \quad D := \bigcup_{j=1}^{m} D_j,$$

and consider a multi-valued function on M

$$U(u) = \prod_{j=1}^{m} P_j(u)^{\alpha_j}, \ \alpha_j \in \mathbb{C} \setminus \mathbb{Z}, \ 1 \le j \le m.$$

K. Aomoto et al., *Theory of Hypergeometric Functions*, Springer Monographs in Mathematics, DOI 10.1007/978-4-431-53938-4_2, © Springer 2011

As was seen in § 1.3 of Chapter 1, for a one-dimensional case and $m = 3$, $P_1 = u, P_2 = 1 - u, P_3 = 1 - ux, \alpha_1 = \alpha - 1, \alpha_2 = \gamma - \alpha - 1, \alpha_3 = -\beta$, we obtain an integral representation of Gauss' hypergeometric function by integrating $U(u)$. When we treat this $U(u)$, one of the ways to eliminate its multi-valuedness and bring it to the ordinary analysis is to lift U to a covering manifold $\widetilde{M}$ of M so that it becomes single-valued, and consider on $\widetilde{M}$. For example, if all the α_j's are $\frac{1}{2}$, $\widetilde{M}$ is the double cover of $\mathbb{C}^n$ ramifying at D. But in general, the relation between $\widetilde{M}$ and M is so complicated that it is not suitable for concrete computations besides theoretical questions. Here, in place of lifting U to $\widetilde{M}$, we introduce some quantities taking a "twist" arising from the multi-valuedness into consideration, and analyze it. Here, we use the following symbols:

$\mathcal{A}_M^p$ 　　the sheaf of C^∞ differential p-forms on M,

$\mathcal{A}^p(M)$ 　　the space of C^∞ differential p-forms on M,

$\mathcal{A}_c^p(M)$ 　　the space of C^∞ differential p-forms with compact support on M,

$\mathcal{O}_M$ 　　the sheaf of germs of holomorphic functions on M,

Ω_M^p 　　the sheaf of holomorphic differential p-forms on M,

$\Omega^p(M)$ 　　the space of holomorphic differential p-forms on M.

Notice that

$$\omega = \frac{dU}{U} = \sum_{j=1}^m \alpha_j \frac{dP_j}{P_j} \tag{2.1}$$

is a single-valued holomorphic 1-form on M. To simplify notation, we choose a smooth triangulation K on M and we explain our idea by using it. Our aim is to extend Stokes' theorem for a multi-valued differential form $U\varphi$, $\varphi \in \mathcal{A}^\bullet(M)$ determined by the multi-valued function U, and to find a twisted version of the de Rham theory via this theorem. We refer to [Si] for the basic notion of analytic functions of several variables.

2.1.2 Intuitive Explanation

Let Δ be one of the p-simplexes of K, φ be a smooth p-form on M. To determine the integral of a "multi-valued" p-form $U\varphi$ on Δ, we have to fix the branch of U on Δ. So, by the symbol $\Delta \otimes U_\Delta$, we mean that the pair of Δ and one of the branches U_Δ of U on Δ is fixed. Then, for $\varphi \in \mathcal{A}^p(M)$, the integral $\int_{\Delta \otimes U_\Delta} U \cdot \varphi$ is, by this rule, defined as

$$\int_{\Delta \otimes U_\Delta} U \cdot \varphi := \int_\Delta [\text{the fixed branch } U_\Delta \text{ of } U \text{ on } \Delta] \cdot \varphi.$$

Since U_Δ can be continued analytically on a sufficiently small neighborhood of Δ, on this neighborhood, for a single-valued p-form $U_\Delta \cdot \varphi$ and a p-simplex Δ, the ordinary Stokes theorem holds and takes the following form:

For $\varphi \in \mathcal{A}^{p-1}(M)$, one has

$$\int_\Delta d(U_\Delta \cdot \varphi) = \int_{\partial\Delta} U_\Delta \cdot \varphi. \tag{2.2}$$

On the other hand, since we have

$$d(U_\Delta \cdot \varphi) = dU_\Delta \wedge \varphi + U_\Delta d\varphi = U_\Delta \left(d\varphi + \frac{dU_\Delta}{U_\Delta} \wedge \varphi \right)$$

on Δ, and $\omega = \frac{dU}{U}$ is a single-valued holomorphic 1-form, by setting

$$\nabla_\omega \varphi := d\varphi + \omega \wedge \varphi, \tag{2.3}$$

the above formula can be rewritten as

$$d(U_\Delta \cdot \varphi) = U_\Delta \nabla_\omega \varphi.$$

(2.3) can be regarded as the defining equation of the covariant differential operator ∇_ω associated to the connection form ω. Clearly, ∇_ω satisfies $\nabla_\omega \cdot \nabla_\omega = 0$, and in such a case, ∇_ω is said to be integrable or to define a Gauss–Manin connection of rank 1 ([De]). Thus, the left-hand side of (2.2) is rewritten, by using the symbol $\Delta \otimes U_\Delta$, as

$$\int_\Delta d(U_\Delta \cdot \varphi) = \int_{\Delta \otimes U_\Delta} U \cdot \nabla_\omega \varphi.$$

We would like to rewrite the right-hand side of (2.2). When the dimension of Δ is low, we can explain our idea clearly with the aid of a figure, so let us start from a 1-simplex Δ.

2.1.3 One-Dimensional Case

For an oriented 1-simplex Δ, with starting point p and ending point q, and a smooth function φ on M, (2.2) becomes

$$\int_\Delta d(U_\Delta \cdot \varphi) = U_\Delta(q)\varphi(q) - U_\Delta(p)\varphi(p).$$

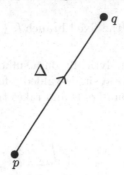

Fig. 2.1

Here, $\partial\Delta = \langle q \rangle - \langle p \rangle$ and $U_\Delta(q)$ is the value of the germ of the function U_q, determined by the branch U_Δ at the boundary q of Δ, at the point q.

Hence, the symbol $\langle q \rangle \otimes U_q$ is determined, and in the above sense, we have

$$U_\Delta(q)\varphi(q) = \text{the integral of } U \cdot \varphi \text{ on } \langle q \rangle \otimes U_q,$$

from which we obtain the formula:

$$\int_{\Delta\otimes U_\Delta} U \cdot \nabla_\omega\varphi = \int_{\langle q\rangle\otimes U_q - \langle p\rangle\otimes U_p} U\cdot\varphi.$$

Since, in the context of a generalized Stokes theorem, the right-hand side should be the boundary of $\Delta\otimes U_\Delta$, we define the "boundary operator" as follows:

$$\partial_\omega(\Delta\otimes U_\Delta) := \langle q\rangle\otimes U_q - \langle p\rangle\otimes U_p.$$

Then, (2.2) takes the following form:

$$\int_{\Delta\otimes U_\Delta} U\cdot\nabla_\omega\varphi = \int_{\partial_\omega(\Delta\otimes U_\Delta)} U\cdot\varphi.$$

2.1.4 Two-Dimensional Case

Let Δ be the 2-simplex with vertices $\langle 1\rangle$, $\langle 2\rangle$, $\langle 3\rangle$ with the orientation given as in Figure 2.2.

For a smooth 1-form φ on M, the right-hand side of (2.2) takes the form:

$$\int_{\partial\Delta} U_\Delta\cdot\varphi = \int_{\langle 12\rangle} U_\Delta\cdot\varphi + \int_{\langle 23\rangle} U_\Delta\cdot\varphi + \int_{\langle 31\rangle} U_\Delta\cdot\varphi.$$

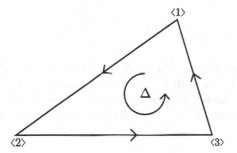

Fig. 2.2

Now, let $U_{\langle 12 \rangle}$ be the branch on the boundary $\langle 12 \rangle$ of Δ determined by U_Δ. By using the symbol $\langle 12 \rangle \otimes U_{\langle 12 \rangle}$, the above formula can be rewritten as follows:

$$\int_{\partial\Delta} U_\Delta \cdot \varphi = \int_{\langle 12 \rangle \otimes U_{\langle 12 \rangle}} U \cdot \varphi + \int_{\langle 23 \rangle \otimes U_{\langle 23 \rangle}} U \cdot \varphi + \int_{\langle 31 \rangle \otimes U_{\langle 31 \rangle}} U \cdot \varphi.$$

Hence, by defining the "boundary" of $\Delta \otimes U_\Delta$ as

$$\partial_\omega(\Delta \otimes U_\Delta) := \langle 12 \rangle \otimes U_{\langle 12 \rangle} + \langle 23 \rangle \otimes U_{\langle 23 \rangle} + \langle 31 \rangle \otimes U_{\langle 31 \rangle},$$

(2.2) can be rewritten as

$$\iint_{\Delta \otimes U_\Delta} U \cdot \nabla_\omega \varphi = \int_{\partial(\Delta \otimes U_\Delta)} U\varphi.$$

2.1.5 Higher-Dimensional Generalization

It is now easy to generalize the examples treated in § 2.1.3 and 2.1.4 to higher-dimensional cases. The boundary operators defined above are the same as those appearing in algebraic topology when one defines homology with coefficients in a local system. Hence, we utilize some symbols used in algebraic topology. Let us explain this below. The differential equation on M

$$\nabla_\omega h - dh + \sum_{j=1}^{m} \alpha_j \frac{dP_j}{P_j} h - 0 \tag{2.4}$$

admits a general solution which can be formally expressed as

$$h = c \prod_{j=1}^{m} P_j^{-\alpha_j}, \quad c \in \mathbb{C} \tag{2.5}$$

and the space generated by the local solutions of (2.4) is of dimension 1. Cover M by a sufficiently fine locally finite open cover $M = \cup U_\nu$ and fix a single-valued non-zero solution h_ν of (2.4) on each U_ν. Since, for $\mu \neq \nu$, h_μ and h_ν are solutions of the same differential equation (2.4) on non-empty $U_\mu \cap U_\nu$, setting

$$h_\mu(u) = g_{\mu\nu}(u)h_\nu(u), \quad u \in U_\mu \cap U_\nu, \tag{2.6}$$

$g_{\mu\nu}$ becomes a constant on $U_\mu \cap U_\nu$. As a solution $h(u)$ on $U_\mu \cap U_\nu$ is expressed in two ways as $h = \xi_\mu h_\mu = \xi_\nu h_\nu$, $\xi_\mu, \xi_\nu \in \mathbb{C}$, we obtain $\xi_\mu = g_{\mu\nu}^{-1}\xi_\nu$ by (2.6). Hence, the set of all local solutions of (2.4) defines a flat line bundle $\mathcal{L}_\omega$ obtained by gluing the fibers $\mathbb{C}$ by the transition functions $\{g_{\mu\nu}^{-1}\}$. Denote the flat line bundle obtained by the transition functions $\{g_{\mu\nu}\}$ by $\mathcal{L}_\omega^\vee$ and call it the dual line bundle of $\mathcal{L}_\omega$. By (2.6), $h_\mu^{-1}(u)$ becomes a local section of $\mathcal{L}_\omega^\vee$. By (2.4) and (2.5), this $\mathcal{L}_\omega^\vee$ can be regarded as the flat line bundle generated by the set of all local solutions of

$$\nabla_{-\omega} h = dh - \sum_{j=1}^{m} \alpha_j \frac{dP_j}{P_j} h = 0, \tag{2.7}$$

and $U(u) = \prod P_j^{\alpha_j}$ is its local section. Below, we use the terminology "local system of rank 1" in the same sense as "flat line bundle". Hence, we call $\mathcal{L}_\omega^\vee$ by the dual local system of $\mathcal{L}_\omega$. Thus, we have seen that the boundary operators which appeared in § 2.1.3 and 2.1.4 coincide with those of chain groups $C_1(K, \mathcal{L}_\omega^\vee)$ and $C_2(K, \mathcal{L}_\omega^\vee)$ with coefficients in $\mathcal{L}_\omega^\vee$. Below, for higher-dimensional cases, using M in place of a simplicial complex K, we denote the chain group by $C_q(M, \mathcal{L}_\omega^\vee)$.

2.1.6 Twisted Homology Group

Summarizing what we discussed above, we formulate them as follows:

$$C_p(M, \mathcal{L}_\omega^\vee) = \left\{ \begin{array}{l} \text{for a } p\text{-simplex } \Delta \text{ of } M, \\ \text{the complex vector space} \\ \text{with basis } \Delta \otimes U_\Delta \end{array} \right\}$$

is called the p-dimensional twisted chain group. As a generalization of § 2.1.3 and 2.1.4, we define the boundary operator

$$\partial_\omega : C_p(M, \mathcal{L}_\omega^\vee) \longrightarrow C_{p-1}(M, \mathcal{L}_\omega^\vee)$$

as follows: for a p-simplex $\Delta = \langle 012 \cdots p \rangle$,

$$\partial_\omega(\Delta \otimes U_\Delta) := \sum_{j=0}^{P} (-1)^j \langle 01 \cdots \widehat{j} \cdots p \rangle \otimes U_{\langle 01 \cdots \widehat{j} \cdots p \rangle}.$$

Here, $\widehat{j}$ means to remove j and $U_{\langle 01 \cdots \widehat{j} \cdots p \rangle}$ is the branch determined by the branch U_Δ at the boundary $\langle 01 \cdots \widehat{j} \cdots p \rangle$ of Δ. Similarly to the ordinary homology theory, one can show the following lemma:

Lemma 2.1. $\partial_\omega \circ \partial_\omega = 0$.

By this lemma, one can define the quotient vector space

$$H_p(M, \mathcal{L}_\omega^\vee) := \{\text{Ker } \partial_\omega : C_p(M, \mathcal{L}_\omega^\vee) \to C_{p-1}(M, \mathcal{L}_\omega^\vee)\}/\partial_\omega C_{p+1}(M, \mathcal{L}_\omega^\vee),$$

called the p-dimensional twisted homology group, and the element of Ker ∂_ω is called a twisted cycle.

Here, we provide an easy example.

Example 2.1. Suppose that $\alpha_j \notin \mathbb{Z}$ and consider the multi-valued function $U(u) = u^{\alpha_1}(u-1)^{\alpha_2}$ on $M = \mathbb{P}^1 \setminus \{0, 1, \infty\}$. Set $\omega = dU/U$.

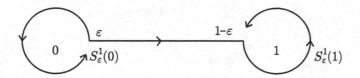

Fig. 2.3

The starting points of the circles are fixed as in Figure 2.3, and the orientation is fixed in the anti-clockwise direction on $\mathbb{C}$. By

$$\partial_\omega S_\varepsilon^1(0) = (c_1 - 1)\langle \varepsilon \rangle, \quad \partial_\omega S_\varepsilon^1(1) = (c_2 - 1)\langle 1 - \varepsilon \rangle,$$

$$\partial_\omega[\varepsilon, 1 - \varepsilon] = \langle 1 - \varepsilon \rangle - \langle \varepsilon \rangle \quad (\text{set } c_j = \exp(2\pi\sqrt{-1}\alpha_j)),$$

the element

$$\Delta(\omega) = \frac{1}{c_1 - 1} S_\omega^1(0) + [\varepsilon, 1 - \varepsilon] - \frac{1}{c_2 - 1} S_\varepsilon^1(1)$$

satisfies $\partial_\omega \Delta(\omega) = 0$, i.e., we have

$$[\Delta(\omega)] \in H_1(M, \mathcal{L}_\omega^\vee).$$

This twisted cycle is called the regularization of the open interval $(0,1)$ and it provides us Hadamard's finite part ([Had], pp. 133–158) of a divergent integral. For detail, see § 3.2.

2.1.7 Locally Finite Twisted Homology Group

Since M is not compact, it requires infinitely many simplices for its triangulation. Besides $C_p(M, \mathcal{L}_\omega^\vee)$ corresponding to ordinary homology, we consider an infinite chain group which is locally finite, and set

$$C_p^{lf}(M, \mathcal{L}_\omega^\vee) = \{\Sigma c_\Delta \Delta \otimes U_\Delta | \Delta\text{'s are locally finite}\}.$$

By Lemma 2.1, $(C_\bullet^{lf}(M, \mathcal{L}_\omega^\vee), \partial_\omega)$ forms a complex, and its homology group is denoted by $H_\bullet^{lf}(M, \mathcal{L}_\omega^\vee)$ and is called a locally finite homology group. For a p-simplex Δ of M, a branch U_Δ of U on Δ, and $\varphi \in \mathcal{A}^{p-1}(M)$, the right-hand side of (2.2) is rewritten as

$$\int_{\partial\Delta} U_\Delta \cdot \varphi = \sum_{i=0}^p (-1)^i \int_{\langle 0 \cdots \widehat{i} \cdots p\rangle} U_\Delta \cdot \varphi$$

$$= \sum_{i=0}^p (-1)^i \int_{\langle 0 \cdots \widehat{i} \cdots p\rangle \otimes U_{\langle 0 \cdots \widehat{i} \cdots p\rangle}} U \cdot \varphi$$

$$= \int_{\partial_\omega(\Delta \otimes U_\Delta)} U \cdot \varphi,$$

and hence, (2.2) takes the following form:

$$\int_{\Delta \otimes U} U \cdot \nabla_\omega \varphi = \int_{\partial_\omega(\Delta \otimes U_\Delta)} U \cdot \varphi.$$

For a general twisted chain, we extend the definition $\mathbb{C}$-linearly and we obtain the following form of Stokes theorem:

Theorem 2.1 (Stokes theorem).

1. For $\sigma \in C_p(M, \mathcal{L}_\omega^\vee)$ and $\varphi \in \mathcal{A}^{p-1}(M)$, we have

$$\int_\sigma U \cdot \nabla_\omega \varphi = \int_{\partial_\omega \sigma} U \cdot \varphi.$$

2. For $\tau \in C_p^{lf}(M, \mathcal{L}_\omega^\vee)$ and $\varphi \in \mathcal{A}_c^{p-1}(M)$, we have

$$\int_\tau U \cdot \nabla_\omega \varphi = \int_{\partial_\omega \tau} U \cdot \varphi.$$

Modeled on the de Rham theory, we would like to construct its twisted version. Thus, by Theorem 2.1, we should establish relations between the cohomology of

$$(\mathcal{A}^\bullet(M), \nabla_\omega), \quad (\mathcal{A}_c^\bullet(M), \nabla_\omega)$$

and the twisted homology groups $H_\bullet(M, \mathcal{L}_\omega^\vee)$, $H_\bullet^{lf}(M, \mathcal{L}_\omega^\vee)$. This is the aim of the following section. See [Stee] for the basic notion of local system.

2.2 Review of the de Rham Theory and the Twisted de Rham Theory

We will introduce a system of equations $E(n+1, m+1; \alpha)$ in § 3.4 which is a direct generalization of the classical hypergeometric differential equations. All of its linearly independent solutions can be expressed as integrals. In Chapter 4, we will consider hypergeometric difference equations, and there, the basic fact is that the cohomology of the complex $(\mathcal{A}^\bullet(M), \nabla_\omega)$ and the twisted homology $H_\bullet(M, \mathcal{L}_\omega^\vee)$ are dual to each other, which is summarized in Lemma 2.9. Although the authors could prove it only by a roundabout way as is stated below, we have provided a proof of the above-mentioned fact by using any result that could be applied, since the results of this section will be essentially used in Chapter 3 and 4. First, we recall the de Rham theory which corresponds to the case $\omega = 0$, and after that we study the objective relation by modifying the former argument appropriately.

2.2.1 Preliminary from Homological Algebra

We take the following fact of homological algebra as our starting point.

Lemma 2.2. *For a chain complex $C_\bullet$ of $\mathbb{C}$-vector spaces, there exists a natural isomorphism*

$$H^p(\mathrm{Hom}_\mathbb{C}(C_\bullet, \mathbb{C})) \xrightarrow{\sim} \mathrm{Hom}_\mathbb{C}(H_p(C_\bullet), \mathbb{C}).$$

Proof. From the chain complex

$$C_\bullet : \cdots \longrightarrow C_{n+1} \xrightarrow{\partial_{n+1}} C_n \xrightarrow{\partial_n} C_{n-1} \longrightarrow \cdots \longrightarrow C_1 \xrightarrow{\partial_1} C_0 \longrightarrow 0,$$

we introduce the new complexes $Z_\bullet, \overline{B}_\bullet$ as follows:

$$Z_n = \operatorname{Ker} \partial_n, \quad n = 1, 2, \ldots, Z_0 = C_0,$$

the boundary operator is the zero-map;

$$\overline{B}_n = \operatorname{Im} \partial_n, \quad n = 1, 2, \ldots, \overline{B}_0 = 0,$$

the boundary operator is the zero-map.

Then, the inclusion $i_n : Z_n \longrightarrow C_n$ and the boundary operator $\partial_n : C_n \longrightarrow \overline{B}_n$ gives us the short exact sequence of chain complexes

$$0 \longrightarrow Z_\bullet \xrightarrow{i_\sharp} C_\bullet \xrightarrow{\partial_\sharp} \overline{B}_\bullet \longrightarrow 0.$$

For a $\mathbb{C}$-vector space E, we denote its dual space $\operatorname{Hom}_{\mathbb{C}}(E, \mathbb{C})$ by E^*. Since $E \longrightarrow E^*$ is a contravariant exact functor,

$$0 \longrightarrow \overline{B}_\bullet{}^* \xrightarrow{\partial^\sharp} C_\bullet{}^* \xrightarrow{i^\sharp} Z_\bullet{}^* \longrightarrow 0$$

is again a short exact sequence. Passing to an exact cohomology sequence, we obtain

$$\longrightarrow H^{n-1}(Z_\bullet{}^*) \xrightarrow{\delta^*_{(n-1)}} H^n(\overline{B}_\bullet{}^*) \xrightarrow{\partial^*_n} H^n(C_\bullet{}^*) \longrightarrow$$

$$\xrightarrow{i^*_n} H^n(Z_\bullet{}^*) \xrightarrow{\delta^*_{(n)}} H^{n+1}(\overline{B}_\bullet{}^*) \longrightarrow \cdots,$$

where we abbreviate $(i^\sharp)^*_n, (\partial^\sharp)^\sharp_n$ by i^*_n, ∂^*_n, respectively. As the boundary operators of $Z_\bullet, \overline{B}_\bullet$ are zero-maps, we have

$$H^{n-1}(Z_\bullet{}^*) = Z^*_{n-1},$$

$$H^n(\overline{B}_\bullet{}^*) = B^*_{n-1}, \quad H^{n+1}(\overline{B}_\bullet{}^*) = B^*_n,$$

and $\delta^*_{(n)}$ coincides with the transposition

$$j^*_n : Z^*_n \longrightarrow B^*_n$$

of the inclusion map $j_n : B_n \longrightarrow Z_n$, where we set $B_n = \operatorname{Im} \partial_{n+1} \subset Z_n$. As the objects we consider are vector spaces over $\mathbb{C}$, the subspace B_n of Z_n is a direct summand, and hence j^*_n is surjective. From this, we have

$$\operatorname{Im} \partial^*_n \cong H^n(\overline{B}_\bullet{}^*)/\operatorname{Im} \delta^*_{(n-1)} = 0$$

which implies that i^*_n is injective. Hence, we obtain

$$H^n(C_\bullet^*) \simeq \operatorname{Im} i^*_n \simeq \operatorname{Ker} \delta^*_{(n)}$$

$$= \left\{ \varphi \in Z^*_n \mid j^*_n(\varphi) = 0 \right\}.$$

On the other hand, by taking the dual of the short exact sequence of the vector spaces

$$0 \longrightarrow B_n \longrightarrow Z_n \longrightarrow H_n(C_\bullet) \longrightarrow 0,$$

we again obtain a short exact sequence

$$0 \longrightarrow \mathrm{Hom}_\mathbb{C}(H_n(C_\bullet), \mathbb{C}) \longrightarrow Z_n^* \xrightarrow{j_n^*} B_n^* \longrightarrow 0.$$

From this, we obtain

$$H^n(C_\bullet{}^*) \simeq \mathrm{Hom}_\mathbb{C}(H_n(C_\bullet), \mathbb{C})$$

and this is exacly what we were looking for.

Here, taking a singular chain complex of M over $\mathbb{C}$ as $C_\bullet$, we obtain the universal coefficient theorem of cohomology:

Lemma 2.3. *There is a natural isomorphism*

$$H^p(M, \mathbb{C}) \xrightarrow{\sim} \mathrm{Hom}_\mathbb{C}(H_p(M, \mathbb{C}), \mathbb{C}). \qquad (2.8)$$

Via this isomorphism, the cohomology group $H^p(M, \mathbb{C})$ can be regarded as the dual of the homology group $H_p(M, \mathbb{C})$. Hence, the value $\langle [\sigma], [\varphi] \rangle$ of a cohomology class $[\varphi] \in H^p(M, \mathbb{C})$ at a homology class $[\sigma] \in H_p(M, \mathbb{C})$ is determined and the bilinear form

$$H_p(M, \mathbb{C}) \times H^p(M, \mathbb{C}) \longrightarrow \mathbb{C} \qquad (2.9)$$

$$([\sigma], [\varphi]) \longrightarrow \langle [\sigma], [\varphi] \rangle$$

is defined. Lemma 2.3 asserts that this bilinear form is non-degenerate.

2.2.2 Current

Expressing the above bilinear form concretely as the integral $\int_\sigma \varphi$ of a C^∞ p-form $\varphi \in \mathcal{A}^p(M)$ over a cycle σ is the key to the de Rham theory. For this purpose, de Rham introduced a very big space which contains both C^∞ differential forms and smooth singular chains, calling its element a current, and showed that the following bilinear form is non-degenerate:

$$H_p(M, \mathbb{C}) \times \frac{\{\varphi \in \mathcal{A}^p(M) | d\varphi = 0\}}{d\mathcal{A}^{p-1}(M)} \longrightarrow \mathbb{C}$$

$$([\sigma], \quad \varphi \mod d\mathcal{A}^{p-1}(M)) \longrightarrow \int_\sigma \varphi.$$

Here, let us briefly explain about currents. We remark that, by definition, M is an open subset of $\mathbb{C}^n$. We introduce the C^∞-topology on the set $\mathcal{A}_c^0(M)$

of all C^∞ functions on M with compact support as follows. A sequence φ_m, $m = 1, 2, \ldots$ of $\mathcal{A}_c^0(M)$ is said to converge to 0 in C^∞-topology if it satisfies the following conditions: there exists a compact subset $K \subset M$ such that supp $\varphi_m \subset K$, $m = 1, 2, \ldots$, and for any multi-index $\alpha = (\alpha_1, \ldots, \alpha_n) \in \mathbb{Z}_{\geq 0}^n$, $\frac{\partial^{|\alpha|}\varphi_m}{\partial u_1^{\alpha_1}\cdots\partial u_n^{\alpha_n}}$ converges uniformly to 0 on K, where we set $|\alpha| = \sum_{i=1}^m \alpha_i$.

Since the space $\mathcal{A}_c^p(M)$ of C^∞ p-forms on M with compact support is isomorphic to the direct product of $\binom{n}{p}$-copies of $\mathcal{A}_c^0(M)$, we introduce the topology of the product space $\mathcal{A}_c^0(M)^{\binom{n}{p}}$ on $\mathcal{A}_c^p(M)$ and call it the C^∞-topology on $\mathcal{A}_c^p(M)$. Now, T is called a p-dimensional (degree $(2n-p)$) current if T is a continuous linear form on $\mathcal{A}_c^p(M)$ with respect to the C^∞-topology. Then, a C^∞ $(2n-p)$-form $\alpha \in \mathcal{A}^{2n-p}(M)$ defines a p-dimensional current by

$$T_\alpha : \varphi \in \mathcal{A}_c^p(M) \longrightarrow \int_M \alpha \wedge \varphi \in \mathbb{C},$$

and a smooth locally finite p-chain $\sigma \in C_p^{lf}(M, \mathbb{C})$ also defines a p-dimensional current by

$$T_\sigma : \varphi \in \mathcal{A}_c^p(M) \longrightarrow \int_\sigma \varphi \in \mathbb{C}.$$

We denote the space of degree p $((2n-p)$-dimensional) currents on M by $\mathcal{K}^p(M)$. By the above consideration, we see that

$$\mathcal{A}^p(M) \subset \mathcal{K}^p(M), \quad C_p^{lf}(M, \mathbb{C}) \subset \mathcal{K}^{2n-p}(M). \tag{2.10}$$

For a p-dimensional current T, we set

$$dT(\varphi) = (-1)^{p+1}T(d\varphi), \quad \varphi \in \mathcal{A}_c^{2n-p-1}(M).$$

By Stokes' theorem, one can easily verify $dT_\alpha = T_{d\alpha}$ for $\alpha \in \mathcal{A}^p(M)$ and $T_{\partial\sigma} = (-1)^{p+1}dT_\sigma$ for $\sigma \in C_p^{lf}(M, \mathbb{C})$. Hence, (2.10) defines the inclusions of complexes:

$$(\mathcal{A}^\bullet(M), d) \subset (\mathcal{K}^\bullet(M), d),$$

$$(C_\bullet^{lf}(M), \pm\partial) \subset (\mathcal{K}^\bullet(M), d).$$

The de Rham theory asserts that the above inclusions of complexes induce the isomorphisms between cohomologies:

$$H^p(\mathcal{A}^\bullet(M), d) \xrightarrow{\sim} H^p(\mathcal{K}^\bullet(M), d), \tag{2.11}$$

$$H_p^{lf}(M, \mathbb{C}) \xrightarrow{\sim} H^{2n-p}(\mathcal{K}^\bullet(M), d). \tag{2.12}$$

2.2.3 Current with Compact Support

Let D be an open subset of M. A p-dimensional current T is said to be zero on D if, for any p-form $\varphi \in A_c^p(M)$ whose support is included in D, one always has $T(\varphi) = 0$. Then, the maximal open subset on which $T = 0$ can be defined and its complement is called the support of T. The support of T is a closed subset of M. We denote the space of degree p currents on M with compact support by $\mathcal{K}_c^p(M)$. By an argument similar to that above, we obtain the inclusions of complexes:

$$(\mathcal{A}_c^\bullet(M), d) \subset (\mathcal{K}_c^\bullet(M), d),$$

$$(C_\bullet(M, \mathbb{C}), \pm\partial) \subset (\mathcal{K}_c^\bullet(M), d).$$

The de Rham theory asserts that the above inclusions of complexes induce the isomorphisms between cohomologies:

$$H^p(\mathcal{A}_c^\bullet(M), d) \xrightarrow{\sim} H^p(\mathcal{K}_c^\bullet(M), d), \tag{2.13}$$

$$H_p(M, \mathbb{C}) \xrightarrow{\sim} H^{2n-p}(\mathcal{K}_c^\bullet(M), d). \tag{2.14}$$

2.2.4 Sheaf Cohomology

Let us explain that the isomorphisms (2.11), (2.13) can be understood uniformly by using sheaf theory. We denote the sheaf of germs of degree p currents on M by $\mathcal{K}_M^p$. It is known that the following three complexes of sheaves are, indeed, exact:

$$0 \longrightarrow \mathbb{C} \longrightarrow \mathcal{A}_M^0 \xrightarrow{d} \mathcal{A}_M^1 \xrightarrow{d} \cdots \longrightarrow \mathcal{A}_M^{2n} \longrightarrow 0, \tag{2.15}$$

$$0 \longrightarrow \mathbb{C} \longrightarrow \mathcal{K}_M^0 \xrightarrow{d} \mathcal{K}_M^1 \xrightarrow{d} \cdots \longrightarrow \mathcal{K}_M^{2n} \longrightarrow 0, \tag{2.16}$$

$$0 \longrightarrow \mathbb{C} \longrightarrow \Omega_M^0 \xrightarrow{d} \Omega_M^1 \xrightarrow{d} \cdots \longrightarrow \Omega_M^n \longrightarrow 0. \tag{2.17}$$

Let $\{\rho_\nu(u)\}$ be a C^∞ partition of unity of associated to a locally finite open cover $\mathcal{U} = \{U_\nu\}$ of M: supp $\rho_\nu \subset\subset U_\nu$, $\Sigma\rho_\nu(u) = 1$. For sections $\varphi \in \Gamma(U_\nu, \mathcal{A}_M^p)$, $T \in \Gamma(U_\nu, \mathcal{K}_M^p)$ over U_ν, one has $\rho_\nu \cdot \varphi \in \Gamma(U_\nu, \mathcal{A}_M^p)$, $\rho_\nu T \in \Gamma(U_\nu, \mathcal{K}_M^p)$ from which one can show

$$H^q(M, \mathcal{A}_M^p) = 0, \quad q \geq 1, \tag{2.18}$$

$$H^q(M, \mathcal{K}_M^p) = 0, \quad q \geq 1. \tag{2.19}$$

Moreover, since M is an open subset of $\mathbb{C}^n$, as it is the complement of an algebraic hypersurface of codimension 1 in $\mathbb{C}^n$, it is a Stein manifold, hence, one may conclude $H^q(M, \mathcal{O}_M) = 0$, $q \geq 1$. (This is a particular case of the fact called Theorem B for Stein manifolds. For example, see [Gra-Rem].) On the other hand, $\Omega_M^p \simeq \mathcal{O}_M^{\binom{n}{p}}$ implies

$$H^q(M, \Omega_M^p) = 0, \quad q \geq 1. \tag{2.20}$$

From these facts, one can derive the following three isomorphisms:

$$H^p(M, \mathbb{C}) \simeq \{\varphi \in \Omega^p(M) | d\varphi = 0\}/d\Omega^{p-1}(M) \tag{2.21}$$

$$\xrightarrow{\sim} \{\varphi \in \mathcal{A}^p(M) | d\varphi = 0\}/d\mathcal{A}^{p-1}(M)$$

$$\xrightarrow{\sim} \{T \in \mathcal{K}^p(M) | dT = 0\}/d\mathcal{K}^{p-1}(M).$$

This can be shown as follows. Since the proofs are all similar, we show the first isomorphism for $p = 2$. By (2.17), we obtain short exact sequences of sheaves

$$0 \longrightarrow \mathbb{C} \longrightarrow \Omega_M^0 \xrightarrow{d} d\Omega_M^0 \longrightarrow 0,$$

$$0 \longrightarrow d\Omega_M^0 \longrightarrow \Omega_M^1 \xrightarrow{d} d\Omega_M^1 \longrightarrow 0.$$

Passing to exact cohomology sequences, by (2.20), we obtain the exact sequences:

$$0 \longrightarrow H^1(M, d\Omega_M^0) \longrightarrow H^2(M, \mathbb{C}) \longrightarrow 0,$$

$$H^0(M, \Omega_M^1) \xrightarrow{d} H^0(M, d\Omega_M^1) \longrightarrow H^1(M, d\Omega_M^0) \longrightarrow 0.$$

By $H^0(M, d\Omega_M^1) = \{\varphi \in \Omega^2(M) | d\varphi = 0\}$, we obtain

$$H^2(M, \mathbb{C}) \simeq H^1(M, d\Omega_M^0)$$

$$\simeq \{\varphi \in \Omega^2(M) | d\varphi = 0\}/d\Omega^1(M).$$

For the other cases, it is enough to discuss them in exactly the same way as above.

By (2.21), we have shown the isomorphism (2.11). Now, the isomorphism

$$H^p(M, \mathbb{C}) \xrightarrow{\sim} H^p(\mathcal{K}^\bullet(M), d) \xleftarrow{\sim} H_{2n-p}^{lf}(M, \mathbb{C}),$$

obtained by combining the above isomorphism with (2.12), is the Poincaré duality. We remark that one can derive the isomorphism (2.12) from one of the above isomorphisms of cohomologies and the Poincaré duality. Instead of extending naïvely the argument of [deR2] to the twisted de Rham theory, although it might be an indirect proof, we show a twisted version of (2.12) in § 2.2.11 by applying the above remark and some results from algebraic topology on (co)homology with coefficients in a local system.

2.2.5 The Case of Compact Support

We denote the cohomology of M over $\mathbb{C}$ with compact support by $H_c^p(M, \mathbb{C})$. Since it is known that the cohomology with compact support of sheaves $\mathcal{A}_M^p$ and $\mathcal{K}_M^p$ vanish for positive degree (see, e.g., [God] p.159, p.174), from (2.15), (2.16), one can derive the isomorphisms:

$$H_c^p(M, \mathbb{C}) \xrightarrow{\sim} \{\varphi \in \mathcal{A}_c^p(M) | d\varphi = 0\}/d\mathcal{A}_c^{p-1}(M)$$

$$\xrightarrow{\sim} \{T \in \mathcal{K}_c^p(M) | dT = 0\}/d\mathcal{K}_c^{p-1}(M)$$

in exactly the same way as in § 2.2.4. The isomorphism, obtained by combining this isomorphism with (2.14),

$$H_c^p(M, \mathbb{C}) \xrightarrow{\sim} H^p(\mathcal{K}_c^\bullet(M), d) \xleftarrow{\sim} H_{2n-p}(M, \mathbb{C})$$

is nothing but the Poincaré duality. Here, the same remark as at the end of § 2.2.4 also applies.

2.2.6 De Rham's Theorem

We summarize the above results in the following lemma:

Lemma 2.4. *The following isomorphisms exist:*

(1) $H^p(M, \mathbb{C}) \simeq \begin{Bmatrix} H^p(\mathcal{A}^\bullet(M), d) \\ \uparrow \wr \\ H^p(\Omega^\bullet(M), d) \end{Bmatrix} \xrightarrow{\sim} H^p(\mathcal{K}^\bullet(M), d) \xleftarrow{\sim} H_{2n-p}^{lf}(M, \mathbb{C}).$

(2) $H_c^p(M, \mathbb{C}) \simeq H^p(\mathcal{A}_c^\bullet(M), d) \xrightarrow{\sim} H^p(\mathcal{K}_c^\bullet(M), d) \xleftarrow{\sim} H_{2n-p}(M, \mathbb{C}).$

Remark 2.1. Since M is an affine algebraic variety, besides the C^∞ de Rham complex $(\mathcal{A}^\bullet(M), d)$ and the analytic de Rham complex $(\Omega^\bullet(M), d)$ which describe $H^p(M, \mathbb{C})$, one can also consider the algebraic de Rham complex $(\Omega^\bullet(*D), d)$, where $\Omega^p(*D)$ is the space of meromorphic p-forms which are holomorphic on M and which admit poles along D. Looking at Lemma 2.4 (1), the reader may wonder whether one can describe $H^p(M, \mathbb{C})$ by $(\Omega^\bullet(*D), d)$ or not. This problem will be discussed in § 2.4. See [B-T] about more details.

2.2.7 Duality

By using an expression of $H^p(M,\mathbb{C})$ obtained in Lemma 2.4 and (2.9), combining this, smooth singular chains and C^∞ p-forms from the viewpoint of a current, we obtain:

Lemma 2.5. *The following bilinear forms are non-degenerate:*

(1) $H_p(M,\mathbb{C}) \times H^p(\mathcal{A}^\bullet(M),d) \longrightarrow \mathbb{C}$

$$([\sigma],[\varphi]) \longrightarrow \int_\sigma \varphi,$$

(2) $H^{2n-p}(\mathcal{A}_c^\bullet(M),d) \times H^p(\mathcal{A}^\bullet(M),d) \longrightarrow \mathbb{C}$

$$([\alpha],[\beta]) \longrightarrow \int_M \alpha \wedge \beta.$$

These are an outline of the de Rham theory. Next, rewriting these results step by step, we will look for the corresponding results for the twisted de Rham theory.

2.2.8 Integration over a Simplex

Denote by $\Delta^p = \langle e_0 \cdots e_p \rangle$ the standard p-simplex having vertices $e_i = (0,\ldots,\overset{i+1}{1},\ldots,0) \in \mathbb{R}^{p+1}$, $0 \le i \le p$, and by σ a smooth singular p-simplex on M, i.e., a C^∞-map $\sigma : \Delta^p \to M$. Let $S^p(M,\mathcal{L}_\omega)$ be the set of all functions associating each singular p-simplex σ on M to an element $u(\sigma)$ of $(\mathcal{L}_\omega)_{\sigma(e_0)}$. The coboundary operator $\delta : S^p(M,\mathcal{L}_\omega) \to S^{p+1}(M,\mathcal{L}_\omega)$ is defined, for each singular $(p+1)$-simplex τ, by

$$(\delta u)(\tau) := \gamma_*^{-1} u(\partial_0 \tau) + \sum_{i=1}^{p+1} (-1)^i u(\partial_i \tau),$$

where γ_* is the isomorphism $(\mathcal{L}_\omega)_{\tau(e_0)} \overset{\sim}{\to} (\mathcal{L}_\omega)_{\tau(e_1)}$ determined by the path γ obtained by restricting τ to the segment $\langle e_0 e_1 \rangle$ and $\partial_i \tau$ is the singular p-simplex $\tau|_{\langle e_0 \cdots \widehat{e}_i \cdots e_{p+1} \rangle}$. One can show $\delta^2 = 0$ by which the singular cochain complex $S^\bullet(M,\mathcal{L}_\omega)$ on M with coefficients in $\mathcal{L}_\omega$ is defined. The cohomology of this cochain complex is denoted by $H^\bullet(M,\mathcal{L}_\omega)$ and is called the singular cohomology of M with coefficients in $\mathcal{L}_\omega$. Now, for a C^∞ p-form φ on M, φ defines an element $I(\varphi)$ of $S^p(M,\mathcal{L}_\omega)$ via an integral: for each singular p-simplex σ, we set

$$I(\varphi)(\sigma) := \int_\sigma U \cdot \varphi.$$

Indeed, we show that it takes a value in $(\mathcal{L}_\omega)_{\sigma(e_0)}$. With the notation used in § 2.1.5, we suppose $\sigma(\Delta^p) \subset U_{\mu \cap} U_\nu$. With the branch h_μ^{-1} on U_μ, this integral becomes $\int_\sigma h_\mu^{-1} \cdot \varphi$, and with h_ν^{-1} on U_ν, it becomes $\int_\sigma h_\nu^{-1} \cdot \varphi$. By $h_\mu = g_{\mu\nu} h_\nu$, $g_{\mu\nu} \in \mathbb{C}$, we have

$$\int_\sigma h_\mu^{-1} \cdot \varphi = g_{\mu\nu}^{-1} \int_\sigma h_\nu^{-1} \cdot \varphi$$

by which $I(\varphi)(\sigma)$ can be regarded as an element of $(\mathcal{L}_\omega)_{\sigma(e_0)}$. By Theorem 2.1 and the definition of δ,

$$I : (\mathcal{A}^\bullet(M), \nabla_\omega) \longrightarrow (S^\bullet(M, \mathcal{L}_\omega), \delta)$$

becomes a cochain map, hence, passing to their cohomologies, we obtain the map

$$I_* : H^\bullet(\mathcal{A}^\bullet(M), \nabla_\omega) \longrightarrow H^\bullet(M, \mathcal{L}_\omega).$$

Later in § 2.2.10, this is shown to be an isomorphism.

Let $\mathcal{L}_\omega$ be the rank 1 local system on M defined in § 2.1.5 , and we denote the singular chain complex on M with coefficients in the dual local system $\mathcal{L}_\omega^\vee$ by $S_\bullet(M, \mathcal{L}_\omega^\vee)$. Notice that there is a natural isomorphism ([Sp])

$$S^\bullet(M, \mathrm{Hom}_\mathbb{C}(\mathcal{L}_\omega^\vee, \mathbb{C})) \simeq \mathrm{Hom}_\mathbb{C}(S_\bullet(M, \mathcal{L}_\omega^\vee), \mathbb{C}).$$

Applying Lemma 2.2 to $C_\bullet = S_\bullet(M, \mathcal{L}_\omega^\vee)$, the above remark implies the universal coefficient theorem for cohomology with coefficients in a local system:

Lemma 2.6. *There is a natural isomorphism:*

$$H^p(M, \mathcal{L}_\omega) \simeq \mathrm{Hom}_\mathbb{C}(H_p(M, \mathcal{L}_\omega^\vee), \mathbb{C}).$$

Via this isomorphism, the cohomology group $H^p(M, \mathcal{L}_\omega)$ can be regarded as the dual space of the homology group $H_p(M, \mathcal{L}_\omega^\vee)$. Hence, the value $\langle [\sigma], [\varphi] \rangle$ of a cohomology class $[\varphi] \in H^p(M, \mathcal{L}_\omega)$ at a homology class $[\sigma] \in H_p(M, \mathcal{L}_\omega^\vee)$ is determined which defines the bilinear form:

$$H_p(M, \mathcal{L}_\omega^\vee) \times H^p(M, \mathcal{L}_\omega) \longrightarrow \mathbb{C} \qquad (2.22)$$

$$([\sigma], [\varphi]) \longrightarrow \langle [\sigma], [\varphi] \rangle.$$

The above lemma says that this bilinear form is non-degenerate.

2.2.9 Twisted Chain

To adapt what we have discussed in § 2.2.2 to the twisted de Rham theory, we should formulate how to regard a twisted chain as a current. Given a smooth triangulation of M, let Δ be a smooth oriented p-simplex. For a branch U_Δ of the multivalued function U on Δ, we symbolized it as $\Delta \otimes U_\Delta$. $\Delta \otimes U_\Delta$ defines a $T_{\Delta \otimes U_\Delta}$ as follows:

$$(T_{\Delta \otimes U_\Delta}, \varphi) := \int_{\Delta \otimes U_\Delta} U \cdot \varphi \qquad (\varphi \in \mathcal{A}_c^p(M)).$$

For any locally finite twisted p-chain $\Sigma c_\Delta \Delta \otimes U_\Delta$, we extend the above definition additively and obtain the inclusion

$$C_p^{lf}(M, \mathcal{L}_\omega^\vee) \subset \mathcal{K}^{2n-p}(M).$$

As the support of the current $T_{\Delta \otimes U_\Delta}$ is Δ, each element of $C_p(M, \mathcal{L}_\omega^\vee)$ defines a current with compact support, and we obtain the inclusion

$$C_p(M, \mathcal{L}_\omega^\vee) \subset \mathcal{K}_c^{2n-p}(M).$$

Next, let us see which operator on the complex of currents corresponds to the boundary operator ∂_ω of the twisted chain complex. For a locally finite twisted p-chain $\sigma = \Sigma c_\Delta \Delta \otimes U_\Delta$ and $\varphi \in \mathcal{A}_c^{p-1}(M)$, we have

$$(T_{\partial_\omega \sigma}, \varphi) = \int_{\partial_\omega \sigma} U\varphi = \Sigma c_\Delta \int_{\partial_\omega (\Delta \otimes U_\Delta)} U\varphi$$

$$= \Sigma c_\Delta \int_{\Delta \otimes U_\Delta} U\nabla_\omega \varphi$$

$$= (T_\sigma, \nabla_\omega \varphi)$$

$$= (T_\sigma, d\varphi + \omega \wedge \varphi)$$

$$= ((-1)^{p+1} dT_\sigma, \varphi) + (T_\sigma \wedge \omega, \varphi)$$

$$= (-1)^{p+1}(dT_\sigma - \omega \wedge T_\sigma, \varphi)$$

$$= (-1)^{p+1}(\nabla_{-\omega} T_\sigma, \varphi).$$

Hence, we obtain the following lemma:

Lemma 2.7. $T_{\partial_\omega \sigma} = (-1)^{p+1}\nabla_{-\omega} T_\sigma.$
Here, σ is either an element of $C_p(M, \mathcal{L}_\omega^\vee)$ or $C_p^{lf}(M, \mathcal{L}_\omega^\vee).$

By this lemma, we obtain the inclusions of complexes:

$$(C_\bullet^{lf}(M, \mathcal{L}_\omega^\vee), \pm\partial_\omega) \subset (\mathcal{K}^\bullet(M), \nabla_{-\omega}), \qquad (2.23)$$

$$(C_\bullet(M, \mathcal{L}_\omega^\vee), \pm\partial_\omega) \subset (\mathcal{K}_c^\bullet(M), \nabla_{-\omega}). \qquad (2.24)$$

On the other hand, we clearly have the inclusions of complexes

$$(\Omega^\bullet(M), \nabla_\omega) \subset (\mathcal{A}^\bullet(M), \nabla_\omega) \subset (\mathcal{K}^\bullet(M), \nabla_\omega),$$

$$(\mathcal{A}_c^\bullet(M), \nabla_\omega) \subset (\mathcal{K}_c^\bullet(M), \nabla_\omega).$$

We would like to compare the cohomology of these complexes. For that, let us construct a twisted version of § 2.2.4 modeled on the de Rham theory.

2.2.10 Twisted Version of § 2.2.4

First, we remark the following fact. With the notation used in § 2.1.5, for a holomorphic p-form φ on $U_\mu \cap U_\nu$, we have $h_\mu^{-1}\varphi = g_{\mu\nu}^{-1} \cdot h_\nu^{-1}\varphi$. This means that, fixing a branch of U, denoted by U for simplicity, on a sufficiently small simply connected neighborhood of each point of M, $U\varphi$ is a holomorphic p-form with values in $\mathcal{L}_\omega$ and we have $U\varphi \in \Omega_M^p \otimes_{\mathbb{C}} \mathcal{L}_\omega$. Since U is not zero on each point of M, $\Omega_M^p \xrightarrow{\sim} \Omega_M^p \otimes_{\mathbb{C}} \mathcal{L}_\omega, \varphi \mapsto U\varphi$ is an isomorphism. Moreover, by $d(U\varphi) = U\nabla_\omega\varphi$, we obtain the commutative diagram:

$$
\begin{array}{ccc}
\Omega_M^p & \xrightarrow{\nabla_\omega} & \Omega_M^{p+1} \\
\wr \downarrow & & \downarrow \wr \\
\Omega_M^p \otimes_{\mathbb{C}} \mathcal{L}_\omega & \xrightarrow{d} & \Omega_M^{p+1} \otimes_{\mathbb{C}} \mathcal{L}_\omega
\end{array}
$$

Since the transition functions $\{g_{\mu\nu}^{-1}\}$ of $\mathcal{L}_\omega$ are constant, we can take the tensor product of (2.17) and $\mathcal{L}_\omega$ over $\mathbb{C}$ and the following sequence becomes exact:

$$0 \longrightarrow \mathcal{L}_\omega \longrightarrow \Omega_M^0 \otimes_{\mathbb{C}} \mathcal{L}_\omega \xrightarrow{d} \Omega_M^1 \otimes_{\mathbb{C}} \mathcal{L}_\omega \longrightarrow \cdots\cdots.$$

Hence, combining this with the above commutative diagram, we obtain the exact sequence of sheaves:

$$0 \to \mathcal{L}_\omega \to \Omega_M^0 \xrightarrow{\nabla_\omega} \Omega_M^1 \to \cdots \xrightarrow{\nabla_\omega} \Omega_M^n \to 0. \qquad (2.25)$$

Similarly, we also obtain

$$0 \to \mathcal{L}_\omega \to \mathcal{A}_M^0 \overset{\nabla_\varphi}{\to} \mathcal{A}_M^1 \to \cdots \overset{\nabla_\varphi}{\to} \mathcal{A}_M^{2n} \to 0, \tag{2.26}$$

$$0 \to \mathcal{L}_\omega \to \mathcal{K}_M^0 \overset{\nabla_\varphi}{\to} \mathcal{K}_M^1 \to \cdots \overset{\nabla_\varphi}{\to} \mathcal{K}_M^{2n} \to 0. \tag{2.27}$$

After the above three exact sequences, the rest of arguments work in exactly the same way as in § 2.2.4 and 2.2.5. (This is an advantage of the sheaf theory!) Thus, we have the following isomorphisms:

$$H^p(M, \mathcal{L}_\omega) \simeq \{\varphi \in \Omega^p(M) | \nabla_\omega \varphi = 0\} / \nabla_\omega \Omega^{p-1}(M) \tag{2.28}$$

$$\simeq \{\varphi \in \mathcal{A}^p(M) | \nabla_\omega \varphi = 0\} / \nabla_\omega \mathcal{A}^{p-1}(M)$$

$$\simeq \{T \in \mathcal{K}^p(M) | \nabla_\omega \varphi = 0\} / \nabla_\omega \mathcal{K}^{p-1}(M),$$

$$H_c^p(M, \mathcal{L}_\omega) \simeq \{\varphi \in \mathcal{A}_c^p(M) | \nabla_\omega \varphi = 0\} / \nabla_\omega \mathcal{A}_c^{p-1}(M) \tag{2.29}$$

$$\simeq \{T \in \mathcal{K}_c^p(M) | \nabla_\omega \varphi = 0\} / \nabla_\omega \mathcal{K}_c^{p-1}(M).$$

2.2.11 Poincaré Duality

Combining the homomorphisms of cohomologies induced from the inclusions of complexes (2.23) and (2.24) with the isomorphisms (2.28) and (2.29) (for (2.23) and (2.24), we replace $\mathcal{L}_\omega^\vee$ by $\mathcal{L}_\omega$), we obtain

$$H^p(M, \mathcal{L}_\omega) \xrightarrow[\;(2.28)\;]{\sim} H^p(\mathcal{K}^\bullet(M), \nabla_\omega) \tag{2.30}$$

$$\text{P.D.} \searrow^{\sim} \qquad \uparrow (2.23)$$

$$H_{2n-p}^{lf}(M, \mathcal{L}_\omega),$$

$$H_c^p(M, \mathcal{L}_\omega) \xrightarrow[\;(2.29)\;]{\sim} H^p(\mathcal{K}_c^\bullet(M), \nabla_\omega) \tag{2.31}$$

$$\text{P.D.} \searrow^{\sim} \qquad \uparrow (2.24)$$

$$H_{2n-p}(M, \mathcal{L}_\omega).$$

Here, P.D. denotes the Poincaré duality. Since the composition of maps (2.23) and (2.28) in (2.30) is just the isomorphism of the Poincaré duality, we obtain the isomorphism ([Br]):

$$H_{2n-p}^{lf}(M, \mathcal{L}_\omega) \longrightarrow H^p(\mathcal{K}^\bullet(M), \nabla_\omega). \tag{2.32}$$

Similarly, we also have

$$H_{2n-p}(M, \mathcal{L}_\omega) \longrightarrow H^p(\mathcal{K}_c^\bullet(M), \nabla_\omega). \tag{2.33}$$

Let us summarize these results in the following lemma:

Lemma 2.8. *The following isomorphisms exist:*

$$1.\ H^p(M, \mathcal{L}_\omega) \simeq \begin{Bmatrix} H^p(\mathcal{A}^\bullet(M), \nabla_\omega) \\ \uparrow \wr \\ H^p(\Omega^\bullet(M), \nabla_\omega) \end{Bmatrix} \xrightarrow{\sim} H^p(\mathcal{K}^\bullet(M), \nabla_\omega) \xleftarrow{\sim} H^{lf}_{2n-p}(M, \mathcal{L}_\omega),$$

$$2.\ H^p_c(M, \mathcal{L}_\omega) \simeq H^p(\mathcal{A}_c^\bullet(M), \nabla_\omega) \xrightarrow{\sim} H^p(\mathcal{K}_c^\bullet(M), \nabla_\omega) \xleftarrow{\sim} H_{2n-p}(M, \mathcal{L}_\omega).$$

2.2.12 Reformulation

From the non-degenerate bilinear form (2.22), the above isomorphisms and the viewpoint of twisted chains as currents, we obtain the following lemma:

Lemma 2.9. *The following bilinear forms are non-degenerate:*

(1) $H_p(M, \mathcal{L}_\omega^\vee) \times H^p(\mathcal{A}^\bullet(M), \nabla_\omega) \longrightarrow \mathbb{C},$

$$([\sigma], [\varphi]) \qquad \longrightarrow \int_\sigma U \cdot \varphi.$$

(2) $H^p(\mathcal{A}_c^\bullet(M), \nabla_\omega) \times H_p^{lf}(M, \mathcal{L}_\omega^\vee) \longrightarrow \mathbb{C},$

$$([\psi], [\tau]) \qquad \longrightarrow \int_\tau U \cdot \psi.$$

(3) $H^{2n-p}(\mathcal{A}_c^\bullet(M), \nabla_{-\omega}) \times H^p(\mathcal{A}^\bullet(M), \nabla_\omega) \longrightarrow \mathbb{C},$

$$([\alpha], [\beta]) \qquad \longrightarrow \int_M \alpha \wedge \beta.$$

(4) $H_p(M, \mathcal{L}_\omega^\vee) \times H^{lf}_{2n-p}(M, \mathcal{L}_\omega) \longrightarrow \mathbb{C}$

$$([\sigma], [\tau]) \qquad \longrightarrow \text{intersection number } [\upsilon] \cdot [\tau].$$

Remark 2.2. The intersection theory of twisted cycles stated in (4) plays an important role in the global theory of hypergeometric functions of type $(n+1, m+1)$ explained in Chapter 3.

2.2.13 Comparison of Cohomologies

Giving a sense of twisted homology groups to the reader, here we state two facts comparing them with ordinary homology groups. One says that the alternating sum $\sum_{p=0}^{2n}(-1)^p \dim H_p(M, \mathcal{L}_\omega^\vee)$ always coincides with the Euler characteristic of M. This fact and the vanishing theorem of twisted cohomology stated in § 2.8 are important facts that provide us a way to calculate the dimension of $H^n(M, \mathcal{L}_\omega)$ which appears in integral representations of hypergeometric functions. An explicit formula of $\chi(M)$ for a special class of M will be explained in § 2.2.14.

By Hironaka's theorem on resolution of singularities, M can be embedded in a projective variety $\overline{M}$ in such a way that $\overline{M} \setminus M =: D$ is a normal crossing divisor; we may assume that $\overline{M} \setminus M$ is normal crossing. Introduce a Riemannian metric on $\overline{M}$ and let $T_\varepsilon(D)$ be the tubular neighborhood of D formed by the points whose distance with D is less than ε. Then, $M_\varepsilon := \overline{M} \setminus T_\varepsilon(D)$ is compact and we have the formula:

$$H_p(M, \mathbb{C}) = \lim_{\varepsilon \to 0} H_p(M_\varepsilon, \mathbb{C}).$$

Triangulating M_ε to a finite simplicial complex K_ε, as is well-known, one has the formula:

$$\chi(K_\varepsilon) = \sum_{p=0}^{2n}(-1)^p \dim C_p(K_\varepsilon, \mathbb{C}) \tag{2.34}$$

for the Euler characteristic $\chi(K_\varepsilon) := \sum_{p=0}^{2n}(-1)^p \dim H_p(K_\varepsilon, \mathbb{C})$. Denoting the p-simplices of K_ε by $\sigma_1^p, \ldots, \sigma_{\lambda_p}^p$, by the definition of twisted chain group, we have

$$C_p(K_\varepsilon, \mathcal{L}_\omega^\vee) = \bigoplus_{i=1}^{\lambda_p} \sigma_i^p \otimes \mathcal{L}_{\omega, <\sigma_i^p>}^\vee, \tag{2.35}$$

where $\langle \sigma_i^p \rangle$ is the barycenter of σ_i^p and $\mathcal{L}_{\omega, \langle \sigma_i^p \rangle}^\vee$ is the fiber of the local system $\mathcal{L}_\omega^\vee$ at the point $\langle \sigma_i^p \rangle$. Since the twisted homology group is the homology group of the complex

$$0 \longrightarrow C_{2n}(K_\varepsilon, \mathcal{L}_\omega^\vee) \xrightarrow{\partial_\omega} \cdots\cdots\cdots \xrightarrow{\partial_\omega} C_0(K_\varepsilon, \mathcal{L}_\omega^\vee) \longrightarrow 0,$$

we obtain

$$\sum_{p=0}^{2n}(-1)^p \dim H_p(K_\varepsilon, \mathcal{L}_\omega^\vee) = \sum_{p=0}^{2n}(-1)^p \dim C_p(K_\varepsilon, \mathcal{L}_\omega^\vee) \qquad (2.36)$$

$$= \sum_{p=0}^{2n}(-1)^p \lambda_p$$

$$= \chi(K_\varepsilon) \qquad \text{(by (2.34))}.$$

By the construction of $|K_\varepsilon| = M_\varepsilon$, we have

$$H_p(M, \mathcal{L}_\omega^\vee) = \lim_{\varepsilon \to 0} H_p(K_\varepsilon, \mathcal{L}_\omega^\vee).$$

Hence, by (2.36) and Lemma 2.6, we obtain the formula:

Theorem 2.2.

$$\sum_{p=0}^{2n}(-1)^p \dim H^p(M, \mathcal{L}_\omega) = \chi(M).$$

Recall that the twisted homology group $H_p(K_\varepsilon, \mathcal{L}_\omega^\vee)$ is defined, by the homomorphisms

$$C_{p+1}(K_\varepsilon, \mathcal{L}_\omega^\vee) \xrightarrow{\partial_\omega^{p+1}} C_p(K_\varepsilon, \mathcal{L}_\omega^\vee) \xrightarrow{\partial_\omega^p} C_{p-1}(K_\varepsilon, \mathcal{L}_\omega^\vee),$$

as the quotient $\mathrm{Ker}\, \partial_\omega^p / \mathrm{Im}\, \partial_\omega^{p+1}$. If we calculate the matrix representations of each $\partial_\omega^{p+1}, \partial_\omega^p$ using (2.35), for a general $\alpha = (\alpha_1, \ldots, \alpha_m)$, we have

$$\mathrm{rank}\, \partial^p \le \mathrm{rank}\, \partial_\omega^p. \qquad (2.37)$$

This is because ∂^p corresponds to ∂_ω^p when $\alpha_1 = \cdots = \alpha_m = 0$ in $\omega = \sum_{j=1}^{m} \alpha_j \frac{dP_j}{P_j}$, but ∂_ω^p is represented by a matrix with coefficients given by polynomials in $\exp(\pm 2\pi\sqrt{-1}\alpha_j)$, $1 \le j \le m$. If $r = \mathrm{rank}\, \partial^p$, then, in the matrix representation of ∂^p, there is an $r \times r$ minor which is non-zero. Hence, around a small neighborhood of $\alpha = 0$, this minor never vanishes, and the rank of ∂_ω^p is in general at least r for such α. $\mathrm{rank}\, \partial^{p+1} \le \mathrm{rank}\, \partial_\omega^{p+1}$ can be shown similarly. On the other hand, since we have

$$\dim \mathrm{Ker}\, \partial_\omega^p = \dim C_p(K_\varepsilon, \mathcal{L}_\omega^\vee) - \mathrm{rank}\, \partial_\omega^p$$

$$\le \lambda_p - \mathrm{rank}\, \partial^p = \dim \mathrm{Ker}\, \partial^p$$

by (2.37), it follows that

$$\dim H_p(K_\varepsilon, \mathcal{L}_\omega^\vee) \le \dim H_p(K_\varepsilon, \mathbb{C}).$$

Proposition 2.1. *For almost all* $\alpha = (\alpha_1, \ldots, \alpha_m)$, *we have*

$$\dim H_p(M, \mathcal{L}_\omega^\vee) \leq \dim H_p(M, \mathbb{C}), \quad 0 \leq p \leq 2n.$$

Remark 2.3. In fact, D. Cohen [Co] showed that the inequality in Proposition 2.1 holds for any α.

2.2.14 Computation of the Euler Characteristic

When the divisor $D_j = \{P_j = 0\}$ defined by a homogeneous polynomial P_j of degree l_j has a particular form, we explain a formula which gives $\chi(M)$ explicitly. First, we introduce some notions. An $(n-s)$-dimensional subvariety V of $\mathbb{P}^n$ is said to be a complete intersection if the homogeneous ideal $I(V)$ of $\mathbb{C}[v_0, \ldots, v_n]$ defining V is generated by s homogeneous polynomials $f_1(v), \ldots, f_s(v)$. Now, if each $f_i(v)$ is of degree a_i, V is represented, by the hypersurface F_i of degree a_i defined by $\{f_i = 0\}$, $1 \leq i \leq s$, as $V = F_{1 \cap \cdots \cap} F_s$. Then, as an application of the Riemann–Roch theorem, the following fact can be shown:

Lemma 2.10. *Fix a positive integer r and let V^n, $n = 0, 1, \ldots$ be a complete intersection subvariety in $\mathbb{P}^{n+r}$ which is represented, by r hypersurfaces F_i of degree a_i, $1 \leq i \leq r$, as the intersection $V = F_{1 \cap \cdots \cap} F_r$. Then, one has*

$$\sum_{n=0}^{\infty} \chi(V^n) z^n = \frac{1}{(1-z)^2} \prod_{i=1}^{r} \frac{a_i}{1 + (a_i - 1)z}. \tag{2.38}$$

Proof. Specializing the formula in Theorem 22.1.1 in Appendix 1 of [Hir] to $k = 0$ and $y = -1$, the result follows.

Let H_0 be the hyperplane at infinity of $\mathbb{P}^n$, and denote the hypersurface of $\mathbb{P}^n$, defined by a divisor $D_j = \{P_j = 0\}$ of $\mathbb{C}^n$, by $\widehat{D}_j$. Here, we impose the following assumption:

Assumption 1. *Setting $\widehat{D}_0 = H_0$, for any $0 \leq s \leq m$ and any $j_1, \ldots, j_s \in \{0, \ldots, m\}$, $\widehat{D}_{j_1 \cap \cdots \cap} \widehat{D}_{j_s}$ is an $(n-s)$-dimensional complete intersection subvariety of $\mathbb{P}^n$. For simplicity, we set $J = \{j_1, \ldots, j_s\}$ and denote this subvariety by $\widehat{D}_J$:*

$$\widehat{D}_J = \widehat{D}_{j_1 \cap \cdots \cap} \widehat{D}_{j_s},$$

we also use the symbol

$$D_J = D_{j_1 \cap \cdots \cap} D_{j_s}.$$

To compute the Euler characteristic $\chi(M)$ of $M = \mathbb{P}^n \setminus \left(H_0 \cup \bigcup_{j=1}^{m} \widehat{D}_j \right)$, let us summarize the additivity of the Euler characteristic in the following lemma.

Lemma 2.11. *For* $J \subset \{1, \ldots, m\}$ *we have*

$$\chi(D_J) = \chi(\widehat{D}_j) - \chi(H_{0\cap}\widehat{D}_J), \tag{2.39}$$

$$\chi\left(\bigcup_{j=1}^{m} D_j\right) = \sum_{j=1}^{m} \chi(D_j) - \sum_{1 \le j_1 < j_2 \le m} \chi(D_{j_1 \cap} D_{j_2}) \tag{2.40}$$

$$+ \cdots + (-1)^{s+1} \sum_{\substack{J \subset \{1, \ldots, m\} \\ |J| = s}} \chi(D_J) + \cdots + (-1)^{m+1} \chi(D_{1\cap \cdots \cap} D_m).$$

Proof. (1) First, we show (2.39). As $D_J = \widehat{D}_J \setminus H_{0\cap}\widehat{D}_J$ and Assumption 1 implies that $\widehat{D}_J$ and $H_{0\cap}\widehat{D}_J$ are compact varieties, one can choose a tubular neighborhood $T_\varepsilon(H_{0\cap}\widehat{D}_J)$ of $H_{0\cap}\widehat{D}_J$ in $\widehat{D}_J$ (cf. § 2.2.13). By the Poincaré duality, we have

$$H_c^p(D_J, \mathbb{C}) \simeq H_{2n-2s-p}(D_J, \mathbb{C}), \quad s = |J|,$$

and by the definition of the cohomology with compact support, we have $H_c^p(D_J, \mathbb{C}) = \lim_{\varepsilon \to 0} H^p(\widehat{D}_J, T_\varepsilon(H_{0\cap}\widehat{D}_J); \mathbb{C})$. On the other hand, since $H_{0\cap}\widehat{D}_J$ is a retraction of $T_\varepsilon(H_{0\cap}\widehat{D}_J)$, the left-hand side of the above formula is equal to $H^p(\widehat{D}_J, H_{0\cap}\widehat{D}_J; \mathbb{C})$. Hence, we have

$$H^p(\widehat{D}_J, H_{0\cap}\widehat{D}_J; \mathbb{C}) \simeq H_{2n-2s-p}(D_J, \mathbb{C}).$$

By making $p = 0, 1, \cdots$ and taking alternating sums, from the exact cohomology sequence associated to the pair $(\widehat{D}_J, H_{0\cap}\widehat{D}_J)$, we obtain

$$\chi(D_J) = \sum_{p=0}^{\infty} (-1)^p \dim H^p(\widehat{D}_J, H_{0\cap}\widehat{D}_J; \mathbb{C})$$

$$= \chi(\widehat{D}_J) - \chi(H_{0\cap}\widehat{D}_J).$$

(2) Second, we show (2.40). First, we prove the case when $m = 2$. The Mayer–Vietoris sequence associated to the inclusions as in Figure 2.4

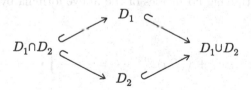

Fig. 2.4

$$\cdots \longrightarrow H_p(D_1 \cap D_2) \longrightarrow H_p(D_1) \oplus H_p(D_2) \longrightarrow H_p(D_1 \cup D_2) \longrightarrow \cdots$$

(cf. [Sp]) (the coefficient $\mathbb{C}$ is omitted) implies $\chi(D_1 \cap D_2) - \{\chi(D_1) + \chi(D_2)\} + \chi(D_1 \cup D_2) = 0$. Hence, we have

$$\chi(D_1 \cup D_2) = \chi(D_1) + \chi(D_2) - \chi(D_1 \cap D_2). \tag{2.41}$$

For $m = 3$, replacing D_2 with $D_2 \cup D_3$, we have

$$\chi(D_1 \cup D_2 \cup D_3) = \chi(D_1) + \chi(D_2 \cup D_3) - \chi(D_1 \cap (D_2 \cup D_3)).$$

The second term in the right-hand side of this formula can be calculated from (2.41). Rewriting the third term as $\chi((D_1 \cap D_2) \cup (D_1 \cap D_3))$, once again by (2.41), the formula (2.40) for $m = 3$ can be obtained. For a general m, one can prove it similarly by induction.

By Assumption 1 and Lemma 2.10, we obtain the formula:

$$\left(\prod_{\sigma=1}^{s} l_{j_\sigma} \right) + \cdots + \chi(\widehat{D}_J) z^{n-s} + \cdots = \frac{1}{(1-z)^2} \prod_{\sigma=1}^{s} \frac{l_{j_\sigma}}{1 + (l_{j_\sigma} - 1)z}. \tag{2.42}$$

On the other hand, again Assumption 1 implies that $H_0 \cap \widehat{D}_J$ is a complete intersection subvariety of $H_0 = \mathbb{P}^{n-1}$ and $H_0 \cap \widehat{D}_j$ is a hypersurface of degree l_j in $\mathbb{P}^{n-1}$. Hence, shifting n to $n-1$ in (2.42), we have

$$\left(\prod_{\sigma=1}^{s} l_{j_\sigma} \right) + \cdots + \chi(H_0 \cap \widehat{D}_J) z^{n-1-s} + \cdots \tag{2.43}$$

$$= \frac{1}{(1-z)^2} \prod_{\sigma=1}^{s} \frac{l_{j_\sigma}}{1 + (l_{j_\sigma} - 1)z}.$$

Subtracting z times both sides of (2.43) from (2.42), and using (2.39), we obtain the formula:

$$\left(\prod_{\sigma=1}^{s} l_{j_\sigma} \right) + \cdots + \chi(D_J) z^{n-s} + \cdots = \frac{1}{1-z} \prod_{\sigma=1}^{s} \frac{l_{j_\sigma}}{1 + (l_{j_\sigma} - 1)z}.$$

Furthermore, multiplying both sides of the above formula by z^s, we obtain:

Lemma 2.12.

$$\left(\prod_{\sigma=1}^{s} l_{j_\sigma} \right) z^s + \cdots + \chi(D_J) z^n + \cdots = \frac{1}{1-z} \prod_{\sigma=1}^{s} \frac{l_{j_\sigma} z}{1 + (l_{j_\sigma} - 1)z}.$$

To compute $\chi(M)$, we remark that, from $M = \mathbb{C}^n \setminus \bigcup_{j=1}^m D_j$ and (2.40), we have

$$\chi(M) = \chi(\mathbb{C}^n) - \chi\left(\bigcup_{j=1}^m D_j\right)$$

$$= 1 + \sum_{s=1}^m \sum_{|J|=s} (-1)^s \chi(D_J).$$

Combining this formula with Lemma 2.12, we obtain

$$1 + \cdots + \chi(M)z^n + \cdots = \sum_{k=0}^\infty z^k + \cdots + \sum_{s=1}^m \sum_{|J|=s} (-1)^s \chi(D_J) z^n + \cdots$$

$$= \frac{1}{1-z} + \frac{1}{1-z} \sum_{|s|=1}^m \sum_{|J|=s} \prod_{j \in J} \frac{-l_j z}{1 + (l_j - 1)z}$$

$$= \frac{1}{1-z} \prod_{j=1}^m \left(1 - \frac{l_j z}{1 + (l_j - 1)z}\right)$$

$$= (1-z)^{m-1} \prod_{j=1}^m \frac{1}{1 + (l_j - 1)z}.$$

Let us summarize these results in the following theorem:

Theorem 2.3. *Let D_j, $1 \le j \le m$ be a divisor determined by a polynomial of degree l_j, and set $M = \mathbb{C}^n \setminus \bigcup_{j=1}^m D_j$. Under Assumption 1 on D_j, $\chi(M)$ is equal to the coefficient of z^n in the expansion of the rational function*

$$(1-z)^{m-1} \prod_{j=1}^m \frac{1}{1 + (l_j - 1)z}$$

at $z = 0$.

For $n = 2$, it is easy to expand this rational function and to calculate its coefficient of z^2. Let us write the result as a corollary:

Corollary 2.1. *Suppose that $M = \mathbb{C}^2 \setminus \bigcup_{j=1}^m D_j$ satisfies the same assumption as above. Then, we have*

$$\chi(M) = \frac{1}{2}(m-1)(m-2) + (m-1)\sum_{j=1}^{m}(l_j - 1)$$

$$+ \sum_{1 \le i < j \le m}(l_i - 1)(l_j - 1) + \sum_{j=1}^{m}(l_j - 1)^2.$$

2.3 Construction of Twisted Cycles (1): One-Dimensional Case

2.3.1 Twisted Cycle Around One Point

Let us construct concretely $m - 1$ independent twisted cycles associated to the multi-valued function

$$U(u) = \prod_{j=1}^{m}(u - x_j)^{\alpha_j}$$

defined on $M = \mathbb{C} \setminus \{x_1, \ldots, x_m\}$, where m points x_j, $1 \le j \le m$ are on the real line satisfying the condition $x_1 < \cdots < x_m$. Setting $\omega = dU/U$, as we have explained in § 2.1.5, the multi-valued function U defines the rank 1 local system $\mathcal{L}_\omega^\vee$. As the branch of $U(u)$, we choose the one that is single-valued on the lower half plane, and satisfies

$$\arg(u - x_j) = \begin{cases} 0 \ (1 \le j \le p) \\ -\pi \ (p+1 \le j \le m) \end{cases} \tag{2.44}$$

on each interval $\Delta_p := (x_p, x_{p+1})$, $1 \le p \le m-1$. Below, we construct twisted cycles with this branch. As for the twisted homology group, the following facts are known ([Sp]):

(1) $H_\bullet(M, \mathcal{L}_\omega^\vee)$ is a homotopy invariant.
(2) The Mayer–Vietoris sequence holds as for the ordinary homology group.

We admit these facts and use them below. First, by (1), it suffices to calculate $H_\bullet(K, \mathcal{L}_\omega^\vee)$ for the one-dimensional simplicial complex K as in Figure 2.5 in place of M. Here, for simplicity, the restriction of the local system $\mathcal{L}_\omega^\vee$ on M to K is expressed by the same symbol. Second, to apply (2), represent K as the union of two subcomplexes: $K = K_1 \cup K_2$,

$$K_1 = \coprod_{j=1}^{m} S_\varepsilon^1(x_j), \quad K_2 = \coprod_{j=1}^{m-1} [x_j + \varepsilon, x_{j+1} - \varepsilon].$$

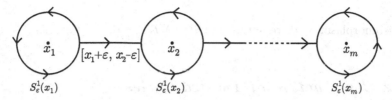

Fig. 2.5

First, let us compute the twisted homology for each circle $S^1_\varepsilon(x_j)$:

Lemma 2.13. *For $\alpha_j \notin \mathbb{Z}$, $1 \leq j \leq m$, we have*

$$H_q(S^1_\varepsilon(x_j), \mathcal{L}^\vee_\omega) = 0, \quad q = 0, 1, \cdots.$$

Proof. Triangulating $S^1_\varepsilon(x_j)$ as in Figure 2.6,

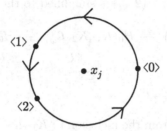

Fig. 2.6

the chain group $C_q(S^1_\varepsilon(x_j), \mathcal{L}^\vee_\omega)$ and the boundary operator ∂_ω look as follows:

$$C_0(S^1_\varepsilon(x_j), \mathcal{L}^\vee_\omega) = \mathbb{C}\langle 0 \rangle + \mathbb{C}\langle 1 \rangle + \mathbb{C}\langle 2 \rangle,$$

$$C_1(S^1_\varepsilon(x_j), \mathcal{L}^\vee_\omega) = \mathbb{C}\langle 01 \rangle + \mathbb{C}\langle 12 \rangle + \mathbb{C}\langle 20 \rangle,$$

$$\partial_\omega \langle 01 \rangle = \langle 1 \rangle - \langle 0 \rangle, \quad \partial_\omega \langle 12 \rangle = \langle 2 \rangle - \langle 0 \rangle, \quad \partial_\omega \langle 20 \rangle = c_j \langle 0 \rangle - \langle 2 \rangle,$$

where we set $c_j = \exp(2\pi\sqrt{-1}\alpha_j)$. In terms of matrices, we have

$$\partial_\omega \begin{pmatrix} \langle 01 \rangle \\ \langle 12 \rangle \\ \langle 20 \rangle \end{pmatrix} - \begin{pmatrix} -1 & 1 & 0 \\ 0 & -1 & 1 \\ c_j & 0 & -1 \end{pmatrix} \begin{pmatrix} \langle 0 \rangle \\ \langle 1 \rangle \\ \langle 2 \rangle \end{pmatrix},$$

and $\det(\partial_\omega) = c_j - 1$ is not zero by the assumption $\alpha_j \notin \mathbb{Z}$. Hence, the map

$$\partial_\omega : C_1(S^1_\varepsilon(x_j), \mathcal{L}^\vee_\omega) \longrightarrow C_0(S^1_\varepsilon(x_j), \mathcal{L}^\vee_\omega)$$

is an isomorphism, and we obtain $H_q(S^1_\varepsilon(x_j), \mathcal{L}^\vee_\omega) = 0$, $q = 0, 1, \ldots$.

2.3.2 Construction of Twisted Cycles

Applying the Mayer–Vietoris sequence to $K = K_1 \cup K_2$, we obtain the exact sequence:

$$0 \longrightarrow H_1(K_1, \mathcal{L}^\vee_\omega) \oplus H_1(K_2, \mathcal{L}^\vee_\omega) \longrightarrow H_1(K, \mathcal{L}^\vee_\omega) \overset{\delta}{\longrightarrow} \qquad (2.45)$$

$$\longrightarrow H_0(K_1 \cap K_2, \mathcal{L}^\vee_\omega) \longrightarrow H_0(K_1, \mathcal{L}^\vee_\omega) \oplus H_0(K_2, \mathcal{L}^\vee_\omega) \longrightarrow H_0(K, \mathcal{L}^\vee_\omega) \longrightarrow 0.$$

K_2 is homotopically equivalent to $m-1$ points, and $K_1 \cap K_2$ to $2m-2$ points. From this and Lemma 2.13, (2.45) is simplified to the exact sequence:

$$0 \longrightarrow H_1(K, \mathcal{L}^\vee_\omega) \overset{\delta}{\longrightarrow} \underset{\substack{\| \wr \\ \mathbb{C}^{2m-2}}}{H_0(K_1 \cap K_2, \mathcal{L}^\vee_\omega)} \overset{i_*}{\longrightarrow} \underset{\substack{\| \wr \\ \mathbb{C}^{m-1}}}{H_0(K_2, \mathcal{L}^\vee_\omega)} \longrightarrow \qquad (2.46)$$

$$\longrightarrow H_0(K, \mathcal{L}^\vee_\omega) \longrightarrow 0.$$

The above i_* is induced from the inclusion $i : K_1 \cap K_2 \longrightarrow K_2$, and we clearly have

$$i_* \left(\sum_{j=1}^{m-1} a_j \langle x_j + \varepsilon \rangle + b_j \langle x_{j+1} - \varepsilon \rangle \right) = \sum_{j=1}^{m-1} (a_j + b_j) \langle x_j + \varepsilon \rangle, \quad a_j, b_j \in \mathbb{C}.$$

Hence, i_* is surjective and we obtain

$$H_0(K, \mathcal{L}^\vee_\omega) = 0,$$

$$\mathrm{Im}\delta = \mathrm{Ker}i_* = \bigoplus_{j=1}^{m-1} \mathbb{C}(\langle x_j + \varepsilon \rangle - \langle x_{j+1} - \varepsilon \rangle).$$

To compute $H_1(K, \mathcal{L}^\vee_\omega)$, one has to calculate the map δ in (2.46) concretely. Coming back to the chain group level, we consider the exact sequences of complexes (below, we omit $\mathcal{L}^\vee_\omega$):

$$
\begin{array}{ccccccccc}
& & 0 & & 0 & & 0 & & \\
& & \downarrow & & \downarrow & & \downarrow & & \\
0 \longrightarrow & C_1(K_1 \cap K_2) & \overset{j_*}{\longrightarrow} & C_1(K_1) \oplus C_1(K_2) & \overset{s}{\longrightarrow} & C_1(K_1 \cup K_2) & \longrightarrow & 0 & \text{(exact)} \\
& \downarrow{\partial_\omega} & & \downarrow{\partial_\omega} & & \downarrow{\partial_\omega} & & & \\
0 \longrightarrow & C_0(K_1 \cap K_2) & \overset{j_*}{\longrightarrow} & C_0(K_1) \oplus C_0(K_2) & \overset{s}{\longrightarrow} & C_0(K_1 \cup K_2) & \longrightarrow & 0 & \text{(exact)}. \\
& \downarrow & & \downarrow & & \downarrow & & & \\
& & 0 & & 0 & & 0 & &
\end{array}
$$

Here, setting $j_1 : K_1 \cap K_2 \longrightarrow K_1$, $j_2 : K_1 \cap K_2 \longrightarrow K_2$, j_* is defined by $j_*(c) := j_{1*}(c) \oplus (-j_{2*}(c))$ for $c \in C_q(K_1 \cap K_2)$, and regarding $C_q(K_i) \subset C_q(K_1 \cup K_2)$, $i = 1, 2$, the map s is defined by

$$
s(c_q \oplus c_q') := c_q + c_q', \quad (c_q \in C_q(K_1), \quad c_q' \in C_q(K_2)).
$$

Let us recall the definition of δ: for $a \in C_1(K_1 \cup K_2)$, $\partial_\omega a = 0$, denote by $a_i \in C_1(K_i)$, $i = 1, 2$ and $a = s(a_1 \oplus a_2)$. By the exactness, there exists $b \in C_0(K_1 \cap K_2)$ such that $j_*(b) = \partial_\omega(a_1 \oplus a_2)$. Recall that δ is the map sending the homology class of a to the homology class of b. Taking $\langle x_j + \varepsilon \rangle - \langle x_{j+1} - \varepsilon \rangle$ from a basis of $\mathrm{Ker}\, i_*$, a simple calculation shows

$$
j_*(\langle x_j + \varepsilon \rangle - \langle x_{j+1} - \varepsilon \rangle)
$$

$$
= \partial_\omega \left(\frac{1}{c_j - 1} S_\varepsilon^1(x_j) - \frac{1}{c_{j+1} - 1} S_\varepsilon^1(x_{j+1}) \right) \oplus \partial_\omega([x_j + \varepsilon, \ x_{j+1} - \varepsilon]),
$$

hence, by setting

$$
\Delta_j(\omega) = \frac{1}{c_j - 1} S_\varepsilon^1(x_j) + [x_j + \varepsilon, \ x_{j+1} - \varepsilon] - \frac{1}{c_{j+1} - 1} S_\varepsilon^1(x_{j+1}), \quad (2.47)
$$

we have $\delta(\Delta_j(\omega)) = \langle x_j + \varepsilon \rangle - \langle x_{j+1} - \varepsilon \rangle$.

Thus, we obtain

$$
H_1(X, \mathcal{L}_\omega^\vee) = \bigoplus_{j=1}^{m-1} \mathbb{C} \cdot \Delta_j(\omega).
$$

Let us summarize this in the following lemma:

Lemma 2.14. *Under the condition* $\alpha_j \notin \mathbb{Z}$, $1 \leq j \leq m$, *we have*

$$
H_0(M, \mathcal{L}_\omega^\vee) = 0,
$$

$$
H_1(M, \mathcal{L}_\omega^\vee) \simeq \bigoplus_{j=1}^{m-1} \mathbb{C} \cdot \Delta_j(\omega),
$$

where $\Delta_j(\omega)$ *is the twisted cycle defined by (2.47).*

2.3.3 Intersection Number (i)

Since we could construct concretely a basis of the twisted homology H_1 $(M, \mathcal{L}_\omega^\vee)$, allowing Lemma 3.2, let us explain intuitively the intersection number between $H_1(M, \mathcal{L}_\omega^\vee)$ and $H_1^{lf}(M, \mathcal{L}_\omega)$ described in Lemma 2.9 (4). If the reader understands the case $m = 2$, the case $m \geq 3$ can be understood similarly, so we restrict ourselves to this case for simplicity. Let $M = \mathbb{C} \setminus \{0, 1\}$ and choose a branch of the multi-valued function $U(u) = u^\alpha(u-1)^\beta, \alpha, \beta, \alpha + \beta \notin \mathbb{Z}$ we have explained in § 2.3.1. Below, we construct the twisted cycles with this branch. By Lemma 2.14, $H_1(M, \mathcal{L}_\omega^\vee)$ is one-dimensional and its basis is given by

$$\Delta(\omega) = \frac{1}{c_1 - 1} S_\varepsilon^1(0) \otimes U_{S_\varepsilon^1(0)} + [\varepsilon, 1 - \varepsilon] \otimes U_{[\varepsilon, 1-\varepsilon]} - \frac{1}{c_2 - 1} S_\varepsilon^1(1) \otimes U_{S_\varepsilon^1(1)},$$

$$c_1 = \exp(2\pi\sqrt{-1}\alpha), \quad c_2 = \exp(2\pi\sqrt{-1}\beta).$$

Here, $S_\varepsilon^1(0)$ is the circle of center 0 and radius ε with starting point ε, which turns in the anticlockwise direction, stopping at an infinitely near point before ε, and $U_{S_\varepsilon^1(0)}$ is the branch obtained by the analytic continuation along $S_\varepsilon^1(0)$ (cf. Figure 2.3), and similarly for the rest.

On the other hand, by Lemma 3.2, $H_1^{lf}(M, \mathcal{L}_\omega)$ is the one-dimensional vector space with a basis $(0, 1) \otimes U_{(0,1)}^{-1}$. Here, $U_{(0,1)}$ is the restriction of the branch determined above to $(0, 1)$. Now, $\Delta := \gamma \otimes U_\gamma^{-1}$ obtained by deforming the interval $(0, 1)$ as in Figure 2.7 is homologous to $(0, 1) \otimes U_{(0,1)}^{-1}$ in $H_1^{lf}(M, \mathcal{L}_\omega)$.

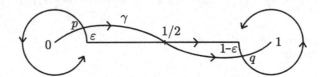

Fig. 2.7

The geometric intersections of $\Delta(\omega)$ and Δ are the three points p, $1/2$, q in Figure 2.7, and their signatures are $-$, $-$, $+$. Since the difference of the branches is canceled by U and U^{-1}, we finally obtain

$$\Delta(\omega) \cdot \Delta = \frac{1}{c_1 - 1} \times (-1) + (-1) + \left(-\frac{1}{c_2 - 1}\right) \times (+1)$$

$$= -\frac{c_1 c_2 - 1}{(c_1 - 1)(c_2 - 1)}.$$

One can compute the intersection matrix for a general m by a similar computation, and the result looks as follows: setting $\Delta_j := (x_j, x_{j+1}) \otimes U_{(x_j,x_{j+1})}^{-1} \in H_1^{lf}(M, \mathcal{L}_\omega)$, $1 \le j \le m-1$, $c_j = \exp(2\pi\sqrt{-1}\alpha_j)$, $d_j := c_j - 1$, $d_{jk} := c_j c_k - 1$, we have

$$
\Delta_j(\omega) \cdot \Delta_k = \begin{cases} c_j/d_j, & j = k+1 \\ -d_{j,j+1}/d_j d_{j+1}, & j = k \\ -1/d_k, & j+1 = k \\ 0, & \text{otherwise.} \end{cases}
$$

Hence, the intersection matrix is given by

$$
\begin{pmatrix} \Delta_1(\omega) \cdot \Delta_1 & \cdots & \Delta_1(\omega) \cdot \Delta_{m-1} \\ \vdots & & \vdots \\ \Delta_{m-1}(\omega) \cdot \Delta_1 & \cdots & \Delta_{m-1}(\omega) \cdot \Delta_{m-1} \end{pmatrix}
$$

$$
= - \begin{pmatrix} \frac{d_{12}}{d_1 d_2} & \frac{-1}{d_2} & 0 \cdots & 0 & 0 \\ \frac{-c_2}{d_2} & \frac{d_{23}}{d_2 d_3} & & 0 & 0 \\ & & & & \vdots \\ & & & & 0 \\ 0 & 0 & & \frac{d_{m-2,m-1}}{d_{m-2}d_{m-1}} & \frac{-1}{d_{m-1}} \\ 0 & 0 & \cdots 0 & \frac{-c_{m-1}}{d_{m-1}} & \frac{d_{m-1,m}}{d_{m-1}d_m} \end{pmatrix}.
$$

Remark 2.4. For an arrangement of hyperplanes in general position and its degenerated case, generalizing this formula, the intersection matrix of $H_n(M, \mathcal{L}_\omega^\vee) \times H_n^{lf}(M, \mathcal{L}_\omega)$ was calculated concretely in [Kit-Yos1, Kit-Yos2]. The research on the monodromy group $\Gamma(n+1, m+1; \alpha)$ of the hypergeometric differential equations $E(n+1, m+1; \alpha)$ stated in § 3.4 is extremely important for the global study of this system of equations (cf. [M-S-T-Y1, M-S-T-Y2]), although this topic will not be touched in this book. The importance of the intersection matrix comes from the fact that this matrix is an invariant of $\Gamma(n+1, m+1; \alpha)$. In the classical research on the uniformization, due to É. Picard, initiated from the research on modular functions, the determination of the intersection matrix by using the Riemann−Hodge period relation for a complex analytic family of algebraic varieties was the core of the research (cf. [Ao12], [Kit-Yos1], [Kit-Yos2], [M-O-Y], [M-S-Y], [Yos4] etc.).

2.4 Comparison Theorem

2.4.1 Algebraic de Rham Complex

Let D be a divisor of the complex affine n-space $\mathbb{C}^n$ with coordinates $(u_1, \ldots, u_n)$ defined by m polynomials $P_j(u)$, $1 \leq j \leq m$, and consider a multi-valued function $U(u) = \prod_{j=1}^{m} P_j(u)^{\alpha_j}$ on the complex manifold $M = \mathbb{C}^n \setminus D$. Since we compare two structures on M, as complex manifold and as algebraic variety, in this section, we denote the former by M^{an} and latter by M^{alg} to make the distinction.

Setting $\omega = dU/U$, the covariant differential operator $\nabla_\omega := d + \omega_\wedge$ with respect to ω defines the differential equation

$$\nabla_\omega h = 0, \qquad h \in \mathcal{O}_{M^{\mathrm{an}}}.$$

Denoting the sheaf of solutions of it by $\mathcal{L}_\omega$, this is a rank 1 local system on M^{an} (cf. § 2.1.5). Regarding M as an affine algebraic variety, besides the analytic sheaves $\mathcal{O}_{M^{\mathrm{an}}}$, $\Omega^p_{M^{\mathrm{an}}}$, one can consider the algebraic structure sheaf $\mathcal{O}_{M^{\mathrm{alg}}}$ on M generated by rational functions that are locally holomorphic on M and the sheaf $\Omega^p_{M^{\mathrm{alg}}}$ of rational p-forms that are locally holomorphic on M. By Lemma 2.8, the twisted cohomology $H^p(M^{\mathrm{an}}, \mathcal{L}_\omega)$ can be computed by using the analytic twisted de Rham complex $(\Omega^\bullet(M^{\mathrm{an}}), \nabla_\omega)$ as follows:

$$H^p(M^{\mathrm{an}}, \mathcal{L}_\omega) \simeq H^p(\Omega^\bullet(M^{\mathrm{an}}), \nabla_\omega). \tag{2.48}$$

For the one-dimensional case, computing the right-hand side of (2.48) by using the partial fraction decomposition of a rational function, we see that one can choose logarithmic 1-forms with poles along D as a representative of cohomology classes. This indicates that one might be able to compute $H^p(M^{\mathrm{an}}, \mathcal{L}_\omega)$ by using the algebraic twisted de Rham complex $(\Omega^\bullet(M^{\mathrm{alg}}), \nabla_\omega)$. This was treated in the most general form for the ordinary de Rham theory (when $\omega = 0$) by Grothendieck [Gro], and was extended for the twisted de Rham theory by Deligne [De].

Advice to the reader In this section, we require the reader to be familiar with hypercohomology with coefficients in a complex of sheaves, which is a generalization of sheaf cohomology and spectral sequence. For those who are not familiar with these topics, the authors hope that he (or she) may admit Theorem 2.5 and pass to § 2.5 where we give concrete computations. In our opinion, the world of hypergeometric functions is a domain where one can use general theories to obtain concrete and explicit results, and it would make no sense otherwise.

2.4.2 Čech Cohomology

Here, we follow the reference [A-B-G] which seems to be accessible for those who work on analysis. There, the above-mentioned result of Grothendieck is explained and some principal results in algebraic geometry are cleverly summarized (without proof) for this purpose. The reader may properly consult it. Here, we define hypercohomology with coefficients in a complex of sheaves by using an open cover of M à la Čech. First, we briefly recall Čech cohomology of a sheaf $\mathcal{S}$: Let $\mathcal{U} = \{U_i\}$ be a locally finite open cover of M, and we denote the set of alternating functions

$$(i_0, \ldots, i_p) \longmapsto \varphi_{i_0 \ldots i_p} \in \Gamma(U_{i_0} \cap \cdots \cap U_{i_p}, \mathcal{S})$$

by $C^p(\mathcal{U}, \mathcal{S})$. Defining $\delta : C^p(\mathcal{U}, \mathcal{S}) \longrightarrow C^{p+1}(\mathcal{U}, \mathcal{S})$ by

$$(\delta\varphi)_{i_0 \cdots i_{p+1}} = \sum_{\nu=0}^{p+1} (-1)^\nu \varphi_{i_0 \cdots \hat{i}_\nu \cdots i_{p+1}}, \tag{2.49}$$

it follows that $\delta^2 = 0$ and $(C^\bullet(\mathcal{U}, \mathcal{S}), \delta)$ becomes a complex which is called a Čech complex. The pth cohomology of this complex

$$H^p(C^\bullet(\mathcal{U}, \mathcal{S}), \delta) = \frac{\mathrm{Ker}\{\delta : C^p(\mathcal{U}, \mathcal{S}) \longrightarrow C^{p+1}(\mathcal{U}, \mathcal{S})\}}{\delta C^{p-1}(\mathcal{U}, \mathcal{S})}$$

is denoted by $H^p(\mathcal{U}, \mathcal{S})$ and is called the Čech cohomology of $\mathcal{U}$ with coefficients in $\mathcal{S}$. Moreover, its limit with respect to the refinement of an open cover is called the pth Čech cohomology of M with coefficients in $\mathcal{S}$, and is denoted by

$$H^p(M, \mathcal{S}) := \varinjlim_{\mathcal{U}} H^p(\mathcal{U}, \mathcal{S}).$$

Here, if we have

$$H^p(U_{i_0} \cap \cdots \cap U_{i_q}, \mathcal{S}) = 0, \quad p = 1, 2, \ldots \tag{2.50}$$

for any q and all $(i_0 \cdots i_q)$, for this open cover, there is an isomorphism

$$H^p(\mathcal{U}, \mathcal{S}) \xrightarrow{\sim} H^p(M, \mathcal{S}), \quad p = 0, 1, \ldots,$$

and we remark that it is not necessary to take the limit.

When covering and a complex of sheaves are given, one can define cohomology group which has a similar property to those of Čech cohomology, called hypercohomology. Here, we will not state it in general form but will explain in a way applicable to our cases.

2.4.3 Hypercohomology

To make the story clear, we consider the twisted de Rham complex $(\Omega_M^\bullet, \nabla_\omega)$ formed by the sheaf Ω_M^p of analytic p-forms on M^{an}

$$0 \longrightarrow \Omega_M^0 \xrightarrow{\nabla_\omega} \Omega_M^1 \longrightarrow \cdots \xrightarrow{\nabla_\omega} \Omega_M^n \longrightarrow 0.$$

Fixing an open cover $\mathcal{U} = \{U_i\}$ of M,

$$C^p(\mathcal{U}, \Omega_M^q), \quad p, q = 0, 1, \ldots$$

becomes a double complex called the twisted Čech−de Rham complex by δ defined for the Čech complex in (2.49) and by ∇_ω, which defines the twisted de Rham complex:

$$C^\bullet(\mathcal{U}, \Omega_M^\bullet) = \left|
\begin{array}{cc}
\vdots & \vdots \\
\uparrow\nabla_\omega & \uparrow\nabla_\omega \\
C^0(\mathcal{U}, \Omega_M^1) \xrightarrow{\delta} C^1(\mathcal{U}, \Omega_M^1) \xrightarrow{\delta} \cdots \cdots \\
\uparrow\nabla_\omega & \uparrow\nabla_\omega \\
C^0(\mathcal{U}, \Omega_M^0) \xrightarrow{\delta} C^1(\mathcal{U}, \Omega_M^0) \xrightarrow{\delta} \cdots \cdots
\end{array}
\right.$$

$$\delta^2 = 0, \quad \nabla_\omega^2 = 0, \quad \delta \cdot \nabla_\omega - \nabla_\omega \cdot \delta = 0.$$

Recollecting this double complex along the anti-diagonal, we set

$$K^r := \bigoplus_{p+q=r} C^p(\mathcal{U}, \Omega_M^q), \quad r = 0, 1, \ldots.$$

On each $C^p(\mathcal{U}, \Omega_M^q)$, the homomorphism $D : K^{p+q} \longrightarrow K^{p+q+1}$ defined by

$$D = \delta + (-1)^p \nabla_\omega$$

satisfies $D^2 = 0$, and $(K^\bullet, D)$ becomes a complex so that one can define the cohomology of this complex.

Definition 2.1. The cohomology $H^p(K^\bullet, D)$ is called the hypercohomology of the open cover $\mathcal{U}$ with coefficients in the twisted de Rham complex $(\Omega_M^\bullet, \nabla_\omega)$, denoted by $\mathbb{H}^p(\mathcal{U}, (\Omega_M^\bullet, \nabla_\omega))$. In addition, the limit with respect to the refinement of $\mathcal{U}$

$$\varinjlim_{\mathcal{U}} \mathbb{H}^p(\mathcal{U}, (\Omega_M^{\bullet}, \nabla_\omega))$$

is denoted by $\mathbb{H}^p(M, (\Omega_M^{\bullet}, \nabla_\omega))$ and is called the hypercohomology of M with coefficients in $(\Omega_M^{\bullet}, \nabla_\omega)$.

For $\mathcal{S} = \Omega_M^q$, $q = 0, 1, \ldots$, when (2.50) holds, it is known that the limiting operation is not necessary and one has

$$\mathbb{H}^p(\mathcal{U}, (\Omega_M^{\bullet}, \nabla_\omega)) \simeq \mathbb{H}^p(M, (\Omega_M^{\bullet}, \nabla_\omega)).$$

2.4.4 Spectral Sequence

The relation between the cohomology $H^{\bullet}(K^{\bullet}, D)$ of the complex induced from the twisted Čech–de Rham complex and the double complex $C^{\bullet}(\mathcal{U}, \Omega_M^{\bullet})$ is not so direct and is related indirectly via a spectral sequence. As the first approximation, there is a spectral sequence which uses the cohomology sheaves $\mathcal{H}^q(\Omega_M, \nabla_\omega)$ of the complex of sheaves $(\Omega_M^{\bullet}, \nabla_\omega)$ as follows:

$${}'E_2^{p,q} = H^p(M, \mathcal{H}^q(\Omega_M^{\bullet}, \nabla_\omega)) \Longrightarrow \mathbb{H}^{p+q}(M, (\Omega_M^{\bullet}, \nabla_\omega)).$$

On the other hand, as the first approximation, there is a spectral sequence which uses δ as follows:

$${}''E_1^{p,q} = H^q(M, \Omega_M^p) \Longrightarrow \mathbb{H}^{p+q}(M, (\Omega_M^{\bullet}, \nabla_\omega)). \tag{2.51}$$

It is clear that what we have mentioned above extends to any complex of sheaves $(\mathcal{S}^{\bullet}, d)$:

$$0 \longrightarrow \mathcal{S}^0 \xrightarrow{d} \mathcal{S}^1 \xrightarrow{d} \cdots \cdots.$$

Moreover, if we have $\mathcal{S}^1 = \mathcal{S}^2 = \cdots = 0$, then it is clear from the definition that the hypercohomology coincides with the ordinary cohomology. Here, we state a fact that can be shown by the functoriality of the spectral sequences.

Lemma 2.15. *Let $(\mathcal{S}_1^{\bullet}, d_1)$, $(\mathcal{S}_2^{\bullet}, d_2)$ be complexes of sheaves, and $j : (\mathcal{S}_1^{\bullet}, d_1)$ $\longrightarrow (\mathcal{S}_2^{\bullet}, d_2)$ be a morphism of complexes. If j induces an isomorphism of cohomology sheaves*

$$j_* : \mathcal{H}^p(\mathcal{S}_1^{\bullet}, d_1) \xrightarrow{\sim} \mathcal{H}^p(\mathcal{S}_2^{\bullet}, d_2), \quad p = 0, 1, \ldots,$$

then j induces an isomorphism of hypercohomologies

$$j_* : \mathbb{H}^p(M, (\mathcal{S}_1^{\bullet}, d_1)) \xrightarrow{\sim} \mathbb{H}^p(M, (\mathcal{S}_2^{\bullet}, d_2)).$$

Proof. As j induces an isomorphism of the ${}'E_2$-term, the rest follows from a general theory of spectral sequences.

2.4.5 Algebraic de Rham Cohomology

For an open set U of $M = \mathbb{C}^n \setminus D$, if there exists a polynomial $Q(u_1, \ldots, u_n)$ such that $U = M \setminus \{Q = 0\}$, U is called an affine open set, and an open cover $\mathcal{U} = \{U_i\}$ of M formed by a finite number of affine open subsets U_i is called an affine open cover of M. For the sheaf $\Omega^\bullet_{M^{\mathrm{alg}}}$, as $\omega = \sum \alpha_j \frac{dP_j}{P_j}$ is a rational 1-form that is homolorphic on M, one can consider the complex of sheaves $(\Omega^\bullet_{M^{\mathrm{alg}}}, \nabla_\omega)$:

$$0 \longrightarrow \Omega^0_{M^{\mathrm{alg}}} \xrightarrow{\nabla_\omega} \Omega^1_{M^{\mathrm{alg}}} \longrightarrow \cdots \xrightarrow{\nabla_\omega} \Omega^n_{M^{\mathrm{alg}}} \longrightarrow 0.$$

Contrary to the analytic $(\Omega^\bullet_{M^{\mathrm{an}}}, \nabla_\omega)$, we remark that the complex $\Omega^p_{M^{\mathrm{alg}}}$, $p \geq 1$ is no more exact. Now, we take an affine open cover of M and consider the double complex $C^\bullet(\mathcal{U}, \Omega^\bullet_{M^{\mathrm{alg}}})$. Then, each $U_{i_0} \cap \cdots \cap U_{i_p}$ is an affine open set and

$$\Gamma(U_{i_0} \cap \cdots \cap U_{i_p}, \Omega^q_{M^{\mathrm{alg}}})$$

is the space of rational q-forms that are holomorphic on $U_{i_0} \cap \cdots \cap U_{i_p}$. Hence, we call this complex the algebraic twisted Čech–de Rham complex. For this complex, we can apply the argument of § 2.4.3 and can define the hyper-cohomology $\mathbb{H}^p(M^{\mathrm{alg}}, (\Omega^\bullet_{M^{\mathrm{alg}}}, \nabla_\omega))$. This is called the algebraic de Rham cohomology.

2.4.6 Analytic de Rham Cohomology

As was shown in § 2.2.10, the following sequence of sheaves is exact:

$$0 \longrightarrow \mathcal{L}_\omega \longrightarrow \Omega^0_{M^{\mathrm{an}}} \xrightarrow{\nabla_\omega} \Omega^1_{M^{\mathrm{an}}} \longrightarrow \cdots \xrightarrow{\nabla_\omega} \Omega^n_{M^{\mathrm{an}}} \longrightarrow 0. \qquad (2.52)$$

Now, considering $\mathcal{L}_\omega$ as a complex of sheaves concentrating only at the 0th degree

$$0 \longrightarrow \mathcal{L}_\omega \longrightarrow 0 \longrightarrow 0 \longrightarrow \cdots \cdots,$$

(2.52) can be regarded as a homomorphism from this complex to the twisted de Rham complex $(\Omega^\bullet_{M^{\mathrm{an}}}, \nabla_\omega)$:

$$j : \{0 \to \mathcal{L}_\omega \to 0\} \longrightarrow (\Omega^\bullet_{M^{\mathrm{an}}}, \nabla_\omega).$$

Both cohomology sheaves are isomorphic, as (2.52) is exact, and by Lemma 2.15, it induces an isomorphism of hypercohomologies. Since the hypercohomology of the complex of sheaves concentrating only at the 0th degree $\{0 \to \mathcal{L}_\omega \to 0 \to \cdots\}$ coincides with the ordinary cohomology $H^p(M, \mathcal{L}_\omega)$, we obtain the isomorphism:

$$H^p(M, \mathcal{L}_\omega) \simeq \mathbb{H}^p(M, (\Omega^\bullet_{M^{\mathrm{an}}}, \nabla_\omega)). \tag{2.53}$$

2.4.7 Comparison Theorem

Clearly, $\Omega^p_{M^{\mathrm{alg}}}$ can be considered as a subsheaf of $\Omega^p_{M^{\mathrm{an}}}$ of $\mathbb{C}$-modules; passing to the complexes, we obtain a natural homomorphism:

$$(\Omega^\bullet_{M^{\mathrm{alg}}}, \nabla_\omega) \longrightarrow (\Omega^\bullet_{M^{\mathrm{an}}}, \nabla_\omega). \tag{2.54}$$

The Grothendieck–Deligne comparison theorem asserts that both hyperco-homologies are isomorphic.

Theorem 2.4 (Grothendieck–Deligne comparison theorem).

$$\mathbb{H}^p(M, (\Omega^\bullet_{M^{\mathrm{alg}}}, \nabla_\omega)) \xrightarrow{\sim} \mathbb{H}^p(M, (\Omega^\bullet_{M^{\mathrm{an}}}, \nabla_\omega)).$$

$$p = 0, 1, \ldots$$

For its proof, see the reference [De] pp. 98–105, cited in § 2.4.1.

Now, we consider the spectral sequence (2.51) associated to hypercoho-mologies. Take $M = M^{\mathrm{alg}}$. We have

$$''E_2^{p,q} = H^q_{\nabla_\omega}(H^p(M, \Omega^\bullet_M)) \Longrightarrow \mathbb{H}^{p+q}(M, (\Omega^\bullet_M, \nabla_\omega)). \tag{2.55}$$

Here, $H^q_{\nabla_\omega}(H^p(M, \Omega^\bullet_M))$ represents the cohomology of the complex $(H^p(M, \Omega^\bullet_M), H^p(\nabla_\omega))$ induced from the cohomological functor H^p. Since our variety M is an affine variety obtained by removing a finite number of algebraic hypersurfaces from $\mathbb{C}^n$, we have $H^p(M, \mathcal{O}_M) = 0$, $p \geq 1$ and since Ω^q_M is isomorphic to $\mathcal{O}_M^{\binom{n}{q}}$, $H^p(M, \Omega^\bullet_M) = 0$, $p \geq 1$. (This fact is a special case of Theorem B for affine varieties. For its proof, see e.g. [Ser].) Hence, the spectral sequence (2.51) beecomes

$$''E_2^{p,q} = \begin{cases} 0 & (p \geq 1) \\ H^q_{\nabla_\omega}(\Gamma(M, \Omega^\bullet_M)) & (p = 0) \end{cases}$$

and is degenerated. Thus, by a well-known device from spectral sequences, we obtain an isomorphism:

$$H^q_{\nabla_\omega}(\Gamma(M, \Omega^\bullet_M)) \xrightarrow{\sim} \mathbb{H}^q(M, (\Omega^\bullet_M, \nabla_\omega)). \tag{2.56}$$

2.4.8 Reformulation

Since $\varphi \in \Gamma(M^{\mathrm{alg}}, \Omega^p_{M^{\mathrm{alg}}})$ is a rational p-form which is holomorphic on M, one can say that this is a rational p-form which admits at most poles only along the divisor D. In this sense, we simplify the symbol $\Gamma(M^{\mathrm{alg}}, \Omega^p_{M^{\mathrm{alg}}})$ and use the symbol $\Omega^p(*D)$, here and after:

$$\Omega^p(*D)$$

:=the space of rational p-forms which may have at most poles only along D.

With this symbol, by (2.53), the comparison theorem and (2.56), we obtain the following theorem:

Theorem 2.5. *For $p = 0, 1, \cdots$, there is an isomorphism:*

$$H^p(M, \mathcal{L}_\omega) \simeq H^p(\Omega^\bullet(*D), \nabla_\omega).$$

In research on integral representations of hypergeometric functions, Theorem 2.5 reduces a reflection on the topological quantity $H^p(M, \mathcal{L}_\omega)$ to an algebraic computation and provides us a theoretical background that allows us to relate a complicated algebraic computation to an intuitive topological quantity, e.g., the Euler characteristic $\chi(M)$. With appropriate restrictions on the divisor D, we will show in § 2.5–2.9 that one can perform this algebraic computation.

Remark 2.5. The computation of $H^p(\Omega^\bullet(*D), \nabla_\omega)$ is performed by elementary algebraic computations of rational functions and the partial derivatives. When the polynomial $P(u)$ defining the divisor D contains parameters depending rationally, by $\nabla_\omega = d + \sum_{j=1}^m \alpha_j \frac{dP_j}{P_j} \wedge$, the final result naturally and rationally depends on these parameters and the exponents $\alpha_1, \ldots, \alpha_m$. The form $U\varphi$ actually belongs to a wider class of multivalued functions, i.e., the category of Nilsson class discussed in [An], [Le]. See also [Pha1] for a topological aspect.

2.5 de Rham-Saito Lemma and Representation of Logarithmic Differential Forms

2.5.1 Logarithmic Differential Forms

By a comparison theorem, computation of a twisted cohomology is reduced to that of $H^p(\Omega^\bullet(*D), \nabla_\omega)$, but if we can find a subcomplex of $\Omega^\bullet(*D)$ stable under ∇_ω whose cohomology is isomorphic to $H^p(\Omega^\bullet(*D), \nabla_\omega)$, we may further simplify the computation. In this section, we will introduce such a subcomplex. Here, for the reader's convenience, we summarize the symbols used frequently in § 2.5–2.9:

$\Omega^\bullet(\mathbb{C}^n)$ = the space of p-forms with polynomial coefficients

$$= \{\varphi = \sum a_{i_1 \cdots i_p}(u) du_{i_1} \wedge \cdots \wedge du_{i_p} | a_{i_1 \cdots i_p} \in \mathbb{C}[u]\},$$

$$U(u) = \prod_{j=1}^{m} P_j(u)^{\alpha_j},$$

$$D := \cup\{P_j = 0\}, \quad P = P_1 \cdots P_m,$$

$$M = \mathbb{C}^n \setminus D,$$

$$\omega = dU/U = \sum_{j=1}^{m} \alpha_j \frac{dP_j}{P_j}, \quad \nabla_\omega = d + \omega\wedge,$$

$\mathcal{L}_\omega$: the local system of rank 1 generated by the solutions of $\nabla_\omega h = 0$,

$$h \in \mathcal{O}_M,$$

$\Omega^p(*D)$: the space of rational p-forms which may have at most poles only

along D,

$\Omega^p(\log D)$: the space of logarithmic p-forms along D.

Following [Sai2], we introduce a complex of logarithmic differential forms along D as a subcomplex of $\Omega^\bullet(*D)$ stable under ∇_ω.

Definition 2.2. A rational p-form $\varphi \in \Omega^p(*D)$ is called a logarithmic p-form along D if it satisfies the condition

$$P\varphi \in \Omega^p(\mathbb{C}^n), \quad dP \wedge \varphi \in \Omega^{p+1}(\mathbb{C}). \tag{2.57}$$

The space of logarithmic p-forms is denoted by $\Omega^p(\log D)$.

We remark that, denoting the irreducible factorization of P by $P = Q_1^{\nu_1} \cdots Q_s^{\nu_s}$, $\nu_l \in \mathbb{Z}_{>0}$, the above condition can be rewritten as follows:

$$P\varphi = \alpha \in \Omega^p(\mathbb{C}^n), \quad \frac{dQ_l}{Q_l} \wedge \alpha \in \Omega^{p+1}(\mathbb{C}^n), \quad 1 \le l \le s. \tag{2.58}$$

Now, let us see that $\Omega^\bullet(\log D)$ is stable under ∇_ω. By (2.57), we have $Pd\varphi = d(P\varphi) - dP \wedge \varphi \in \Omega^{p+1}(\mathbb{C}^n)$ and $d(dP \wedge \varphi) = -dP \wedge d\varphi \in \Omega^{p+2}(\mathbb{C}^n)$, hence, $d\varphi \in \Omega^{p+1}(\log D)$. Next, we show $\frac{dP_j}{P_j} \wedge \varphi \in \Omega^{p+1}(\log D)$. Denoting the irreducible factorization of each P_j by $P_j = \prod Q_l^{\mu_l}$, $\mu_l \in \mathbb{Z}_{\ge 0}$, we have

$$\psi := \frac{dP_j}{P_j} \wedge \varphi = \sum \mu_l \frac{dQ_l}{Q_l} \wedge \varphi,$$

hence, by (2.58), we obtain

$$\beta := P\psi = \sum \mu_l \frac{dQ_l}{Q_l} \wedge \alpha \in \Omega^{p+1}(\mathbb{C}^n).$$

For $1 \le k \le s$, we have

$$\gamma := \frac{dQ_k}{Q_k} \wedge \beta = \sum_{l \ne k} \mu_l \frac{dQ_k}{Q_k} \wedge \frac{dQ_l}{Q_l} \wedge \alpha,$$

but since Q_k and Q_l are different irreducible polynomials, $\{Q_k = 0\}$ and $\{Q_l = 0\}$ intersect only at an algebraic set of codimension at least 2. By (2.58), γ is a holomorphic $(p+2)$-form at each point of $\{Q_k = 0\} \setminus \{Q_l = 0\}$, and similarly on $\{Q_l = 0\} \setminus \{Q_k = 0\}$. Hence, γ is a rational $(p+2)$-form which is holomorphic except for an algebraic set of codimension at least 2. Since a pole of a rational $(p+2)$-form is necessarily of codimension 1, γ has no pole in $\mathbb{C}^n$ which means $\gamma \in \Omega^{p+2}(\mathbb{C}^n)$. Thus, we obtain the following lemma:

Lemma 2.16. $\Omega^\bullet(\log D)$ *is stable under d and $\frac{dP_j}{P_j}\wedge$, $1 \le j \le m$. Hence, it is stable under ∇_ω, and $(\Omega^\bullet(\log D), \nabla_\omega)$ becomes a subcomplex of the twisted de Rham complex $(\Omega^\bullet(*D), \nabla_\omega)$.*

For later use in § 2.9, here we prove a property of logarithmic differential forms.

Lemma 2.17. *Suppose that in $P = P_1 \cdots P_m$, two different P_i and P_j are coprime and that $\alpha_j \ne 1$, $1 \le j \le m$. Then, a rational p-form $\varphi \in \Omega^p(*D)$ is a logarithmic differential form if and only if it satisfies the following conditions:*

$$\varphi \in \frac{1}{P}\Omega^p(\mathbb{C}^n), \quad \nabla_\omega\varphi \in \frac{1}{P}\Omega^{p+1}(\mathbb{C}^n). \tag{2.59}$$

Proof. It follows from Lemma 2.16 that $\varphi \in \Omega^p(\log D)$ satisfies (2.59). Let us show the converse. Setting $\varphi = \beta/P$, $\beta \in \Omega^p(\mathbb{C}^n)$, we have

$$\nabla_\omega\varphi = \frac{1}{P}\left\{ d\beta + \sum_{j=1}^m (\alpha_j - 1)\frac{dP_j}{P_j} \wedge \beta \right\}.$$

By (2.59), it follows that $\sum_{j=1}^m (\alpha_j - 1)\frac{dP_j}{P_j} \wedge \beta \in \Omega^{p+1}(\mathbb{C}^n)$, but since P_j, $1 \le j \le m$ are corpime, $\{P_j = 0\} \cap \left(\bigcup_{l \ne j}\{P_l = 0\} \right)$ is of codimension at least 2. Hence, we see that each term of the sum in the above formula satisfies $(\alpha_j - 1)\frac{dP_j}{P_j} \wedge \beta \in \Omega^{p+1}(\mathbb{C}^n)$ which implies, by the assumption $\alpha_j \ne 1$, $\frac{dP_j}{P_j} \wedge \beta \in \Omega^{p+1}(\mathbb{C}^n)$. Thus, it satisfies the condition (2.58) for logarithmic differential forms and $\varphi \in \Omega^p(\log D)$ follows.

2.5.2 de Rham–Saito Lemma

To compute the twisted rational de Rham cohomology, one has to express elements of $\Omega^p(\log D)$ with an exterior product of $\frac{dP_j}{P_j}$, $1 \leq j \leq m$ and a differential form with polynomial coefficient. For this purpose, one is required to consider the division of differential forms which is assured by the de Rham–Saito lemma in the title. It uses commutative algebras for its formulation; for the reader who is not familiar with them, we will define and explain the notion and state the facts that will be used.

Let A be a commutative Noetherian ring with unit 1 and $\mathfrak{a}$ its ideal. A sequence of elements $f_1, \ldots, f_t$ of $\mathfrak{a}$ is called a regular sequence if f_1 is not a zero-divisor of A and f_i is not a zero-divisor of $A/(f_1, \ldots, f_{i-1})$ for $2 \leq i \leq t$. The maximal length of the regular sequences in $\mathfrak{a}$ is called the depth of the ideal $\mathfrak{a}$ in A and is denoted by $\mathrm{depth}_{\mathfrak{a}} A$.

Here, for later comprehension, we explain the depth of an ideal $\mathfrak{a}$ with an example. The image of the map $\nu : \mathbb{P}^1 \longrightarrow \mathbb{P}^3$ defined by the homogeneous coordinates of projective space

$$[x_0 : x_1] \longmapsto [x_0^3 : x_0^2 x_1 : x_0 x_1^2 : x_1^3] = [z_0 : z_1 : z_2 : z_3]$$

is a smooth rational curve C, called twisted cubic. C is the intersection of three quadratic surfaces in $\mathbb{P}^3$:

$$Q_1 : F_1(z) := z_0 z_2 - z_1^2 = 0,$$

$$Q_2 : F_2(z) := z_0 z_3 - z_1 z_2 = 0,$$

$$Q_3 : F_3(z) := z_1 z_3 - z_2^2 = 0.$$

Indeed, the inclusion $C \subset Q_1 \cap Q_2 \cap Q_3$ is clear, and conversely, for $z \in Q_1 \cap Q_2 \cap Q_3$, $z_0 = z_3 = 0$ cannot happen. If $z_0 \neq 0$, we have

$$\nu([z_0 : z_1]) = \left[z_0 : z_1 : \frac{z_1^2}{z_0} : \frac{z_1^3}{z_0^2} \right] = z,$$

and similarly for the case $z_3 \neq 0$. Hence, we have $C = Q_1 \cap Q_2 \cap Q_3$. On the other hand, for example, studying $Q_1 \cap Q_2$, we see that

$$(F_1, F_2) = (F_1, F_2, F_3) \cap (z_0, z_1)$$

by computation. Here, (F_1, F_2) is the homogeneous ideal of the polynomial ring $\mathbb{C}[z_0, z_1, z_2, z_3]$ generated by F_1, F_2 and similarly for the other cases. $Q_1 \cap Q_2$ looks as in Figure 2.8.

Hence, C cannot be expressed as the intersection of two quadratic surfaces Q_1 and Q_2. Again, by simple computation, we see that

$$z_0 F_3 = -z_2 F_1 + z_1 F_2 \in (F_1, F_2)$$

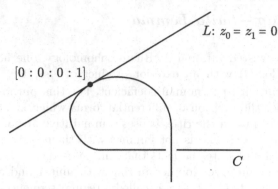

L is tangent to C at $[0:0:0:1]$

Fig. 2.8

and F_3 becomes a zero-divisor in the quotient ring $\mathbb{C}[z_0, z_1, z_2, z_3]/(F_1, F_2)$, which means the sequence F_1, F_2, F_3 is not a regular sequence of the ideal (F_1, F_2, F_3). In this way, a regular sequence $f_1, \ldots, f_t$ in a polynomial ring has an intuitive meaning that the dimensions of each algebraic subset $\{f_1 = 0\}, \{f_1 = f_2 = 0\}, \cdots$ decrease by 1. Since we should treat regular sequences in a quotient ring of a polynomial ring, we have formulated in a general manner.

Using this concept, let us explain the de Rham−Saito lemma concerning the division of differential forms which play an important role later. Let M be a A-free module of rank n with basis $e_1, \ldots, e_n$, and denote the pth exterior power of M by $\wedge^p M$. Notice that $\wedge^0 M = A$ and $\wedge^{-1}M = 0$. Now, for $\omega_1, \ldots, \omega_r \in M$, the element $\omega_1 \wedge \cdots \wedge \omega_r$ is uniquely expressed in the form

$$\omega_1 \wedge \cdots \wedge \omega_r = \sum_{1 \le i_1 < \cdots < i_r \le n} a_{i_1 \cdots i_r} e_{i_1} \wedge \cdots \wedge e_{i_r}, \quad a_{i_1 \cdots i_r} \in A. \quad (2.60)$$

Let $\mathfrak{a}$ be the ideal generated by the coefficients $a_{i_1 \cdots i_r}$, $1 \le i_1 < \cdots < i_r \le n$; we further use the symbols:

$$Z^p := \{\varphi \in \wedge^p M | \omega_1 \wedge \cdots \wedge \omega_r \wedge \varphi = 0\},$$

$$H^p := Z^p / \sum_{k=1}^{r} \omega_k \wedge \wedge^{p-1} M.$$

Since we prove the lemma below by induction, for $r = 0$, we also set $\mathfrak{a} = A$, $Z^p = 0$, and $H^p = 0$ $(p = 0, 1, \cdots)$. We have:

Lemma 2.18 (de Rham−Saito lemma [deR1], [Sai1]).

1. *There exists an integer $\nu \in \mathbb{Z}_{\ge 0}$ such that $\mathfrak{a}^\nu H^p = 0$, $0 \le p \le n$.*
2. *For $0 \le p < \operatorname{depth}_{\mathfrak{a}} A$, one has $H^p = 0$.*

Before going into the proof, we explain localization of a commutative ring A. For example, if we take $A = \mathbb{C}[u_1, \ldots, u_n]$, this can be regarded as the ring of all regular functions (in an algebraic sense) over the n-dimensional affine space $\mathbb{C}^n$. If we choose a non-constant $f \in A$ and consider the affine open subset $U_f := \mathbb{C}^n \setminus \{f = 0\}$, the ring of functions on it becomes

$$A_{(f)} := \{a/f^n | a \in A, \; n = 0, 1, \cdots\}.$$

If we fix a point $u_0 = (u_1^{(0)}, \ldots, u_n^{(0)})$ of $\mathbb{C}^n$, the ring of germs of regular functions at u_0 is expressed as

$$A_{\underline{m}} := \{a/b | a, b \in A, \; b(u_0) \neq 0\},$$

where $\underline{m}$ is the maximal ideal $(u_1 - u_1^{(0)}, \ldots, u_n - u_n^{(0)})$ of A. Notice that $\{b \in A | b(u_0) \neq 0\} = A \setminus \underline{m}$. In this way, on A, the abstraction of the operation passing to the ring of regular functions on an affine open subset or at a point u_0, is called a localization of A.

Definition 2.3. Let A be a commutative ring with 1, and S be a subset of A which is closed under the multiplication and which contains 1 but not 0. (a, s) and (b, t) belonging to $A \times S$ are said to be related if there exists $u \in S$ satisfying $(at - bs)u = 0$. This is an equivalence relation. The equivalence class of (a, s) is denoted by a/s and the set of all equivalence classes is denoted by $S^{-1}A$. Then, as ordinary computation on fractions, one can define the addition, the subtraction and the multiplication on $S^{-1}A$ and they define a new commutative ring. In particular, for $S = \{a^n | n = 0, 1, \ldots, \; \forall \, a^n \neq 0\}$, we denote $S^{-1}A$ by $A_{(a)}$, and for $S = A \setminus \underline{m}$ with a maximal ideal $\underline{m}$ of A, we denote $S^{-1}A$ by $A_{\underline{m}}$.

Proof of Lemma 2.18. (1)   Since A is a Notherian ring, the submodule Z^p of a finitely generated free module M is finitely generated over A. Hence, to show (1), it is sufficient to show that for any $a_{i_1 \cdots i_r} \in \mathfrak{a}$ and $\varphi \in Z^p$, there exists $m \in \mathbb{Z}_{\geq 0}$ such that

$$(a_{i_1 \ldots i_r})^m \varphi \in \sum_{i=1}^{r} \omega_i{}^{\wedge} \wedge^{p-1} M.$$

Consider two cases separately. First, if $a_{i_1 \cdots i_r}$ is nilpotent (i.e., some power of $a_{i_1 \cdots i_r}$ becomes zero), the assertion is clear. Second, if $a := a_{i_1 \cdots i_r}$ is not nilpotent, the set of the powers of a $\{a^k | k \in \mathbb{Z}_{\geq 0}\}$ does not contain 0 and is closed under the multiplication, hence, we can consider the localization $A_{(a)}$ of A and $M_{(a)} := M \otimes_A A_{(a)}$ of M. We remark that, since M is a finitely generated free module, the following isomorphism holds $(\wedge^p M) \otimes_A A_{(a)} \simeq \wedge^p (M \otimes_A A_{(a)}) \simeq \wedge^p M_{(a)}$. We consider the homomorphism induced from a natural homomorphism $u \in A \longrightarrow [u] \in A_{(a)}$

$$\wedge^p M \longrightarrow (\wedge^p M) \otimes_A A_{(a)} \simeq \wedge^p M_{(a)}.$$

$$\varphi \longrightarrow [\varphi]$$

Expanding $[\omega_1] \wedge \cdots \wedge [\omega_r]$ as in (2.60), the coefficient $[a_{i_1 \cdots i_r}] \in A_{(a)}$ of $e_{i_1} \wedge \cdots \wedge e_{i_r}$ is invertible in $A_{(a)}$. Writing

$$\omega_k = \sum_{i=1}^{n} u_{ki} e_i, \quad u_{ki} \in A, \quad 1 \le k \le r,$$

we remark that the $r \times r$ minor of the matrix

$$\begin{pmatrix} u_{11} & \cdots\cdots & u_{1n} \\ \vdots & & \vdots \\ u_{r1} & \cdots\cdots & u_{rn} \end{pmatrix} \in M_{r,n}(A)$$

obtained by removing the i_1th$\cdots$, the i_rth columns is nothing but $a_{i_1 \cdots i_r}$. For simplicity, we consider $a_{12 \cdots r}$. The determinant of the matrix in the relation

$$\begin{pmatrix} \omega_1 \\ \vdots \\ \omega_r \\ e_{r+1} \\ \vdots \\ e_n \end{pmatrix} = \begin{pmatrix} u_{11} & \cdots & u_{1r} & u_{1,r+1} & \cdots & u_{1n} \\ \vdots & & \vdots & \vdots & & \vdots \\ u_{r1} & \cdots & u_{rr} & u_{r,r+1} & \cdots & u_{rn} \\ & & & 1 & & \\ & 0 & & & \ddots & \\ & & & & & 1 \end{pmatrix} \begin{pmatrix} e_1 \\ \vdots \\ e_n \end{pmatrix} \tag{2.61}$$

is $a_{12 \cdots r}$, and if we consider this equation in $M_{(a)}$, its determinant $[a_{1 \cdots r}]$ is invertible and (2.61) can be solved. That is, $[\omega_1], \ldots, [\omega_r], e_{r+1}, \ldots, e_n$ form a basis of $M_{(a)}$. With this base, we express $[\varphi] \in \wedge^p M_{(a)}$ as

$$[\varphi] = \sum a_{JK} [\omega]_J \wedge e_K,$$

where $J = \{j_1, \ldots, j_s\}$, $1 \le j_1 < \cdots < j_s \le r$, $K = \{k_1, \ldots, k_t\}$, $r + 1 \le k_1 < \cdots < k_t \le n$, $s + t = p$, and

$$[\omega]_J = [\omega_{j_1}] \wedge \cdots \wedge [\omega_{j_s}], \quad e_K = e_{k_1} \wedge \cdots \wedge e_{k_t}.$$

The condition $[\omega_1] \wedge \cdots \wedge [\omega_r] \wedge [\varphi] = 0$ is equivalent to the fact that the above expression of $[\varphi]$ does not contain the term corresponding to $|K| = p$. This can be further expressed as

$$[\varphi] = \sum_{k=1}^{r} [\omega_k] \wedge \alpha'_k, \quad \alpha'_k \in \wedge^{p-1} M_{(a)},$$

where α'_k also have an expression $\alpha'_k = \alpha_k/a^m$, $\alpha_k \in \wedge^{p-1} M$, $m \in \mathbb{Z}_{\geq 0}$. Hence, in $\wedge^p M_{(a)}$, we have

$$\left[a^m \varphi - \sum_{k=1}^{r} \omega_k \wedge \alpha_k \right] = 0.$$

By the definition of the localization $A_{(a)}$, there exists $m' \in \mathbb{Z}_{\geq 0}$ such that, in $\wedge^p M$, we have

$$a^{m'} \left\{ a^m \varphi - \sum_{k=1}^{r} \omega_k \wedge \alpha_k \right\} = 0.$$

This is what we should show.

(2) We show the lemma by a double induction on r and p.

(i) For $r = 0$, as we have defined $H^p = 0$ for any p, the assertion is trivial.

(ii) For $p = 0$, as $p = 0 < \text{depth}_a A$, there exists an element $u \in a$ which is not a zero-divisor of A. Now, $\varphi \in Z^0$ implies $\varphi \in A$ and $\varphi \wedge \omega_1 \wedge \cdots \wedge \omega_r = 0$, hence by the definition of a, we have $\varphi u = 0$. Since u is not a zero-divisor of A, we have $\varphi = 0$, i.e., $Z^0 = 0$. Hence, we obtain $H^0 = 0$.

(iii) For $r > 0$ and $0 < p < \text{depth}_a A$, we assume that the assertion holds for $(r - 1, p)$ and $(r, p - 1)$, and we show the assertion for (r, p). By $\text{depth}_a A > 0$, A possesses an element $a \in a$ which is not a zero-divisor. From the proof of (1), first, we can take $m \in \mathbb{Z}_{\geq 0}$ such that $a^m H^p = 0$, but since a^m is not a zero-divisor of A and belongs to a, we may assume that $m = 1$: we remark $(\wedge^p M) \otimes_A (A/aA) \simeq \wedge^p (M \otimes_A (A/aA))$ as $aH^p = 0$ and M is a finitely generated free module. We also remark that the kernel of a natural surjection

$$\wedge^p M \longrightarrow (\wedge^p M) \otimes_A (A/aA)$$

$$\omega \longrightarrow \overline{\omega}$$

is $a \wedge^p M$. For $\varphi \in Z^p$, by $aH^p = 0$, we have an expression

$$a\varphi = \sum_{k-1}^{r} \omega_k \wedge \alpha_k, \quad \alpha_k \in \wedge^{p-1} M, \quad 1 \leq k \leq r, \tag{2.62}$$

and from the above remark, we have

$$\sum_{k=1}^{r} \overline{\omega}_k \wedge \overline{\alpha}_k = 0.$$

From this, we obtain

$$\overline{\omega}_1 \wedge \cdots \wedge \overline{\omega}_r \wedge \overline{\alpha}_k = (-1)^{r-k} \overline{\omega}_1 \wedge \cdots \wedge \widehat{\overline{\omega}_k} \wedge \cdots \wedge \overline{\omega}_r \wedge \left(\sum \overline{\omega}_k \wedge \overline{\alpha}_k \right)$$

$$= 0.$$

Expanding $\overline{\omega}_1 \wedge \cdots \wedge \overline{\omega}_r$ as in (2.60), the ideal of A/aA generated by their coefficients is $\mathfrak{a}/aA$, and by the definition of regular sequence, we have

$$\mathrm{depth}_{\mathfrak{a}/aA}(A/aA) = \mathrm{depth}_{\mathfrak{a}} A - 1 > p - 1 \geq 0.$$

Hence, by the induction hypothesis for $(r, p-1)$, $\overline{\alpha}_k$ can be expressed as

$$\overline{\alpha}_k = \sum_{l=1}^{r} \overline{\omega}_l \wedge \overline{\beta}_{kl}, \quad \beta_{kl} \in \wedge^{p-2} M, \quad 1 \leq k \leq r.$$

Further, from the above remark, we also have

$$\alpha_k - \sum_{l=1}^{r} \omega_l \wedge \beta_{kl} = a \gamma_k, \quad \gamma_k \in \wedge^{p-2} M.$$

By this formula, (2.62) becomes

$$a \left(\varphi - \sum_{k=1}^{r} \omega_k \wedge \gamma_k \right) = \sum_{k \neq l} \omega_k \wedge \omega_l \wedge \beta_{kl}.$$

Each term of the right-hand side contains different ω_k and ω_l which means

$$a \left(\varphi - \sum_{k=1}^{r} \omega_k \wedge \gamma_k \right) \wedge \omega_2 \wedge \cdots \wedge \omega_r = 0,$$

and since a is not a zero-divisor of A, this further implies

$$\left(\varphi - \sum_{k=1}^{r} \omega_k \wedge \gamma_k \right) \wedge \omega_2 \wedge \cdots \wedge \omega_r = 0.$$

For the ideal $\mathfrak{a}'$ generated by the coefficients of the expansion of $\omega_2 \wedge \cdots \wedge \omega_r$, it follows from a simple computation that $\mathfrak{a} \subset \mathfrak{a}'$ from which, by the definition of depth, we obtain $p < \mathrm{depth}_{\mathfrak{a}} A \leq \mathrm{depth}_{\mathfrak{a}'} A$. Hence, with the case of $(r-1, p)$, we obtain an expression

$$\varphi - \sum_{k=1}^{r} \omega_k \wedge \gamma_k = \sum_{k=2}^{r} \omega_k \wedge \gamma_k', \quad \gamma_k' \in \wedge^{p-1} M,$$

and this is what we are looking for.

2.5.3 Representation of Logarithmic Differential Forms (i)

Let us rephrase the above lemma in a suitable form for us. We consider non-constant homogeneous polynomials $\overline{P}_j(u)$, $1 \leq j \leq m$ with variables $u_1, \ldots, u_n$. To avoid the complexity of symbols, let $(d\overline{P}_{j_1} \wedge \cdots \wedge d\overline{P}_{j_r}, \overline{P}_{j_1}, \cdots, \overline{P}_{j_r})$ be the ideal of $\mathbb{C}[u_1, \ldots, u_n]$ generated by all $r \times r$-minors

$$\frac{\partial(\overline{P}_{j_1}, \ldots, \overline{P}_{j_r})}{\partial(u_{i_1}, \ldots, u_{i_r})}, \quad 1 \leq i_1 < \cdots < i_r \leq n$$

of the Jacobi matrix

$$\begin{pmatrix} \frac{\partial \overline{P}_{j_1}}{\partial u_1} & \cdots \cdots & \frac{\partial \overline{P}_{j_1}}{\partial u_n} \\ \vdots & & \vdots \\ \frac{\partial \overline{P}_{j_r}}{\partial u_1} & \cdots \cdots & \frac{\partial \overline{P}_{j_r}}{\partial u_n} \end{pmatrix}$$

and $\overline{P}_{j_1}, \ldots, \overline{P}_{j_r}$. This symbol takes the expansion of the exterior product $d\overline{P}_{j_1} \wedge \cdots \wedge d\overline{P}_{j_r}$ with respect to the basis $du_{i_1} \wedge \cdots \wedge du_{i_r}, 1 \leq i_1 < \cdots < i_r \leq n$

$$d\overline{P}_{j_1} \wedge \cdots \wedge d\overline{P}_{j_r} = \sum \frac{\partial(\overline{P}_{j_1}, \ldots, \overline{P}_{j_r})}{\partial(u_{i_1}, \ldots, u_{i_r})} du_{i_1} \wedge \cdots \wedge du_{i_r}$$

into account and is practical as it simplifies formulas.

Here, we explain some notation and terminologies that will be used later. The set of all prime ideals of a commutative ring A is denoted by $\mathrm{Spec}(A)$. Starting from a prime ideal $\mathfrak{p} = \mathfrak{p}_0$, the maximal length k of the sequences $\mathfrak{p}_0 \supsetneq \mathfrak{p}_1 \supsetneq \cdots \supsetneq \mathfrak{p}_k$ of prime ideals in A is called the height of $\mathfrak{p}$ and is denoted by height $\mathfrak{p}$. For any ideal I of A, we set

$$\text{height } I = \inf\{\text{height } \mathfrak{p} | \mathfrak{p} \supset I\}.$$

We remark that, for a ring of polynomials, this corresponds to the codimension $\mathrm{codim} V(I)$ of the algebraic set $V(I)$ defined by I.

Assumption 2. *(1) For any $1 \leq r \leq \min\{m, n-1\}$ and $1 \leq j_1 < \cdots < j_r \leq m$, the algebraic subset defined by the ideal $\{d\overline{P}_{j_1} \wedge \cdots \wedge d\overline{P}_{j_r}, \overline{P}_{j_1}, \cdots, \overline{P}_{j_r}\}$ consists of finitely many points of $\mathbb{C}^n$. That is, in the language of commutative algebras, one has*

$$\text{height}(d\overline{P}_{j_1} \wedge \cdots \wedge d\overline{P}_{j_r}, \overline{P}_{j_1}, \cdots, \overline{P}_{j_r}) \geq n. \tag{2.63}$$

(2) For any $1 \leq s \leq \min\{m, n\}$ and $1 \leq j_1 < \cdots < j_s \leq m$, $\overline{P}_{j_1}, \ldots, \overline{P}_{j_s}$ forms a regular sequence.

Remark 2.6. In this book, one of the most important cases satisfying Assumption 2 is an arrangement of m hyperplanes in $\mathbb{C}^n$ such that the arrangement of $m + 1$ hyperplanes in $\mathbb{P}^n$ obtained by adding the hyperplane at infinity is in general position (cf. § 2.9.1 for its precise definition).

We reformulate Lemma 2.18 to the following form so that we can use the division of differential forms. This plays an essential role in representing logarithmic differential forms. We owe the proof given here to T. Yamazaki.

Lemma 2.19. *Under Assumption 2, for $\psi \in \Omega^p(\mathbb{C}^n)$ such that $r + p \leq n - 1$, if the condition*

$$d\overline{P}_{j_1} \wedge \cdots \wedge d\overline{P}_{j_r} \wedge \psi \equiv 0 \pmod{\overline{P}_{j_1}, \cdots, \overline{P}_{j_r}}$$

is satisfied, then we have

$$\psi \equiv 0 \pmod{d\overline{P}_{j_1}, \ldots, d\overline{P}_{j_r}, \overline{P}_{j_1}, \ldots, \overline{P}_{j_r}}.$$

Proof. We use the symbols:

$$B := \mathbb{C}[u_1, \ldots, u_n], \quad I := (d\overline{P}_{j_1} \wedge \cdots \wedge d\overline{P}_{j_r}, \overline{P}_{j_1}, \ldots, \overline{P}_{j_r}),$$

$$A := B/(\overline{P}_{j_1}, \ldots, \overline{P}_{j_r}), \quad \mathfrak{a} := I/(\overline{P}_{j_1}, \ldots, \overline{P}_{j_r}),$$

$$\pi : B \longrightarrow A \quad \text{(a natural homomorphism)}.$$

As $\mathfrak{a}$ is an ideal of A and $I = \pi^{-1}(\mathfrak{a})$, we have $B/I \simeq A/\mathfrak{a}$. Applying Lemma 2.18 to our A, $\mathfrak{a}$, and $d\overline{P}_{j_1}, \ldots, d\overline{P}_{j_r}$ here, we see that it is sufficient to show this lemma for either $\mathfrak{a} = A$ or $\text{depth}_{\mathfrak{a}} A = n - r$. Let us consider two cases separately. First, if $I = B$, then we have $\mathfrak{a} = A$ and Lemma 2.18 (1) implies $H^p = 0$, $0 \leq p \leq n$. Second, if $I \neq B$, by Assumption 2 (1), we have height $I = n$. Here, we recall the known fact:

$$\text{depth}_{\mathfrak{a}} A = \inf\{\text{depth} A_{\underline{m}} | \underline{m} \in V(\mathfrak{a})\}, \tag{2.64}$$

where $V(\mathfrak{a})$ is the set $\{\underline{m} \in \text{Spec}(A) | \mathfrak{a} \subset \underline{m}\}$ of all prime ideals of A containing $\mathfrak{a}$, $A_{\underline{m}}$ is the localization of A at the prime ideal $\underline{m}$, and $\text{depth} A_{\underline{m}}$ is the depth of the maximal ideal $\underline{m} A_{\underline{m}}$ of the local ring $A_{\underline{m}}$ (cf. [Matsu], p.105, for its proof). Since height $I = n$, we have $\dim A/\mathfrak{a} = \dim B/I = 0$ which implies that the prime ideal $\mathfrak{a} \subset \underline{m}$ of A is a maximal ideal. $\underline{n} := \pi^{-1}(\underline{m})$ is a maximal ideal of B, and for its localization $B_{\underline{n}}$, there is an isomorphism

$$A_{\underline{m}} \simeq B_{\underline{n}}/(\overline{P}_{j_1}, \ldots, \overline{P}_{j_r}) B_{\underline{n}}. \tag{2.65}$$

By Assumption 2 (2), since the sequence $\overline{P}_{j_1}, \ldots, \overline{P}_{j_r}$ is regular in the ring B, it is also regular in the localization $B_{\underline{n}}$. But since $B_{\underline{n}}$ is the regular local ring of germs of regular functions at the point of $\mathbb{C}^n$ corresponding to the maximal ideal $\underline{n}$, we have $\text{depth} B_{\underline{n}} = \dim B_{\underline{n}} = n$. Hence, $A_{\underline{m}}$ is a Cohen–Macaulay local ring, i.e., it is a local ring satisfying

$$\text{depth} A_{\underline{m}} = \dim A_{\underline{m}}.$$

Moreover, the formula $\operatorname{depth} A_{\underline{m}} = \operatorname{depth} B_{\underline{n}} - r$ is known (cf. [Matsu], p.107). Hence, we obtain

$$\operatorname{depth} A_{\underline{m}} = \dim B_{\underline{n}} - r,$$

$$= n - r$$

and this together with (2.64) imply $\operatorname{depth}_a A = n - r$.

With these preparations, we prove the following proposition which gives a concrete expression of logarithmic differential forms.

Proposition 2.2 (Representation of logarithmic differential forms).
Suppose that m homogeneous polynomials $\overline{P}_j(u) \in \mathbb{C}[u_1, \dots, u_n]$ satisfy Assumption 2. If $\psi \in \Omega^p(\mathbb{C}^n)$ for $0 \le p \le n - 2$ satisfies the condition

$$d\overline{P}_j \wedge \psi \equiv 0 \pmod{\overline{P}_j}, \quad 1 \le j \le m, \tag{2.66}$$

ψ can be expressed in the form:

$$\psi = \overline{P}_1 \cdots \overline{P}_m \left\{ \psi_0 + \sum_{j=1}^{m} \frac{d\overline{P}_j}{\overline{P}_j} \wedge \psi_j + \cdots \right. \tag{2.67}$$

$$\cdots + \left. \sum_{1 \le j_1 < \cdots < j_p \le m} \frac{d\overline{P}_{j_1}}{\overline{P}_{j_1}} \wedge \cdots \wedge \frac{d\overline{P}_{j_p}}{\overline{P}_{j_p}} \psi_{j_1 \cdots j_p} \right\},$$

$$\psi_{j_1 \cdots j_\nu} \in \Omega^{p-\nu}(\mathbb{C}^n), \quad 1 \le j_1 < \cdots < j_\nu \le m, \quad 0 \le \nu \le p.$$

Proof. Let us show this by induction on m. For $m = 1$, by the condition (2.66), Lemma 2.19 implies that for $\psi \in \Omega^p(\mathbb{C}^n)$ with $p \le n - 2$, we have

$$\psi \equiv 0 \pmod{d\overline{P}_1, \overline{P}_1}.$$

Hence, there exist $\psi_0 \in \Omega^p(\mathbb{C}^n)$ and $\psi_1 \in \Omega^{p-1}(\mathbb{C}^n)$ such that

$$\psi = \overline{P}_1 \psi_0 + d\overline{P}_1 \wedge \psi_1 = \overline{P}_1 \left\{ \psi_0 + \frac{d\overline{P}_1}{\overline{P}_1} \wedge \psi_1 \right\},$$

which is what we should prove. Suppose that this proposition is proved for m polynomials, and we assume that $\psi \in \Omega^p(\mathbb{C}^n)$ with $0 \le p \le n - 2$ satisfies $d\overline{P}_j \wedge \psi \equiv 0 \pmod{\overline{P}_j}$, $1 \le j \le m + 1$. By induction hypothesis, ψ can be expressed in the form (2.67). Let $N \in \mathbb{Z}_{\ge 0}$ be the maximal integer satisfying $\psi_{j_1 \cdots j_N} \ne 0$. By induction on N, we show that ψ can be expressed in the form similar to (2.67) for $m + 1$ $\overline{P}_1, \dots, \overline{P}_{m+1}$. For $N = 0$, we have $\psi = \overline{P}_1 \cdots \overline{P}_m \psi_0$. On the other hand, since we have $d\overline{P}_{m+1} \wedge \psi \equiv 0 \pmod{\overline{P}_{m+1}}$, it follows that

$$\overline{P}_1 \cdots \overline{P}_m d\overline{P}_{m+1} \wedge \psi_0 \equiv 0 \pmod{\overline{P}_{m+1}}.$$

As $n \geq 2$, by Assumption 2 (2), $\{\overline{P}_j, \overline{P}_{m+1}\}$ for $1 \leq j \leq m$ form regular sequences, we obtain $d\overline{P}_{m+1} \wedge \psi_0 \equiv 0 \pmod{\overline{P}_{m+1}}$. Applying Lemma 2.19 to this formula, we have $\psi_0 \equiv 0 \pmod{d\overline{P}_{m+1}, \overline{P}_{m+1}}$ which implies that ψ_0 can be expressed as

$$\psi_0 = \overline{P}_{m+1}\alpha + d\overline{P}_{m+1} \wedge \beta, \quad \alpha \in \Omega^p(\mathbb{C}^n), \quad \beta \in \Omega^{p-1}(\mathbb{C}^n).$$

Hence, ψ can be expressed as

$$\psi = \overline{P}_1 \cdots \overline{P}_{m+1} \left\{ \alpha + \frac{d\overline{P}_{m+1}}{\overline{P}_{m+1}} \wedge \beta \right\},$$

and the proposition for $N = 0$ is proved. Next, suppose that the proposition is proved for $N - 1$, and let us prove it for N. First, we remark that, by the choice of N, we have $\psi_{j_1 \cdots j_\nu} = 0$ for $\nu \geq N + 1$. For simplicity, if we consider the case with the index $(12 \cdots N)$, by (2.67), we have

$$\psi - \overline{P}_{N+1} \cdots \overline{P}_m d\overline{P}_1 \wedge \cdots \wedge d\overline{P}_N \wedge \psi_{1 \dots N} \equiv 0 \tag{2.68}$$

$$(\text{mod } \overline{P}_1, \cdots, \overline{P}_N).$$

As $N \leq p \leq n - 2$, Assumption 2 (2) implies that, for any $N + 1 \leq j \leq m$, $\{\overline{P}_1, \ldots, \overline{P}_N, \overline{P}_j, \overline{P}_{m+1}\}$ forms a regular sequence. On the other hand, as we have $d\overline{P}_{m+1} \wedge \psi \equiv 0 \pmod{\overline{P}_{m+1}}$ by assumption, this together with (2.68) implies

$$d\overline{P}_1 \wedge \cdots \wedge d\overline{P}_N \wedge d\overline{P}_{m+1} \wedge \psi_{1 \dots N} \equiv 0 \pmod{\overline{P}_1, \cdots, \overline{P}_N, \overline{P}_{m+1}}.$$

By $\psi_{1 \dots N} \in \Omega^{p-N}(\mathbb{C}^n)$ and $N \leq p \leq n - 2$, we obtain from Lemma 2.19

$$\psi_{1 \dots N} \equiv 0 \pmod{d\overline{P}_1, \ldots, d\overline{P}_N, d\overline{P}_{m+1}, \overline{P}_1, \ldots, \overline{P}_N, \overline{P}_{m+1}},$$

and $\psi_{1 \dots N}$ can be expressed in the form:

$$\psi_{1 \dots N} = \sum_{j=1}^{N} \overline{P}_j \alpha_{1 \dots N; \, j} + \overline{P}_{m+1} \alpha_{1 \dots N; \, m+1}$$

$$+ \sum_{j=1}^{N} d\overline{P}_j \wedge \beta_{1 \dots N; \, j} + d\overline{P}_{m+1} \wedge \beta_{1 \dots N; \, m+1},$$

$$\alpha_{1 \dots N; \, j} \in \Omega^{p-N}(\mathbb{C}^n), \quad \beta_{1 \dots N; \, j} \in \Omega^{p-N-1}(\mathbb{C}^n).$$

Similarly, we have

$$\psi_{j_1 \cdots j_N} = \sum_{k=1}^{N} \overline{P}_{j_k} \alpha_{j_1 \cdots j_N;\, j_k} + \overline{P}_{m+1} \alpha_{j_1 \cdots j_N;\, m+1} \tag{2.69}$$

$$+ \sum_{k=1}^{N} d\overline{P}_{j_k} \wedge \beta_{j_1 \cdots j_N;\, j_k} + d\overline{P}_{m+1} \wedge \beta_{j_1 \cdots j_N;\, m+1},$$

$$\alpha_{j_1 \cdots j_N;\, j_k} \in \Omega^{p-N}(\mathbb{C}^n), \quad \beta_{j_1 \cdots j_N;\, j_k} \in \Omega^{p-N-1}(\mathbb{C}^n).$$

Replacing (2.67) with (2.69), we obtain

$$\psi = \overline{P}_1 \cdots \overline{P}_m \left\{ \sum_{\nu=0}^{N-1} \sum_{1 \le j_1 < \cdots < j_\nu \le m} \frac{d\overline{P}_{j_1}}{\overline{P}_{j_1}} \wedge \cdots \wedge \frac{d\overline{P}_{j_\nu}}{\overline{P}_{j_\nu}} \wedge \psi_{j_1 \cdots j_\nu} \right\} \tag{2.70}$$

$$+ \overline{P}_1 \cdots \overline{P}_m \left\{ \sum_{1 \le j_1 < \cdots < j_\nu \le m} \frac{d\overline{P}_{j_1}}{\overline{P}_{j_1}} \wedge \cdots \wedge \frac{d\overline{P}_{j_N}}{\overline{P}_{j_N}} \right.$$

$$\left. \wedge \left(\sum_{k=1}^{N} \overline{P}_{j_k} \alpha_{j_1 \cdots j_N;\, j_k} + \overline{P}_{m+1} \alpha_{j_1 \cdots j_N;\, m+1} \right) \right\}$$

$$+ \overline{P}_1 \cdots \overline{P}_m \left\{ \sum_{1 \le j_1 < \cdots < j_N \le m} \frac{d\overline{P}_{j_1}}{\overline{P}_{j_1}} \wedge \cdots \wedge \frac{d\overline{P}_{j_N}}{\overline{P}_{j_N}} \right.$$

$$\left. \wedge \left(\sum_{k=1}^{N} d\overline{P}_{j_k} \wedge \beta_{j_1 \cdots j_N;\, j_k} + d\overline{P}_{m+1} \wedge \beta_{j_1 \cdots j_N;\, m+1} \right) \right\}.$$

Notice that in the right-hand side of the above formula, the last { } becomes

$$\sum_{1 \le j_1 < \cdots < j_N \le m} \frac{d\overline{P}_{j_1}}{\overline{P}_{j_1}} \wedge \cdots \wedge \frac{d\overline{P}_{j_N}}{\overline{P}_{j_N}} \wedge d\overline{P}_{m+1} \wedge \beta_{j_1 \cdots j_N;\, m+1}.$$

Now, in the right-hand side of the above formula, picking up the terms containing $\overline{P}_{m+1}$, we set

$$\eta := \psi - \overline{P}_1 \cdots \overline{P}_{m+1} \left\{ \sum_{1 \le j_1 < \cdots < j_N \le m} \frac{d\overline{P}_{j_1}}{\overline{P}_{j_1}} \wedge \cdots \right. \tag{2.71}$$

$$\wedge \frac{d\overline{P}_{j_N}}{\overline{P}_{j_N}} \wedge \alpha_{j_1 \cdots j_N;\, m+1}$$

$$\left. + \sum_{1 \le j_1 < \cdots < j_N \le m} \frac{d\overline{P}_{j_1}}{\overline{P}_{j_1}} \wedge \cdots \wedge \frac{d\overline{P}_{j_N}}{\overline{P}_{j_N}} \wedge \frac{d\overline{P}_{m+1}}{\overline{P}_{m+1}} \wedge \beta_{j_1 \cdots j_N;\, m+1} \right\}.$$

Comparing this with (2.70), we see that η is expressed in the form:

$$\eta = \overline{P}_1 \cdots \overline{P}_m \left\{ \sum_{\nu=0}^{N-1} \sum_{1 \le j_1 < \cdots < j_\nu \le m} \frac{d\overline{P}_{j_1}}{\overline{P}_{j_1}} \wedge \cdots \wedge \frac{d\overline{P}_{j_\nu}}{\overline{P}_{j_\nu}} \wedge \eta_{j_1 \cdots j_\nu} \right\}, \quad (2.72)$$

$$\eta_{j_1 \cdots j_\nu} \in \Omega^{p-\nu}(\mathbb{C}^n).$$

From the conditions $d\overline{P}_j \wedge \psi \equiv 0 \pmod{\overline{P}_j}$, $1 \le j \le m+1$ and (2.71), one can easily check $d\overline{P}_j \wedge \eta \equiv 0 \pmod{\overline{P}_j}$, $1 \le j \le m+1$, and from the form of (2.72), one can apply induction to η, hence, η can be expressed as

$$\eta = \overline{P}_1 \cdots \overline{P}_{m+1} \left\{ \tilde{\eta}_0 + \sum_{j=1}^{m+1} \frac{d\overline{P}_j}{\overline{P}_j} \wedge \tilde{\eta}_j + \cdots \right.$$

$$\left. \cdots + \sum_{1 \le j_1 < \cdots < j_p \le m+1} \frac{d\overline{P}_{j_1}}{\overline{P}_{j_1}} \wedge \cdots \wedge \frac{d\overline{P}_{j_p}}{\overline{P}_{j_p}} \wedge \tilde{\eta}_{j_1 \cdots j_p} \right\}.$$

Rewriting (2.71) with this formula, we see that η is expressed in the form we are looking for and the induction is terminated.

Remark 2.7. For $\varphi \in \Omega^p(\log D)$ with $0 \le p \le n-2$, by definition, one can express as $\varphi = \psi/P$, $\psi \in \Omega^p(\mathbb{C}^n)$, and ψ for $1 \le j \le m$ satisfies $\frac{d\overline{P}_j}{\overline{P}_j} \wedge \psi \in \Omega^{p+1}(\mathbb{C}^n)$. Hence, ψ satisfies the assumption of Proposition 2.2. Under Assumption 2 on the divisor D, φ is expressed in the form of the parenthesis of the right-hand side of (2.67). In the discussion given below, the reader should notice that it is essential to express elements of $\Omega^p(\log D)$ in this concrete form.

2.6 Vanishing of Twisted Cohomology for Homogeneous Case

2.6.1 Basic Operators

For m homogeneous polynomials $\overline{P}_j(u)$, $1 \le j \le m$ in n variables $u = (u_1, \ldots, u_n)$, we use the same notation as in § 2.5: let $\overline{D} = \cup\{\overline{P}_j = 0\}$ be the divisor defined by $\overline{P} = \overline{P}_1 \cdots \overline{P}_m$, and set $\overline{U}(u) = \prod_{j=1}^m \overline{P}_j(u)^{\alpha_j}$, $\overline{\omega} = d\overline{U}/\overline{U}$, $\nabla_{\overline{\omega}} = d + \overline{\omega}\wedge$. To compute the cohomologies of the twisted de Rham complex $(\Omega^\bullet(*\overline{D}), \nabla_{\overline{\omega}})$ and its subcomplex $(\Omega^\bullet(\log \overline{D}), \nabla_{\overline{\omega}})$, we first recall some basic operators acting on $\Omega^\bullet(*\overline{D})$.

A linear map θ of vector spaces on $\Omega^\bullet(*D)$ such that there exists an even integer r satisfying $\theta(\Omega^p(*D)) \subset \Omega^{p+r}(*D)$, $0 \le p \le n$, and

$$\theta(\alpha \wedge \beta) = \theta(\alpha) \wedge \beta + \alpha \wedge \theta(\beta), \quad \alpha, \beta \in \Omega^\bullet(*D)$$

is called a derivation of degree r. A linear map δ such that there exists an odd integer r satisfying $\delta(\Omega^p(*D)) \subset \Omega^{p+r}(*D)$, $0 \le p \le n$, and

$$\delta(\alpha \wedge \beta) = \delta\alpha \wedge \beta + (-1)^p\alpha \wedge \delta\beta, \quad \alpha \in \Omega^p(*D), \quad \beta \in \Omega^\bullet(*D)$$

is called an anti-derivation of degree r. For example, the exterior derivative d is an anti-derivation of degree 1. We remark that since $\Omega^\bullet(*D)$ is generated by $\Omega^0(*D)$ and $du_1, \dots, du_n$, a derivation θ and an anti-derivation δ are uniquely determined by their values on $\Omega^0(*D)$ and $du_1, \dots, du_n$. Now, let $v = \sum_{i=1}^n u_i \frac{\partial}{\partial u_i}$ be the Euler vector field. We introduce a derivation and an anti-derivation determined uniquely by the two conditions:

1. The Lie derivative L_v with respect to the Euler vector field v: L_v is a derivation of degree 0 defined for a rational function $f \in \Omega^0(*D)$ by $L_v(f) = v(f)$ and for du_i by $L_v(du_i) = du_i$, $1 \le i \le n$.
2. The interior product i_v with respect to the Euler vector field v: i_v is an anti-derivation of degree -1 defined for $f \in \Omega^0(*D)$ by $i_v(f) = 0$ and for du_i by $i_v(du_i) = u_i$, $1 \le i \le n$.

As we see below, $\Omega^\bullet(*\overline{D})$ and $\Omega^\bullet(\log \overline{D})$ are stable under the action of i_v and L_v. First, since $\varphi \in \Omega^p(*\overline{D})$ is the sum of elements of the form

$$\alpha := a(u)du_{i_1} \wedge \cdots \wedge du_{i_p} / \overline{P}_1^{\nu_1} \cdots \overline{P}_m^{\nu_m}, \ a \in \mathbb{C}[u], \ \nu_j \in \mathbb{Z}_{\ge 0},$$

it suffices to show $i_v(\alpha)$, $L_v(\alpha) \in \Omega^\bullet(*\overline{D})$. By a simple computation, we have

$$i_v(\alpha) = \frac{a(u)}{\overline{P}_1^{\nu_1} \cdots \overline{P}_m^{\nu_m}} \sum_{k=1}^p (-1)^{k-1} u_{i_k} du_{i_1} \wedge \cdots \wedge \widehat{du_{i_k}} \wedge \cdots \wedge du_{i_p}$$

which implies $i_v(\alpha) \in \Omega^{p-1}(*D)$. Similarly, notice that $L_v(a/\overline{P}_1^{\nu_1} \cdots \overline{P}_m^{\nu_m}) = $ a polynomial$/(\overline{P}_1^{\nu_1} \cdots \overline{P}_m^{\nu_m})^2$. We have

$$L_v(\alpha) = L_v\left(\frac{a}{\overline{P}_1^{\nu_1} \cdots \overline{P}_m^{\nu_m}}\right) du_{i_1} \wedge \cdots \wedge du_{i_p}$$

$$+ \frac{a}{\overline{P}_1^{\nu_1} \cdots \overline{P}_m^{\nu_m}} \sum_{k=1}^p du_{i_1} \wedge \cdots \wedge du_{i_k} \wedge \cdots \wedge du_{i_p},$$

which implies $L_v(\alpha) \in \Omega^p(*D)$. Next, let us show that $\Omega^\bullet(\log \overline{D})$ is stable under i_v and L_v. For $\varphi \in \Omega^p(\log \overline{D})$, we have $\overline{P}\varphi = \alpha \in \Omega^p(\mathbb{C}^n)$ and $\frac{d\overline{P}}{\overline{P}} \wedge \alpha \in \Omega^{p+1}(\mathbb{C}^n)$. Notice that since $\overline{P}$ is a homogeneous polynomial, the Euler identity implies $i_v(d\overline{P}) = v(\overline{P}) = (\deg \overline{P})\overline{P}$. Hence, we have

$$\overline{P}i_v(\varphi) = i_v(\alpha) \in \Omega^{p-1}(\mathbb{C}^n),$$

$$\Omega^p(\mathbb{C}^n) \ni i_v\left(\frac{d\overline{P}}{\overline{P}} \wedge \alpha\right) = (\deg \overline{P})\alpha - \frac{d\overline{P}}{\overline{P}} \wedge i_v(\alpha).$$

From the second formula, we see that $\frac{d\overline{P}}{\overline{P}} \wedge i_v(\alpha) \in \Omega^p(\mathbb{C}^n)$, hence, the first formula together with (2.57) implies $i_v(\varphi) \in \Omega^{p-1}(\log D)$. Finally, let us show $L_v(\varphi) \in \Omega^p(\log \overline{D})$. Notice that $L_v(\overline{P}) = (\deg \overline{P})\overline{P}$, $L_v(d\overline{P}) = (deg\overline{P})d\overline{P}$. Applying L_v to the both sides of (2.57), we obtain

$$(\deg \overline{P})\overline{P}\varphi + \overline{P}L_v(\varphi) \in \Omega^p(\mathbb{C}^n),$$

$$(\deg \overline{P})d\overline{P} \wedge \varphi + d\overline{P} \wedge L_v(\varphi) \in \Omega^{p+1}(\mathbb{C}^n).$$

By $\overline{P}\varphi$, $d\overline{P} \wedge \varphi \in \Omega^\bullet(\mathbb{C}^n)$, this implies $\overline{P}L_v(\varphi) \in \Omega^p(\mathbb{C}^n)$, $d\overline{P} \wedge L_v(\varphi) \in \Omega^{p+1}(\mathbb{C}^n)$, which means $L_v(\varphi) \in \Omega^p(\log \overline{D})$.

2.6.2 Homotopy Formula

For two operators A, B acting on $\Omega^\bullet(*\overline{D})$, we define their commutator $[A, B]$ by $[A, B] = AB - BA$. Then, by the fact that the degree of a derivation is even and that of an anti-derivation is odd, simple calculations show:

(1)   If δ_1, δ_2 are anti-derivatives, then $\delta_1\delta_2 + \delta_2\delta_1$ is a derivative.
(2)   If θ_1, θ_2 are derivatives, then so is $[\theta_1, \theta_2]$.
(3)   If θ is a derivative and δ is an anti-derivative, then $[\theta, \delta]$ is an anti-derivative.

From (1)–(3) and the fact that a derivation and an anti-derivation are determined by their actions on $\Omega^0(*D)$ and $du_1, \ldots, du_n$, the following relations can be easily checked:

$$[L_v, d] = 0, \quad [L_v, i_v] = 0, \quad [L_v, \overline{\omega}\wedge] = 0, \quad [L_v, \nabla_{\overline{\omega}}] = 0. \quad (2.73)$$

In fact, the first two formulas of (2.73) follow from the remark mentioned above, the third formula follows immediately from $L_v(d\overline{P}_j/\overline{P}_j) = 0$, and this together with the first formula implies the last one. Moreover, the following homotopy formula is known:

$$d \circ i_v + i_v \circ d = L_v.$$

This can be shown as follows: Since d and i_v are anti-derivations of degree 1 and -1 respectively, the (1) above implies that the left-hand side of this formula is a derivation of degree 0. Since the action of the both sides on $\Omega^0(*D)$ and du_i coincide by simple calculation, we conclude the equality.

Let us deform this homotopy formula in a way suitable for our purpose. Set $l_j = \deg \overline{P}_j$, $1 \le j \le m$. By simple computation, we have

$$i_v(\overline{\omega}) = \sum_{j=1}^{m} l_j \alpha_j,$$

which together with the above homotopy formula imply the homotopy formula of $\nabla_{\overline{\omega}}$:

$$\nabla_{\overline{\omega}} \circ i_v + i_v \circ \nabla_{\overline{\omega}} = L_v + \sum_{j=1}^{m} l_j \alpha_j, \tag{2.74}$$

where

$$\overline{\omega} = \sum_{j=1}^{m} \alpha_j \frac{d\overline{P}_j}{\overline{P}_j}, \quad \deg \overline{P}_j = l_j.$$

2.6.3 Eigenspace Decomposition

The plan of the remaining part of this section is as follows. With all these preparations, we first decompose $\Omega^\bullet(*\overline{D})$ to the direct sum of the μ-th homogeneous components $\Omega^\bullet(*\overline{D})_\mu$, $\mu \in \mathbb{Z}$, and show that each component $\Omega^\bullet(*\overline{D})_\mu$ is stable by $\nabla_{\overline{\omega}}$ and $(\Omega^\bullet(*\overline{D})_\mu, \nabla_{\overline{\omega}})$ becomes a subcomplex of the twisted de Rham complex. Next, by this fact, we reduce the vanishing of $H^\bullet(\Omega^\bullet(*\overline{D}), \nabla_{\overline{\omega}})$ to the vanishing of this subcomplex. Finally, we derive the vanishing of the cohomology of the subcomplex $(\Omega^\bullet(*\overline{D})_\mu, \nabla_{\overline{\omega}})$ with the aid of the homotpy formula (2.74).

First, we should define a p-form $\varphi \in \Omega^p(*\overline{D})$ to be homogeneous of degree μ. When $g(u)$ is a homogeneous polynomial of degree μ, by the Euler formula, we have $L_v(g) = \mu g$. From this and $L_v(du_i) = du_i$, we have

$$L_v(g(u)du_{i_1} \wedge \cdots \wedge du_{i_p}) = (\mu + p)g(u)du_{i_1} \wedge \cdots \wedge du_{i_p}$$

by simple computations. Hence, if $\psi \in \Omega^p(\mathbb{C}^n)$ satisfies $L_v(\psi) = \mu\psi$, we say that ψ is homogeneous of degree μ. When ψ has the form

$$\psi = \sum_{1 \le i_1 < \cdots < i_p \le n} \psi_{i_1 \cdots i_p}(u) du_{i_1} \wedge \cdots \wedge du_{i_p},$$

ψ is homogeneous of degree μ if and only if each coefficient $\psi_{i_1 \cdots i_p}(u)$ is homogeneous of degree $(\mu - p)$. When an inhomogeneous $\psi \in \Omega^p(\mathbb{C}^n)$ is expressed as the sum of homogeneous forms of degree at most μ, if the coefficient of the highest degree μ is not zero, ψ is said to be of degree μ. Notice that since the polynomial $\overline{P} = \overline{P}_1 \cdots \overline{P}_m$ is homogeneous of degree $l = \sum_{j=1}^{m} l_j$, we have $L_v(1/\overline{P}) = -l/\overline{P}$. For a rational p-form $\varphi \in \Omega^p(*\overline{D})$ there exists $k \in \mathbb{Z}_{>0}$ such that $\varphi = \beta/\overline{P}^k$, $\beta \in \Omega^p(\mathbb{C}^n)$. If we define ψ to be homogeneous of degree μ when β is homogeneous of degree $\mu + kl$, this does not depend on the expression $\beta/\overline{P}^k$, and by the above remark it can be characterized by $L_v(\varphi) = \mu\varphi$. That is, a homogeneous rational p-form $\varphi \in \Omega^p(*\overline{D})$ of degree μ is an element of the eigenspace of the action of the linear operator L_v on $\Omega^p(*\overline{D})$ with eigenvalue μ. Setting

$$\Omega^p(*\overline{D})_\mu := \{\varphi \in \Omega^p(*\overline{D}) | L_v\varphi = \mu\varphi\},$$

$\Omega^p(*\overline{D})$ decomposes into the direct sum of the eigenspaces of L_v, and we have

$$\Omega^p(*\overline{D}) = \bigoplus_{\mu \in \mathbb{Z}} \Omega^p(*\overline{D})_\mu.$$

On the other hand, as was shown in § 2.6.1, $\Omega^p(\log \overline{D})$ is stable under the action of L_v, hence it admits a similar decomposition to the direct sum: setting

$$\Omega^p(\log \overline{D})_\mu := \Omega^p(\log \overline{D}) \cap \Omega^p(*\overline{D})_\mu,$$

we have

$$\Omega^p(\log \overline{D}) = \bigoplus_{\mu \in \mathbb{Z}} \Omega^p(\log \overline{D})_\mu.$$

2.6.4 Vanishing Theorem (i)

By the above decomposition, fundamental commutation relations (2.73) and the homotopy formula (2.74) for $\nabla_{\overline{\omega}}$, we can formally derive the vanishing of the cohomology. First, L_v and $\nabla_{\overline{\omega}}$ commute by (2.73), $\nabla_{\overline{\omega}}$ acts on each $\Omega^\bullet(*\overline{D})_\mu$ and $\Omega^\bullet(\log \overline{D})_\mu$, and we have the decompositions as complexes:

$$(\Omega^\bullet(*\overline{D}), \nabla_{\overline{\omega}}) = \bigoplus_{\mu \in \mathbb{Z}} (\Omega^\bullet(*\overline{D})_\mu, \nabla_{\overline{\omega}}),$$

$$(\Omega^\bullet(\log \overline{D}), \nabla_{\overline{\omega}}) = \bigoplus_{\mu \in \mathbb{Z}} (\Omega^\bullet(\log \overline{D})_\mu, \nabla_{\overline{\omega}}).$$

Hence, passing to their cohomology, we obtain the decompositions:

$$H^p(\Omega^\bullet(*\overline{D}), \nabla_{\overline{\omega}}) = \bigoplus_{\mu \in \mathbb{Z}} H^p(\Omega^\bullet(*\overline{D})_\mu, \nabla_{\overline{\omega}}),$$

$$H^p(\Omega^\bullet(\log \overline{D}), \nabla_{\overline{\omega}}) = \bigoplus_{\mu \in \mathbb{Z}} H^p(\Omega^\bullet(\log \overline{D})_\mu, \nabla_{\overline{\omega}}).$$

Thus, to show the vanishing of the cohomologies, it suffices to show the vanishing of each homogeneous component of the cohomologies. Suppose that $\varphi \in \Omega^p(*\overline{D})_\mu$ satisfies $\nabla_{\overline{\omega}}\varphi = 0$. By the homotopy formula (2.74) for $\nabla_{\overline{\omega}}$, we have

$$\nabla_{\overline{\omega}} \cdot i_v(\varphi) = \left(\mu + \sum_{j=1}^m l_j \alpha_j \right) \varphi,$$

and since L_v and i_v commute by (2.73), we obtain $i_v(\varphi) \in \Omega^{p-1}(*\overline{D})_\mu$. Hence, if $\mu + \sum_{j=1}^m l_j \alpha_j \neq 0$, it follows that $\varphi = \nabla_{\overline{\omega}}(i_v(\varphi)/(\mu + \sum l_j \alpha_j))$ which implies $H^p(\Omega^\bullet(*\overline{D})_\mu, \nabla_{\overline{\omega}}) = 0$. A similar argument also applies to $\Omega^\bullet(\log \overline{D})_\mu$. Notice that we also have

$$\Omega^\bullet(\log \overline{D})_\mu = 0 \qquad (\mu < -l).$$

Summarizing these, we obtain the following theorem:

Theorem 2.6. *(1) If $\sum_{j=1}^m l_j \alpha_j \notin \mathbb{Z}$, for any p, we have $H^p(\Omega^\bullet(*\overline{D})_\mu, \nabla_{\overline{\omega}}) = 0$, in particular, $H^p(\Omega^\bullet(*\overline{D}), \nabla_{\overline{\omega}}) = 0$.*
(2) If $\sum_{j=1}^m l_j \alpha_j \neq l, l-1, l-2, \cdots$, for any p, we have $H^p(\Omega^\bullet(\log \overline{D})_\mu, \nabla_{\overline{\omega}}) = 0$, in particular, $H^p(\Omega(\log \overline{D}), \nabla_{\overline{\omega}}) = 0$. Here, we set $l = \sum_{j=1}^m l_j = \deg \overline{P}$.

2.7 Filtration of Logarithmic Complex

For the imhomogeneous case, we introduce a filtration by degree on the logarithmic complex, and compare the cohomology of the associated graded complex with that of the cohomology of logarithmic complex for the homogeneous case. To study this difference, Proposition 2.2 proved in § 2.5, which gives a representation of logarithmic differential forms, will be essentially used.

2.7.1 Filtration

Denote the highest homogeneous component of non-constant m polynomials $P_j(u)$, $1 \leq j \leq m$ by $\overline{P}_j(u)$ and its degree by l_j, respectively. In this section, for P_j we use the symbols defined in § 2.5.1, and for $\overline{P}_j$ we use those defined in § 2.6.1, without notice. Let us define a filtration on $\Omega^p(\log D)$. By definition, $\varphi \in \Omega^p(\log D)$ can be expressed uniquely as $\varphi = \alpha/P$, $\alpha \in \Omega^p(\mathbb{C}^n)$. As the degree of P is $l = \sum_{j=1}^m l_j$, we formally define the degree of $1/P$ by $-l$, and when the degree of α (cf. § 2.6.3) is $\mu + l$, we say that φ is of degree μ and denote it by $\deg \varphi = \mu$. Now, we define:

Definition 2.4. Let $F_\mu \Omega^p(\log D)$ be the space of logarithmic p-forms, which may have poles along D, of degree at most μ:

$$F_\mu \Omega^p(\log D) := \{\varphi \in \Omega^p(\log D) \mid \deg \varphi \leq \mu\}.$$

The sequence $F_\mu \Omega^p(\log D)$, $\mu \geq -l + p$ of subspaces of $\Omega^p(\log D)$ is called an increasing filtration.

Now, let us show that this filtration is compatible with the covariant differential operator ∇_ω, i.e.,

$$\nabla_\omega F_\mu \Omega^p(\log D) \subset F_\mu \Omega^{p+1}(\log D). \tag{2.75}$$

For $\varphi \in F_\mu \Omega^p(\log D)$, by definition, we have $\varphi = \alpha/P$, $\alpha \in \Omega^p(\mathbb{C}^n)$, $\deg \alpha \leq \mu + l$, and $\frac{dP}{P} \wedge \alpha := \beta \in \Omega^{p+1}(\mathbb{C}^n)$. Since $d\varphi = (d\alpha - \beta)/P$ and the degree of $d\alpha - \beta \in \Omega^{p+1}(\mathbb{C}^n)$ is at most $\mu + l$, we have $d\varphi \in F_\mu \Omega^{p+1}(\log D)$. In the formula $\omega \wedge \varphi \in \Omega^{p+1}(\log D)$, since the degree of $\omega = \sum \alpha_j \frac{dP_j}{P_j}$ is 0, we see that the degree of $\omega \wedge \varphi$ is at most μ. Hence, $\nabla_\omega \varphi \in F_\mu \Omega^{p+1}(\log D)$ is proved.

By (2.75), $(F_\mu \Omega(\log D), \nabla_\omega)$, $\mu \geq -l$ $(\Omega(\log D), \nabla_\omega)$ is an increasing sequence of subcomplexes, and defines a filtration on the complex $(\Omega^\bullet(\log D), \nabla_\omega)$. The associated graded complex with respect to this filtration is the complex $(Gr_\mu^F \Omega^\bullet(\log D), Gr_\mu^F(\nabla_\omega))$, $\mu \geq -l$ defined by

$$Gr_\mu^F \Omega^\bullet(\log D) := F_\mu \Omega^\bullet(\log D)/F_{\mu-1}\Omega^\bullet(\log D)$$

with the differential $Gr_\mu^F(\nabla_\omega)$ naturally induced from ∇_ω.

2.7.2 Comparison with Homogeneous Case

Now, to compare $(Gr_\mu^F \Omega^\bullet(\log D), Gr_\mu^F(\nabla_\omega))$ defined above with $(\Omega^\bullet(\log \overline{D})_\mu, \nabla_{\overline{\omega}})$ defined in § 2.6.3, let us define a natural homomorphism

$$\sigma_\mu^p : Gr_\mu^F \Omega^p(\log D) \longrightarrow \Omega^p(\log \overline{D})_\mu$$

as follows: by definition, $\varphi \in F_\mu \Omega^p(\log D)$ is expressed as $\varphi = \alpha/P$, $\alpha \in \Omega^p(\mathbb{C}^n)$ and $\deg \alpha \leq \mu + l$. Let $\overline{\alpha}$ be the homogeneous component of α of degree $\mu + l$, and set $\overline{\varphi} = \overline{\alpha}/\overline{P}$. First, let us show that $\overline{\varphi} \in \Omega^p(\log \overline{D})_\mu$. Since φ is a logarithmic differential form, we have $\frac{dP}{P} \wedge \alpha = \beta \in \Omega^{p+1}(\mathbb{C}^n)$ and $\deg P = l$, $\deg \alpha \leq \mu + l$ imply $\deg \beta \leq \mu + l$. Taking the homogeneous components of both sides $dP \wedge \alpha = P\beta$ of degree $\mu + 2l$, we have $d\overline{P} \wedge \overline{\alpha} = \overline{P}\,\overline{\beta}$, where $\overline{\beta}$ is the homogeneous component of β of degree $\mu + l$, which implies $\overline{\varphi} \in \Omega^p(\log \overline{D})$. On the other hand, since $\overline{\varphi}$ is 0 or homogeneous of degree μ by construction, we obtain $\overline{\varphi} \in \Omega^p(\log \overline{D})_\mu$.

Now, associating $\varphi \in F_\mu \Omega^p(\log D)$ to $\overline{\varphi} \in \Omega^p(\log \overline{D})_\mu$, it clearly becomes a homomorphism. Since the kernel of this map is $F_{\mu-1}\Omega^p(\log D)$, it induces the map σ_μ^p we are looking for, which is, in addition, injective. Next, let us show that it defines a homomorphism of complexes

$$\sigma_\mu^\bullet : (Gr_\mu^F \Omega^\bullet(\log D), Gr_\mu^F(\nabla_\omega)) \longrightarrow (\Omega^\bullet(\log \overline{D})_\mu, \nabla_{\overline{\omega}})$$

by collecting σ_μ^p, $0 \leq p \leq n$. Here, for $\varphi \in F_\mu \Omega^p(\log D)$, we set

$$[\varphi] := \varphi \bmod F_{\mu-1}\Omega^p(\log D), \quad \overline{\varphi} := \sigma_\mu^p([\varphi]).$$

Then, we have

$$\left(\nabla_{\overline{\omega}} \circ \sigma_\mu^p\right)([\varphi]) = d\overline{\varphi} + \sum \alpha_j \frac{d\overline{P}_j}{\overline{P}_j} \wedge \overline{\varphi}$$

$$= \overline{\left(d\varphi + \sum \alpha_j \frac{dP_j}{P_j} \wedge \varphi\right)}$$

$$= \left(\sigma_\mu^{p+1} \circ Gr_\mu^F(\nabla_\omega)\right)([\varphi]),$$

which implies that $\sigma_\mu^\bullet$ is a homomorphism of complexes.

2.7.3 Isomorphism

As a homomorphism σ_μ^p is injective, setting

$$N^p(\log D)_\mu := \Omega^p(\log \overline{D})_\mu / \mathrm{Im}\sigma_\mu^p,$$

we obtain the short exact sequence of complexes:

$$0 \longrightarrow Gr_\mu^F \Omega^\bullet(\log D) \xrightarrow{\sigma_\mu^\bullet} \Omega^\bullet(\log \overline{D})_\mu \longrightarrow N^\bullet(\log D)_\mu \longrightarrow 0. \quad (2.76)$$

Then, we have the following important lemma.

Lemma 2.20. *If the highest homogeneous components $\overline{P}_j$ of m polynomials P_j, $1 \leq j \leq m$ of degree l_j satisfy Assumption 2, then σ_μ^p is an isomorphism for $p \neq n-1$. Hence, we have $N^p(\log D)_\mu = 0$.*

Proof. It suffices to show that σ_μ^p is surjective for $p \neq n-1$. Suppose that $0 \leq p \leq n-2$. By Assumption 2, $\overline{P}_j$'s are relatively prime to each other, and by (2.58), $\overline{\varphi} \in \Omega^p(\log \overline{D})_\mu$ is expressed as follows:

$$\overline{\varphi} = \overline{\alpha}/\overline{P}, \quad \overline{\alpha} \in \Omega^p(\mathbb{C}^n), \quad \overline{\alpha} \text{ is homogeneous of degree } \mu + l,$$

$$\frac{d\overline{P}_j}{\overline{P}_j} \wedge \overline{\alpha} \in \Omega^{p+1}(\mathbb{C}^n), \quad 1 \le j \le m.$$

Hence, $\overline{\alpha}$ satisfies the condition of Proposition 2.2 and is expressed as follows:

$$\overline{\alpha} = \overline{P}_1 \cdots \overline{P}_m \left\{ \overline{\alpha}_0 + \sum_{j=1}^{m} \frac{d\overline{P}_j}{\overline{P}_j} \wedge \overline{\alpha}_j + \cdots \right.$$

$$\left. + \sum_{1 \le j_1 < \cdots < j_p \le m} \frac{d\overline{P}_{j_1}}{\overline{P}_{j_1}} \wedge \cdots \wedge \frac{d\overline{P}_{j_p}}{\overline{P}_{j_p}} \overline{\alpha}_{j_1 \cdots j_p} \right\},$$

$$\overline{\alpha}_{j_1 \cdots j_\nu} \in \Omega^{p-\nu}(\mathbb{C}^n) : \text{ homogeneous of degree } \mu.$$

Setting

$$\varphi := \overline{\alpha}_0 + \sum_{j=1}^{m} \frac{dP_j}{P_j} \wedge \overline{\alpha}_j + \cdots$$

$$+ \sum_{1 \le j_1 < \cdots < j_p \le m} \frac{dP_{j_1}}{P_{j_1}} \wedge \cdots \wedge \frac{dP_{j_p}}{P_{j_p}} \overline{\alpha}_{j_1 \cdots j_p},$$

by the definition of σ_μ^p, we have $\sigma_\mu^p([\varphi]) = \overline{\varphi}$. For $p = n$, by the definition of logarithmic differential form, we have $\Omega^n(\log D) = \frac{1}{P}\Omega^n(\mathbb{C}^n)$, $\Omega^n(\log \overline{D}) = \frac{1}{P}\Omega^n(\mathbb{C}^n)$. This together with the definition of the filtration F_μ implies the surjectivity of σ_μ^n.

2.8 Vanishing Theorem of the Twisted Rational de Rham Cohomology

In this section, we assume that for the highest homogeneous components $\overline{P}_j$ of m polynomials P_j, Assumption 2 is always satisfied. Here, under this condition, we show the vanishing theorem of the twisted rational de Rham cohomology, i.e., that they are 0 except for $H^n(\Omega^\bullet(*D), \nabla_\omega)$. By this fact, Theorem 2.2 and Theorem 2.5, we obtain

$$\dim H^n(\Omega(*D), \nabla_\omega) = (-1)^n \chi(M)$$

which is extremely important for applications. For some concrete cases, it often happens that the dimension of the left-hand side can be computed rather easily by the right-hand side (cf. § 2.2.14). The proof proceeds according to [Kit-No]. See also [Cho].

2.8.1 *Vanishing of Logarithmic de Rham Cohomology*

We use the symbols defined in the previous sections. When $\sum_{j=1}^{m} l_j \alpha_j \neq l, l-1, \cdots$, Theorem 2.6 (2) implies that $H^p(\Omega^\bullet(\log \overline{D}), \nabla_{\overline{\omega}}) = 0$ for any p. Applying this result to the exact cohomology sequence associated to the short exact sequence of complexes (2.76), we obtain an isomorphism

$$H^{p-1}(N^\bullet(\log D)_\mu) \xrightarrow{\sim} H^p(Gr_\mu^F \Omega^\bullet(\log D), Gr_\mu^F(\nabla_\omega)). \qquad (2.77)$$

By Lemma 2.20, we see that $N^p(\log D)_\mu$ is 0 for $p \neq n-1$ which implies

$$H^p(Gr_\mu^F \Omega^\bullet(\log D), Gr_\mu^F(\nabla_\omega)) = 0 \quad (p \neq n). \qquad (2.78)$$

From this, we obtain the following result.

Theorem 2.7. *Suppose that the highest homogeneous components $\overline{P}_j$ of m polynomials P_j, $1 \leq j \leq m$ satisfy Assumption 2 and $\sum_{j=1}^{m} l_j \alpha_j \neq l, l-1, \cdots$. Then, we have*

$$H^p(\Omega^\bullet(\log D), \nabla_\omega) = 0 \quad (p \neq n).$$

Proof. Let $p \neq n$ and take $\varphi \in \Omega^p(\log D)$ satisfying $\nabla_\omega \varphi = 0$. Let the degree of φ be μ and denote the element of $Gr_\mu^F \Omega^p(\log D)$ corresponding to φ by $[\varphi]$. We have $Gr_\mu^F(\nabla_\omega)([\varphi]) = 0$. By (2.78), we can choose $\psi_\mu \in \Omega^{p-1}(\log D)$, $\deg \psi_\mu \leq \mu$ in such a way that

$$\varphi_{\mu-1} := \varphi - \nabla_\omega \psi_\mu \in F_{\mu-1}\Omega^p(\log D)$$

holds. Since we have $\nabla_\omega \varphi_{\mu-1} = \nabla_\omega \varphi - \nabla_\omega^2 \psi_\mu = 0$, we can apply the same argument to $\varphi_{\mu-1}$. Repeating this, by noting that $F_{p-l-1}\Omega^p(\log D) = 0$, we can choose $\psi_\nu \in F_\nu \Omega^p(\log D)$, $p - l \leq \nu \leq \mu$ in such a way that $\varphi = \sum_{\nu=p-l}^{\mu} \nabla_\omega \psi_\nu$ holds. This is what we should show. $\qquad\qquad\blacksquare$

2.8.2 *Vanishing of Algebraic de Rham Cohomology*

To show that the twisted rational de Rham cohomology $H^p(\Omega^\bullet(*D), \nabla_\omega)$ vanishes for $p \neq n$, we state the following facts:

(1) $\Omega^\bullet(*D) = \bigcup_{k=1}^{\infty} P^{-k}\Omega^\bullet(\log D)$,

(2) $(P^{-k}\Omega^\bullet(\log D), \nabla_\omega)$ is a subcomplex of $(\Omega^\bullet(*D), \nabla_\omega)$.

Since (1) is evident, let us show (2). For $\varphi = \psi/P^k$, $\psi \in \Omega^p(\log D)$, both $d\psi$ and $\frac{dP_j}{P_j} \wedge \psi$ belong to $\Omega^{p+1}(\log D)$ by Lemma 2.16. On the other hand, since we have

$$\nabla_\omega \varphi = P^{-k}\left[d\psi + \sum_{j=1}^m (\alpha_j - k)\frac{dP_j}{P_j} \wedge \psi\right] \qquad (2.79)$$

by simple computation, with the fact mentioned above, we obtain $\nabla_\omega \varphi \in P^{-k}\Omega^{p+1}(\log D)$. Hence, $(P^{-k}\Omega^\bullet(\log D), \nabla_\omega)$ becomes a complex.

Now, we consider an isomorphism

$$\varepsilon_k^p : \Omega^p(\log D) \xrightarrow{\;\sim\;} P^{-k}\Omega^p(\log D).$$
$$\psi \longmapsto \psi/P^k$$

Denote the covariant differential operator with respect to the connection form

$$\omega(k) = \sum_{j=1}^m (\alpha_j - k)\frac{dP_j}{P_j}$$

by $\nabla_{\omega(k)}$; (2.79) implies $\nabla_\omega \circ \varepsilon_k^p = \varepsilon_k^{p+1} \circ \nabla_{\omega(k)}$. Hence, $\varepsilon_k^\bullet$ defines an isomorphism of complexes

$$\varepsilon_k^\bullet : (\Omega^\bullet(\log D), \nabla_{\omega(k)}) \xrightarrow{\;\sim\;} (P^{-k}\Omega^\bullet(\log D), \nabla_\omega),$$

passing to the cohomologies, we obtain an isomorphism

$$H^p(\Omega^\bullet(\log D), \nabla_{\omega(k)}) \simeq H^p(P^{-k}\Omega^\bullet(\log D), \nabla_\omega). \qquad (2.80)$$

Applying Theorem 2.7 to the left-hand side of the above formula, under the assumption $\sum_{j=1}^m l_j(\alpha_j - k) \neq l, l-1, \cdots$, the left-hand side of (2.80) becomes 0 for $p \neq n$. Hence, under the assumption $\sum_{j=1}^m l_j\alpha_j \notin \mathbb{Z}$, we have shown

$$H^p(P^{-k}\Omega^\bullet(\log D), \nabla_\omega) = 0, \quad k \in \mathbb{Z}_{>0}$$

for $p \neq n$. Since we have $\Omega^\bullet(*D) = \bigcup_{k=1}^\infty P^{-k}\Omega^\bullet(\log D)$, we obtain $H^p(\Omega^\bullet(*D), \nabla_\omega) = 0$ for $p \neq n$. Summarizing these, we obtain:

Theorem 2.8 (Vanishing theorem of twisted rational de Rham cohomology). *Suppose that the highest homogeneous components $\overline{P}_j$ of polynomials P_j, $1 \leq j \leq m$ satisfy Assumption 2 and $\sum_{j=1}^m l_j\alpha_j \notin \mathbb{Z}$. Then, we have*

$$H^p(\Omega^\bullet(*D), \nabla_\omega) = 0 \qquad (p \neq n).$$

In particular, we have

$$H^p(M, \mathcal{L}_\omega) = 0 \qquad (p \neq n).$$

Remark 2.8. By (2.56), Theorem 2.8 signifies the fact that the algebraic de Rham cohomology $\mathbb{H}^p(M, (\Omega^\bullet_{M^{\mathrm{alg}}}, \nabla_\omega))$ vanishes for $p \neq n$.

By Theorem 2.2, we obtain the following corollary.

Corollary 2.2. $\dim H^n(\Omega^\bullet(*D), \nabla_\omega) = (-1)^n \chi(M)$.

When $M = \mathbb{C}^n \setminus \cup D_j$ satisfies Assumption 1 and 2, one can compute $\dim H^n(\Omega^\bullet(*D), \nabla_\omega)$ explicitly by Theorem 2.3 and Corollary 2.2. This is summarized in the following corollary:

Corollary 2.3. If $M = \mathbb{C}^n \setminus \cup D_j$ satisfies Assumption 1 and 2, then the formula holds:

$$\dim H^n(\Omega^\bullet(*D), \nabla_\omega)$$

$$= (-1)^n \left\{ \text{the coefficient of } z^n \text{ in } (1-z)^{m-1} \prod_{j=1}^{m} \frac{1}{1+(l_j-1)z} \right\}.$$

Remark 2.9. The assumption of Corollary 2.3 is satisfied for an arrangement of hyperplanes in $\mathbb{P}^n$ in general position defined in § 2.9.1. In this case, each l_j becomes 1, hence we have

$$\dim H^n(\Omega^\bullet(*D), \nabla_\omega) = \binom{m-1}{n}.$$

This apparently coincides with the results obtained in § 2.9 by direct algebraic computation (cf. Theorem 2.11).

2.8.3 Two-Dimensional Case

When M is of dimension 2, the vanishing theorem of cohomology holds under a weaker assumption than Assumption 2. Below, we explain this. Assuming $\sum_{j=1}^{m} l_j \alpha_j \notin \mathbb{Z}$, by Theorem 2.6 (2) and the exact cohomology sequence associated to (2.76), we obtain

$$H^0(Gr_\mu^F \Omega^\bullet(\log D), Gr_\mu^F(\nabla_\omega)) = 0,$$

$$H^1(Gr_\mu^F \Omega^\bullet(\log D), Gr_\mu^F(\nabla_\omega)) \simeq H^0(N^\bullet(\log D)_\mu). \qquad (2.81)$$

Here, we assume that the irreducible components of $\overline{P} = \overline{P}_1 \cdots \overline{P}_m$ are multiplicity free. It turns out that the same holds for $P = P_1 \cdots P_m$. Then, we can conclude $\Omega^0(\log D) = \Omega^0(\log D) = \mathbb{C}[u_1, u_2]$. In fact, by definition, $f \in \Omega^0(\log D)$ can be expressed as

$$f = a/P, \qquad a \in \mathbb{C}[u_1, u_2],$$

and we also have

$$\frac{dP}{P}a = g_1du_1 + g_2du_2, \quad g_1, g_2 \in \mathbb{C}[u_1, u_2]. \tag{2.82}$$

Notice that, by the assumption on P, P and $\frac{\partial P}{\partial u_i}$ are coprime. (2.82) implies $\frac{\partial P}{\partial u_i}a = Pg_i$, $i = 1, 2$, but, by the above remark, P divides a and hence $f \in \mathbb{C}[u_1, u_2]$.

By this fact and the definition of the filtration on $\Omega^0(\log D)$, we have

$$F_\mu \Omega^0(\log D) = \{\text{polynomials of degree at most } \mu\},$$

and similarly,

$$\Omega^0(\log \overline{D})_\mu = \{\text{homogeneous polynomials of degree } \mu\}.$$

Combining these and returning to the the definition of σ_μ^0, we see that $\sigma_\mu^0 : Gr_\mu^F \Omega^0(\log D) \longrightarrow \Omega^0(\log \overline{D})_\mu$ becomes an isomorphism which implies $N^0(\log D)_\mu = 0$. Hence, by (2.81), we have shown $H^1(Gr_\mu^F \Omega^\bullet(\log D), Gr_\mu^F(\nabla_\omega)) = 0$. Thus, we have

$$H^p(Gr_\mu^F \Omega^\bullet(\log D), Gr_\mu^F(\nabla_\omega)) = 0, \quad p = 0, 1.$$

For the rest of arguments, repeating those of § 2.8.1 and 2.8.2, we conclude $H^p(\Omega^\bullet(*D), \nabla_\omega) = 0$, $p = 0, 1$. Summarizing these, we obtain the following theorem:

Theorem 2.9. *Let $\overline{P}_j$ be the highest homogeneous components of polynomials P_j, $1 \le j \le m$ in two variables. If the irreducible components of the product $\overline{P}_1 \cdots \overline{P}_m$ are multiplicity free and $\sum_{j=1}^m l_j\alpha_j \notin \mathbb{Z}$, we have*

$$H^p(\Omega^\bullet(*D), \nabla_\omega) = 0 \qquad (p \ne 2).$$

By Corollary 2.1, 2.2 and Theorem 2.9, we obtain the following corollary:

Corollary 2.4. *If $M = \mathbb{C}^2 \backslash \bigcup_{j=1}^m D_j$ satisfies Assumption 1 and the assumption of Theorem 2.9, then we have the following formula:*

$$\dim H^2(\Omega^\bullet(*D), \nabla_\omega) = \frac{1}{2}(m-1)(m-2) + (m-1)\sum_{j=1}^m (l_j - 1)$$

$$+ \sum_{1 \le i < j \le m} (l_i - 1)(l_j - 1) + \sum_{j=1}^m (l_j - 1)^2.$$

2.8.4 Example

In the proofs of Theorem 2.8 and 2.9, we showed $N^p(\log D)_\mu = 0$ for $p \ne n-1$ under Assumption 2, but since it suffices to show $H^p(N^\bullet(\log D)_\mu) = 0$, $p \ne$

$n - 1$ by (2.77), we may say that this assumption was too strong. Below, we explain an important example where the twisted cohomology vanishes even if it does not satisfy the assumption of Theorem 2.9. We owe this example to J. Kaneko. The Appell hypergeometric function F_4, discussed in § 3.1.8, will be shown in § 3.3.7 to have the integral representation up to a constant:

$$\iint_{\Delta^2(\omega)} u_1^{\alpha_1} u_2^{\alpha_2} (1 - u_1 - u_2)^{\alpha_3} (u_1 u_2 - x_1 u_1 - x_2 u_2)^{\alpha_4} du_1 \wedge du_2. \quad (2.83)$$

Take for P_1, P_2, P_3, P_4 as $u_1, u_2, 1-u_1-u_2, u_1u_2-x_1u_1-x_2u_2$ respectively, we have $\overline{P_1}\overline{P_2}\overline{P_3}\overline{P_4} = (u_1u_2)^2(-u_1 - u_2)$ and it does not satisfy the assumption of Theorem 2.9.

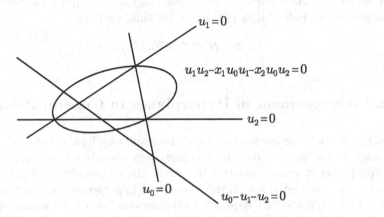

Fig. 2.9

Adding the line at infinity to a curve $P = 0$ in $\mathbb{C}^2$, we regard it as a curve in $\mathbb{P}^2(\mathbb{C})$. Applying the coordinate transformation of $\mathbb{P}^2(\mathbb{C})$

$$v_1 = \frac{-u_1}{1 - u_1 - u_2}, \qquad v_2 = \frac{-u_2}{1 - u_1 - u_2},$$

which sends this curve to itself, by simple computation, the integral (2.83) transforms to the integral:

$$\iint_{\widetilde{\Delta}^2(\varnothing)} \prod_{j=1}^{4} Q_j(v_1, v_2)^{\beta_j} dv_1 \wedge dv_2, \quad (2.84)$$

where

$$Q_1 = v_1, \quad Q_2 = v_2, \quad Q_3 = 1 - v_1 - v_2,$$

$$Q_4 = x_1 v_1 + x_2 v_2 - x_1 v_1^2 + (1 - x_1 - x_2) v_1 v_2 - x_2 v_2^2,$$

$$\beta_1 = \alpha_1, \quad \beta_2 = \alpha_2, \quad \beta_3 = -\alpha_1 - \alpha_2 - \alpha_3 - 2\alpha_4 - 3, \quad \beta_4 = \alpha_4.$$

Here, we have $\overline{Q}_1 \overline{Q}_2 \overline{Q}_3 \overline{Q}_4 = v_1 v_2 (-v_1 - v_2)\{-x_1 v_1^2 + (1 - x_1 - x_2) v_1 v_2 - x_2 v_2^2\}$, and when $(1 - x_1 - x_2)^2 - 4 x_1 x_2 \neq 0$, for general x_1, x_2, this becomes the product of five irreducible polynomials and it satisfies the assumption of Theorem 2.9. Hence, under the assumption $\beta_1 + \beta_2 + \beta_3 + 2\beta_4 = -\alpha_3 - 3 \notin \mathbb{Z}$, we have $H^p(M, \mathcal{L}_\omega) = 0$, $p \neq 2$, which implies $H^p(\Omega(*D), \mathcal{L}_\omega) = 0$, $p \neq 2$ by the comparison theorem.

Remark 2.10. As Figure 2.9 shows, one can easily compute the Euler characteristic of M, indeed it is $\chi(M) = 4$. By this, we have

$$\dim H^2(\Omega^\bullet(*D), \nabla_\omega) = 4.$$

2.9 Arrangement of Hyperplanes in General Position

In this section, we use the tools and the results we have developed up to now to study the twisted de Rham cohomology associated to arrangements of hyperplanes in general position in detail. The results obtained in this section will play an essential role in the research of hypergeometric functions of type $(n+1, m+1)$ stated in Chapter 3. In this section, we always assume $m \geq n+1$.

2.9.1 Vanishing Theorem (ii)

Suppose that the arrangement of m hyperplanes $\{P_j = 0\}$ defined by m polynomials of degree 1

$$P_j(u) = x_{0j} + x_{1j} u_1 + \cdots + x_{nj} u_n, \quad 1 \leq j \leq m$$

in n variables $u_1, \ldots, u_n$ is in general position, that is, any $(n+1) \times (n+1)$ minor of the $(n+1) \times (m+1)$ matrix

$$x = \begin{pmatrix} x_{01} & x_{02} & \cdots\cdots & x_{0m} & 1 \\ x_{11} & x_{12} & \cdots\cdots & x_{1m} & 0 \\ \vdots & & & \vdots \\ x_{n1} & x_{n2} & \cdots\cdots & x_{nm} & 0 \end{pmatrix} \tag{2.85}$$

is not zero. Then, it is clear that any $(n+1) \times (n+1)$ minor of the $n \times m$ matrix

$$\begin{pmatrix} x_{11} & x_{12} & \cdots\cdots & x_{1m} \\ \vdots & \vdots & & \\ x_{n1} & x_{n2} & \cdots\cdots & x_{nm} \end{pmatrix} \qquad (2.86)$$

is not zero. From this, we see that $\overline{P}_1, \ldots, \overline{P}_m$ satisfy Assumption 2, hence, under the assumption $\sum_{j=1}^{m} \alpha_j \notin \mathbb{Z}$, Theorem 2.8 implies

$$H^p(\Omega^\bullet(*D), \nabla_\omega) = 0 \quad (p \neq n). \qquad (2.87)$$

2.9.2 Representation of Logarithmic Differential Forms (ii)

We would like to study $H^n(\Omega^\bullet(*D), \nabla_\omega)$ in detail by using the filtration on $\Omega^\bullet(*D)$ introduced in § 2.7. For this purpose, let us prepare three lemmata. First, we extend the idea of partial fraction expansion for one variable to the case of several variables. Since, any $(n+1) \times (n+1)$ minor of (2.85) is not zero by assumption, for $n+1$ values $P_{j_0}, \ldots, P_{j_n}$, we can solve

$$x_{0j_\nu} + x_{1j_\nu} u_1 + \cdots + x_{nj_\nu} u_n = P_{j_\nu}, \quad 0 \leq \nu \leq n$$

with respect to $1, u_1, \ldots, u_n$, and it can be expressed as linear combinations of $P_{j_0}, \ldots, P_{j_n}$ over $\mathbb{C}$ (or to be precise, if we regard x_{ij} as parameters, then as linear combinations over the field of rational functions $\mathbb{C}(x_{ij})$). With this remark, let us decompose $g(u) = u_1^{\nu_1} \cdots u_n^{\nu_n} / P_1 \cdots P_m$, $\nu = (\nu_1, \cdots, \nu_n) \in \mathbb{Z}_{\geq 0}^n$ into simple fractions. First, when $|\nu| > 0$, assuming $\nu_1 > 0$ for simplicity, there exist constants c_j, $1 \leq j \leq n+1$ such that

$$u_1 = c_1 P_1 + \cdots + c_{n+1} P_{n+1}.$$

Hence, we have

$$g(u) = \sum_{j=1}^{n+1} c_j \frac{u_1^{\nu_1-1} u_2^{\nu_2} \cdots u_n^{\nu_n}}{P_1 \cdots \widehat{P}_j \cdots P_m}.$$

Next, similarly for $|\nu| = 0$, we have

$$1 = \sum_{j=1}^{n+1} c_j P_j$$

for some $c_j \in \mathbb{C}$, from which we obtain

$$g(u) = \frac{1}{P_1 \cdots P_m} = \sum_{j=1}^{n+1} c_j \frac{1}{P_1 \cdots \widehat{P}_j \cdots P_m}.$$

By repeating this operation as far as we can, we finally arrive at the formula:

$$
\frac{u_1^{\nu_1} \cdots u_n^{\nu_n}}{P_1 \cdots P_m} =
\begin{cases}
\displaystyle\sum_J \frac{a_J(u)}{P_{j_1} \cdots P_{j_n}} & \text{(for } |\nu| \geq m - n) \\[2ex]
\displaystyle\sum_J \frac{c_J}{P_{j_1} \cdots P_{j_n}} & \text{(for } |\nu| < m - n),
\end{cases}
$$

where for $J = \{j_1, \ldots, j_n\}$, $1 \leq j_1 < \cdots < j_n \leq m$, $a_J(u)$ is a homogeneous polynomial of degree $|\nu| - m + n$ and $c_J \in \mathbb{C}$.

In the above formula, for $|\nu| > m - n$, we can further reduce it by expressing the polynomial $a_J(u)$ as a polynomial in $P_{j_1}, \ldots, P_{j_n}$. Let us summarize the final formula obtained by this reduction in the following lemma.

Lemma 2.21. *A fraction $a(u)/P_1 \cdots P_m$ with a polynomial $a(u) \in \mathbb{C}[u_1, \cdots u_n]$ of degree μ admits the partial fraction decomposition as follows:*

$$
\frac{a(u)}{P_1 \cdots P_m} =
\begin{cases}
\displaystyle a_0(u) + \sum_{j=1}^{m} a_j(u)\frac{1}{P_j} + \cdots + \\[2ex]
\displaystyle \quad + \sum_{1 \leq j_1 < \cdots < j_n \leq m} \frac{a_{j_1 \cdots j_n}(u)}{P_{j_1} \cdots P_{j_n}} & \text{(for } \mu \geq m - n), \\[2ex]
\displaystyle \sum_J \frac{c_J}{P_{j_1} \cdots P_{j_n}} & \text{(for } \mu < m - n)
\end{cases}
$$

where $a_{j_1 \cdots j_\nu}(u)$ is a polynomial of degree $\mu - m + \nu$ and $c_J \in \mathbb{C}$.

Proposition 2.2 assures us that we can express an element of $\Omega^p(\log D)$, $0 \leq p \leq n - 2$ concretely. For an arrangement of hyperplanes in general position, if the degree of an element of $\Omega^{n-1}(\log D)$ or $\Omega^n(\log D)$ in question is not negative, applying the above lemma, we can obtain a similar expression. To explain this, we simplify the symbol on minors as follows:

$$
x\begin{pmatrix} i_1 \cdots i_\nu \\ j_1 \cdots j_\nu \end{pmatrix} := \det \begin{pmatrix} x_{i_1 j_1} & \cdots\cdots & x_{i_1 j_\nu} \\ \vdots & & \vdots \\ x_{i_\nu j_1} & \cdots\cdots & x_{i_\nu j_\nu} \end{pmatrix}.
$$

First, we consider $\varphi \in \Omega^n(\log D)$ with $\deg \varphi \geq 0$. By definition, we have

$$
\varphi = \frac{a(u)}{P_1 \cdots P_m} du_1 \wedge \cdots \wedge du_n, \quad a(u) \in \mathbb{C}[u], \quad \deg a \geq m - n,
$$

hence, Lemma 2.21 implies that φ is the sum of elements of the form

$$
\frac{a_{j_1 \cdots j_\nu}(u)}{P_{j_1} \cdots P_{j_\nu}} du_1 \wedge \cdots \wedge du_n, \quad \deg a_{j_1 \cdots j_\nu} = \deg a - m + \nu.
$$

Since any $n \times n$ minor of (2.86) is not zero, the rank of $n \times \nu$ matrix

$$\begin{pmatrix} x_{1j_1} & \cdots\cdots & x_{1j_\nu} \\ \vdots & & \vdots \\ x_{nj_1} & \cdots\cdots & x_{nj_\nu} \end{pmatrix}$$

becomes ν and there exists a $\nu\times\nu$ minor which does not vanish. For simplicity, we assume $x\begin{pmatrix} 1 & 2 & \cdots & \nu \\ j_1 & j_2 & \cdots & j_\nu \end{pmatrix} \neq 0$. By simple computation, we have

$$dP_{j_1} \wedge \cdots \wedge dP_{j_\nu} \wedge du_{\nu+1} \wedge \cdots \wedge du_n = x\begin{pmatrix} 1 & 2 & \cdots & \nu \\ j_1 & j_2 & \cdots & j_\nu \end{pmatrix} du_1 \wedge \cdots \wedge du_n$$

from which, by setting $J = \{j_1,\ldots,j_\nu\}$ and using multi-indices, we obtain

$$\frac{a_J(u)}{P_{j_1}\cdots P_{j_\nu}} du_1 \wedge \cdots \wedge du_n$$

$$= \frac{dP_{j_1}}{P_{j_1}} \wedge \cdots \wedge \frac{dP_{j_\nu}}{P_{j_\nu}} \wedge \left\{ \frac{a_J(u)}{x\binom{1\cdots\nu}{J}} du_{\nu+1} \wedge \cdots \wedge du_n \right\}.$$

As $\deg a_J = \deg a - m + \nu$, the degree of $(n-\nu)$-form in $\{\ \}$ is

$$\deg a_J + (n-\nu) = \deg a - m + n = \deg \varphi.$$

Hence, when the degree μ of $\varphi \in \Omega^n(\log D)$ is non-negative, φ can be expressed as the sum of elements of the form

$$\frac{dP_{j_1}}{P_{j_1}} \wedge \cdots \wedge \frac{dP_{j_\nu}}{P_{j_\nu}} \wedge \varphi_{j_1\cdots j_\nu}, \quad \varphi_{j_1\cdots j_\nu} \in \Omega^{n-\nu}(\mathbb{C}^n), \quad \deg \varphi_{j_1\cdots j_\nu} = \mu.$$

Second, we show that a similar expression also exists for $\varphi \in \Omega^{n-1}(\log D)$, $\mu = \deg \varphi \geq 0$. To simplify the notation, we set

$$*du_i := (-1)^{i-1} du_1 \wedge \cdots \wedge \widehat{du_i} \wedge \cdots \wedge du_n.$$

Notice that

$$du_i \wedge *du_i = du_1 \wedge \cdots \wedge du_n.$$

By definition, φ can be expressed as

$$\varphi = \frac{1}{P_1\cdots P_m} \sum_{i=1}^{n} b_i(u) * du_i, \quad b_i \in \mathbb{C}[u],$$

and we have $\deg b_i = \mu + 1 + m - n \geq m - n$. Hence, by Lemma 2.21, we have

$$\frac{b_i(u)}{P_1\cdots P_m} = \sum_J \frac{b_{i,J}(u)}{P_{j_1}\cdots P_{j_n}}, \quad \deg b_{i,J} = \mu + 1,$$

with $J = \{j_1, \ldots, j_n\}$, and φ can be finally expressed as

$$\varphi = \sum_{i=1}^{n} \sum_{J} \frac{b_{i,J}(u)}{P_{j_1} \cdots P_{j_n}} * du_i.$$

In the right-hand side of this formula, since

$$\frac{1}{P_{j_1} \cdots P_{j_n}} \sum_{i=1}^{n} b_{i,J}(u) * du_i$$

has to be a logarithmic differential form in a neighborhood of $\bigcap_{\nu=1}^{n}\{P_{j_\nu} = 0\}$, we obtain

$$\frac{dP_{j_\nu}}{P_{j_\nu}} \wedge \sum_{i=1}^{n} b_{i,J}(u) * du_i \in \Omega^n(\mathbb{C}^n) \tag{2.88}$$

by definition. On the other hand, $x\begin{pmatrix} 1 & 2 & \cdots & n \\ j_1 & j_2 & \cdots & j_n \end{pmatrix} \neq 0$ implies that both $\{du_1, \ldots, du_n\}$ and $\{dP_{j_1}, \ldots, dP_{j_n}\}$ are bases of T^*M. Hence, by a general theory of exterior algebra, the change-of-basis matrix from $\{*du_1, \ldots, *du_n\}$ to $\{dP_{j_1} \wedge \cdots \wedge \widehat{dP_{j_\nu}} \wedge \cdots \wedge dP_{j_n}, 1 \leq \nu \leq n\}$ is given by the matrix whose components are $(n-1) \times (n-1)$ minors of the matrix

$$\begin{pmatrix} x_{1j_1} & \cdots\cdots & x_{1j_n} \\ \vdots & & \vdots \\ x_{nj_1} & \cdots\cdots & x_{nj_n} \end{pmatrix}.$$

Hence, there exists $g_{ij_\nu} \in \mathbb{C}$ (to be precise, $\mathbb{C}(x_{ij})$) such that

$$*du_i = \sum_{\nu=1}^{n} g_{ij_\nu} dP_{j_1} \wedge \cdots \wedge \widehat{dP_{j_\nu}} \wedge \cdots \wedge dP_{j_n}, \quad 1 \leq i \leq n, \quad g_{ij_\nu} \in \mathbb{C}.$$

Hence, $\sum_{i=1}^{n} b_{i,J}(u) * du_i$ is rewritten as

$$\sum_{\nu=1}^{n} (-1)^{\nu-1} c_{j_\nu}(u) dP_{j_1} \wedge \cdots \wedge \widehat{dP}_{j_\nu} \wedge \cdots \wedge dP_{j_n},$$

$$c_{j_\nu} \in \mathbb{C}[u], \qquad \deg c_{j_\nu} = \mu + 1.$$

Hence, (2.88) becomes

$$\frac{1}{P_{j_\nu}} c_{j_\nu}(u) dP_{j_1} \wedge \cdots \wedge dP_{j_n} \in \Omega^n(\mathbb{C}^n)$$

which implies that P_{j_ν} divides c_{j_ν}. Hence, setting $c_{j_\nu} = P_{j_\nu} \tilde{c}_{j_\nu}$, $\tilde{c}_{j_\nu}$ is a polynomial of degree μ. Thus, φ is rewritten as follows:

$$\varphi = \sum_J \sum_{\nu=1}^n (-1)^{\nu-1} \frac{P_{j_\nu} \tilde{c}_{j_\nu}}{P_{j_1} \cdots P_{j_n}} dP_{j_1} \wedge \cdots \wedge \widehat{dP_{j_\nu}} \wedge \cdots \wedge dP_{j_n}$$

$$\tilde{c}_{j_\nu} \in \mathbb{C}[u], \qquad \deg \tilde{c}_{j_\nu} = \mu.$$

The rest is a simple calculation and we obtain a desired expression. Summarizing this, we obtain:

Lemma 2.22. *Suppose that $\varphi \in \Omega^p(\log D)$, $p = n - 1$ or $p = n$ satisfies $\mu := \deg \varphi \geq 0$. Then, φ can be expressed as follows:*

$$\varphi = \varphi_0 + \sum_{j=1}^m \frac{dP_j}{P_j} \wedge \varphi_j + \cdots \tag{2.89}$$

$$+ \sum_{1 \leq j_1 < \cdots < j_p \leq m} \frac{dP_{j_1}}{P_{j_1}} \wedge \cdots \wedge \frac{dP_{j_p}}{P_{j_p}} \varphi_{j_1 \cdots j_p},$$

$$\varphi_{j_1 \cdots j_\nu} \in \Omega^{p-\nu}(\mathbb{C}^n), \qquad \deg \varphi_{j_1 \cdots j_\nu} = \mu.$$

Moreover, when $\varphi \in \Omega^{n-1}(\log D)$, $\deg \varphi = -1$, we have $\varphi = 0$.

Looking at the proof of Lemma 2.22, we see that the assumption $\mu = \deg \varphi \geq 0$ is used to state that $\varphi_{i_1 \cdots i_\nu}$ of (2.89) is of degree μ. If we remove this assumption, we would not have the assertion on the degree of $\varphi_{i_1 \cdots i_\nu}$ but the expression (2.89) itself is still valid. That is:

Corollary 2.5. *For $p = n - 1$ or $p = n$, $\varphi \in \Omega^p(\log D)$ has the expression:*

$$\varphi = \varphi_0 + \sum_{j=1}^m \frac{dP_j}{P_j} \wedge \varphi_j + \cdots \tag{2.90}$$

$$+ \sum_{1 \leq j_1 < \cdots < j_p \leq m} \frac{dP_{j_1}}{P_{j_1}} \wedge \cdots \wedge \frac{dP_{j_p}}{P_{j_p}} \varphi_{j_1 \cdots j_p},$$

$$\varphi_{j_1 \cdots j_\nu} \in \Omega^{p-\nu}(\mathbb{C}^n).$$

2.9.3 Reduction of Poles

As the final preliminary, we show a kind of comparison theorem between the cohomology of the logarithmic complex and the twisted rational de Rham cohomology. This assures that for $p = n - 1, n$, a rational p-forms with poles of higher order can be reduced to that of logarithmic poles in the same cohomology class.

Lemma 2.23. *Suppose that $p = n - 1$ or n , and $\alpha_j \notin \mathbb{Z}_{>0}$, $1 \le j \le m$. Then, those $\varphi \in \Omega^p(*D)$ satisfying $\nabla_\omega \varphi \in (1/P_1 \cdots P_m)\Omega^{p+1}(\mathbb{C}^n)$ can be expressed as*

$$\varphi = \psi + \nabla_\omega \beta, \quad \psi \in \Omega^p(\log D), \quad \beta \in \Omega^{p-1}(*D).$$

Proof. $\varphi \in \Omega^p(*D)$ can be expressed as

$$\varphi = \widetilde{\varphi}/P_1^{k_1} \cdots P_m^{k_m}, \quad \widetilde{\varphi} \in \Omega^p(\mathbb{C}^n), \quad k_j \ge 1, \quad 1 \le j \le m.$$

First, let us show that it suffices to prove: when a k_j is bigger than 1, there exists $\alpha \in \Omega^p(\mathbb{C}^n)$, $\beta \in \Omega^{p-1}(*D)$ such that

$$\varphi = \alpha/P_1^{k_1} \cdots P_j^{k_j-1} \cdots P_m^{k_m} + \nabla_\omega \beta. \tag{2.91}$$

Admitting this, by $\nabla_\omega(\alpha/P_1^{k_1} \cdots P_j^{k_j-1} \cdots P_m^{k_m}) = \nabla_\omega \varphi \in \frac{1}{P}\Omega^{p+1}(\mathbb{C}^n)$, we apply the same argument repeatedly to $\alpha/P_1^{k_1} \cdots P_j^{k_j-1} \cdots P_m^{k_m}$, and we see that φ can be finally expressed as

$$\varphi = \alpha'/P_1 \cdots P_m + \nabla_\omega \beta', \quad \alpha' \in \Omega^p(\mathbb{C}^n), \quad \beta' \in \Omega^{p-1}(*D).$$

On the other hand, since we have $\nabla_\omega(\alpha'/P) = \nabla_\omega \varphi \in \frac{1}{P}\Omega^{p+1}(\mathbb{C}^n)$, we can apply Lemma 2.17 and show $\alpha'/P \in \Omega^p(\log D)$.

Now, let us show (2.91). By computation, we have

$$\nabla_\omega \varphi = (1/\prod_{}^{m} P_j^{k_j})\{d\widetilde{\varphi} + \sum_{j=1}^{m}(\alpha_j - k_j)\frac{dP_j}{P_j} \wedge \widetilde{\varphi}\} \in \frac{1}{P_1 \cdots P_m}\Omega^{p+1}(\mathbb{C}^n),$$

and $k_j \ge 1$, $1 \le j \le m$ implies

$$d\widetilde{\varphi} + \sum_{j=1}^{m}(\alpha_j - k_j)\frac{dP_j}{P_j} \wedge \widetilde{\varphi} \in \Omega^{p+1}(\mathbb{C}^n).$$

As $\widetilde{\varphi} \in \Omega^p(\mathbb{C}^n)$ and P_j's are coprime to each other, we see that $(\alpha_j - k_j)\frac{dP_j}{P_j} \wedge \widetilde{\varphi} \in \Omega^{p+1}(\mathbb{C}^n)$ for each j. By the assumption $\alpha_j \notin \mathbb{Z}_{>0}$, we have

$$\frac{dP_j}{P_j} \wedge \widetilde{\varphi} \in \Omega^{p+1}(\mathbb{C}^n),$$

and we can conclude $\widetilde{\varphi}/P_1 \cdots P_m \in \Omega^p(\log D)$ by (2.58). Here, by Corollary 2.5, we obtain the expression for $\widetilde{\varphi}$:

$$\widetilde{\varphi} = P_1 \cdots P_m \left\{ \widetilde{\varphi}_0 + \sum_{j=1}^{m} \frac{dP_j}{P_j} \wedge \widetilde{\varphi}_j + \cdots \right.$$

$$\left. + \sum_{1 \le j_1 < \cdots < j_p \le m} \frac{dP_{j_1}}{P_{j_1}} \wedge \cdots \wedge \frac{dP_{j_p}}{P_{j_p}} \widetilde{\varphi}_{j_1 \cdots j_p} \right\},$$

$$\widetilde{\varphi}_{j_1 \cdots j_\nu} \in \Omega^{p-\nu}(\mathbb{C}^n).$$

Hence, $\varphi = \widetilde{\varphi}/P_1^{k_1} \cdots P_m^{k_m}$ is the sum of terms of the form

$$[\widetilde{\varphi}]_J / P_1^{k_1} \cdots P_m^{k_m}, \tag{2.92}$$

where we set $[\widetilde{\varphi}]_J := P_1 \cdots P_m \frac{dP_{j_1}}{P_{j_1}} \wedge \cdots \wedge \frac{dP_{j_\nu}}{P_{j_\nu}} \wedge \widetilde{\varphi}_J$, $J = \{j_1, \ldots, j_\nu\}$, $1 \le j_1 < \cdots < j_\nu \le m$. For simplicity, assuming $k_1 > 1$, let us show that we can reduce the order of a pole of (2.92) by 1. We consider two cases separately.

(1) When $1 \notin J$, $[\widetilde{\varphi}]_J$ can be expressed as $P_1 \times$ [a p-form with polynomial coefficients], hence (2.92) can be written as

$$\text{[a } p\text{-form with polynomial coefficients]}/P_1^{k_1-1} P_2^{k_2} \cdots P_m^{k_m}.$$

(2) When $1 \in J$, we assume that $J = \{1, \ldots, \nu\}$, for simplicity. Then, we have

$$[\widetilde{\varphi}]_J = dP_1 \wedge \cdots \wedge dP_\nu \wedge (P_{\nu+1} \cdots P_m \widetilde{\varphi}_J).$$

Setting

$$\xi = dP_2 \wedge \cdots \wedge dP_\nu \wedge (P_{\nu+1} \cdots P_m \widetilde{\varphi}_J)/P_1^{k_1-1} P_2^{k_2} \cdots P_m^{k_m}$$

and calculating $\nabla_\omega \xi$, we easily obtain the formula:

$$\nabla_\omega \xi = (\alpha_1 - k + 1) \frac{[\widetilde{\varphi}]_J}{P_1^{k_1} \cdots P_m^{k_m}}$$

$$+ (-1)^{\nu-1} \frac{dP_2 \wedge \cdots \wedge dP_\nu \wedge d(P_{\nu+1} \cdots P_m \widetilde{\varphi}_J)}{P_1^{k_1-1} P_2^{k_2} \cdots P_m^{k_m}}$$

$$+ \sum_{j=\nu+1}^{m} (\alpha_j - k_j + 1) \frac{dP_j}{P_j} \wedge \frac{dP_2 \wedge \cdots \wedge dP_\nu \wedge (P_{\nu+1} \cdots P_m \widetilde{\varphi}_J)}{P_1^{k_1-1} P_2^{k_2} \cdots P_m^{k_m}}.$$

The second and the third terms of the right-hand side of the above formula apparently have the form $\eta/P_1^{k_1-1} P_2^{k_2} \cdots P_m^{k_m}$, $\eta \in \Omega^p(\mathbb{C}^n)$. On the other hand, since the condition $\alpha_1 \notin \mathbb{Z}_{>0}$ implies $\alpha_1 - k + 1 \ne 0$, (2.92) is finally expressed as

$$\frac{1}{\alpha_1 - k_1 + 1} \cdot \frac{\eta}{P_1^{k_1-1} P_2^{k_2} \cdots P_m^{k_m}} + \nabla_\omega(\xi/(\alpha_1 - k + 1)),$$

and the order of a pole is reduced by 1.

2.9.4 Comparison Theorem

With these preparations, we show that a homomorphism of cohomologies

$$\iota_* : H^p(\Omega^\bullet(\log D), \nabla_\omega) \longrightarrow H^p(\Omega^\bullet(*D), \nabla_\omega), \quad p = 0, 1, \dots, n$$

induced by the natural map $\iota : (\Omega^\bullet(\log D), \nabla_\omega) \longrightarrow (\Omega^\bullet(*D), \nabla_\omega)$ is an isomorphism under the assumption $\alpha_j \notin \mathbb{Z}$ $(1 \leq j \leq m)$, $\sum_{j=1}^m \alpha_j \notin \mathbb{Z}$. First, for $0 \leq p \leq n-1$, by Theorem 2.7 and 2.8, both sides become zero, hence are isomorphic. Second, let us show the case when $p = n$. First, we show that ι_* is surjective. By Lemma 2.23, $\varphi \in \Omega^n(*D)$ is expressed as

$$\varphi = \psi + \nabla_\omega\beta, \quad \psi \in \Omega^n(\log D), \quad \beta \in \Omega^{n-1}(*D)$$

which implies the surjectivity of ι_*. Second, we show that ι_* is injective. Suppose that $\psi \in \Omega^n(\log D)$ can be expressed as

$$\psi = \nabla_\omega\varphi, \quad \varphi \in \Omega^{n-1}(*D).$$

Since we have $\nabla_\omega\varphi = \psi \in \frac{1}{P}\Omega^n(\mathbb{C})$, Lemma 2.23 implies that φ can be expressed as

$$\varphi = \alpha + \nabla_\omega\beta, \quad \alpha \in \Omega^{n-1}(\log D), \quad \beta \in \Omega^{n-2}(*D).$$

Hence, we have

$$\psi = \nabla_\omega\varphi = \nabla_\omega\alpha$$

which implies the injectivity of ι_*. Thus, we obtain the following theorem.

Theorem 2.10. *For an arrangement of hyperplanes in general position, under the assumption $\alpha_j \notin \mathbb{Z}$ $(1 \leq j \leq m)$, $\sum_{j=1}^m \alpha_j \notin \mathbb{Z}$, we have*

$$H^p(\Omega^\bullet(*D), \nabla_\omega) = 0 \quad \text{(for $p \neq n$)},$$

$$H^n(\Omega^\bullet(*D), \nabla_\omega) \simeq H^n(\Omega^\bullet(\log D), \nabla_\omega).$$

2.9.5 Filtration

By the above theorem, it suffices to compute $H^n(\Omega^\bullet(\log D), \nabla_\omega)$ to determine the twisted rational de Rham cohomology. Here, let us study the filtration F_μ on $\Omega^\bullet(\log D)$ introduced in § 2.7 in detail for this case. Simplifying the notation, we use the symbols:

$$\Omega^p(\mathbb{C}^n)_\nu = \{\varphi \in \Omega^p(\mathbb{C}^n) \mid \varphi : \text{homogeneous of degree } \nu\},$$

$$\Omega^p(\mathbb{C}^n)_{\leq \nu} = \{\varphi \in \Omega^p(\mathbb{C}^n) \mid \deg \varphi \leq \nu\},$$

$$\varphi\langle j_1, \ldots, j_p\rangle := \frac{dP_{j_1}}{P_{j_1}} \wedge \cdots \wedge \frac{dP_{j_p}}{P_{j_p}}.$$

Recall that $N^{n-1}(\log D)_\mu$ is defined as the cokernel of the injective homomorphism

$$\sigma_\mu^{n-1} : Gr_\mu^F \Omega^{n-1}(\log D) \longrightarrow \Omega^{n-1}(\log \overline{D})_\mu$$

and by § 2.8.1, we have an isomorphism

$$H^n(Gr_\mu^F \Omega^\bullet(\log D), Gr_\mu^F(\nabla_\omega)) \simeq H^{n-1}(N^\bullet(\log D)_\mu) \qquad (2.93)$$

under the condition $\sum \alpha_j \notin \mathbb{Z}$.

First, notice that the degree of a p-form φ with polynomial coefficients is at least p. By the above remark and Lemma 2.22, $\xi \in F_1\Omega^{n-1}(\log D)$ can be expressed as

$$\xi = \sum \varphi\langle j_1, \ldots, j_{n-1}\rangle \xi_{j_1 \cdots j_{n-1}} + \sum \varphi\langle j_1, \ldots, j_{n-2}\rangle \wedge \xi_{j_1 \cdots j_{n-2}},$$

$$\xi_{j_1 \cdots j_{n-1}} \in \Omega^0(\mathbb{C}^n)_{\leq 1}, \quad \xi_{j_1 \cdots j_{n-2}} \in \Omega^1(\mathbb{C}^n)_{\leq 1}.$$

Then, by the definition of the degree of a logarithmic differential form defined in § 2.7.1, each term $\varphi\langle J\rangle \wedge \xi_J$, $J = \{j_1, \ldots, j_p\}$, $p = n - 1, n$ of the above expression again belongs to $F_1\Omega^{n-1}(\log D)$. Hence, we have

$$F_1\Omega^{n-1}(\log D) = \sum \varphi\langle j_1, \ldots, j_{n-1}\rangle \wedge \Omega^0(\mathbb{C}^n)_{\leq 1}$$

$$+ \sum \varphi\langle j_1, \ldots, j_{n-2}\rangle \wedge \Omega^1(\mathbb{C}^n)_{\leq 1}.$$

Similarly, for $F_\mu\Omega^{n-1}(\log D)$, $\mu \geq 0$, we have

$$F_\mu\Omega^{n-1}(\log D) = \sum_{|J|=0}^{n-1} \sum_J \varphi\langle J\rangle \wedge \Omega^{n-1-|J|}(\mathbb{C}^n)_{\leq \mu}.$$

Here, by using the symbol

$$[p] \wedge \Omega_{\leq \mu}^{n-1-p} := \sum_{|J|=p} \varphi\langle J\rangle \wedge \Omega^{n-1-p}(\mathbb{C}^n)_{\leq \mu},$$

we summarize the above result in Table 9.1.

Table 9.1

F_{-1}	F_0	F_1	F_2	$\cdots$	F_μ
0	$[n-1]\wedge\Omega^0_0$	$[n-1]\wedge\Omega^0_{\leq 1}$	$[n-1]\wedge\Omega^0_{\leq 2}$	$\cdots$	$[n-1]\wedge\Omega^0_{\leq\mu}$
		$[n-2]\wedge\Omega^1_1$	$[n-2]\wedge\Omega^1_{\leq 2}$	$\cdots$	$[n-2]\wedge\Omega^1_{\leq\mu}$
			$[n-3]\wedge\Omega^2_2$	$\cdots$	$[n-3]\wedge\Omega^2_{\leq\mu}$
					$[n-4]\wedge\Omega^3_{\leq\mu}$
					$\vdots$

2.9.6 Basis of Cohomology

Next, let us study $\Omega^{n-1}(\log\overline{D})$. In the case of an arrangement of hyperplanes in general position, $u_1,\ldots,u_n$ are expressed as any n linear combinations of $\overline{P}_j$, $1\leq j\leq m$ over $\mathbb{C}$. By this fact, retracing the proof of Proposition 2.2, we can show the following lemma.

Lemma 2.24. *For $\mu\geq 0$, $\overline{\varphi}\in\Omega^{n-1}(\log\overline{D})_\mu$ can be expressed in the following form:*

$$\overline{\varphi}=\overline{\varphi}_0+\sum_{j=1}^m\frac{d\overline{P}_j}{\overline{P}_j}\wedge\varphi_j+\cdots+ \tag{2.94}$$

$$+\sum_{1\leq j_1<\cdots<j_{n-1}\leq m}\frac{d\overline{P}_{j_1}}{\overline{P}_{j_1}}\wedge\cdots\wedge\frac{d\overline{P}_{j_{n-1}}}{\overline{P}_{j_{n-1}}}\overline{\varphi}_{j_1\cdots j_{n-1}},$$

$$\overline{\varphi}_{j_1\cdots j_\nu}\in\Omega^{n-1-\nu}(\mathbb{C}^n)_\mu.$$

Proof. Since this lemma can be proved along almost the same lines as in the proof of Proposition 2.2, here we only state important points. As $\overline{\varphi}\in\Omega^{n-1}(\log\overline{D})_\mu$ is expressed as $\overline{\varphi}=\psi/\overline{P}_1\cdots\overline{P}_m$, $\psi\in\Omega^{n-1}(\mathbb{C}^n)_{\mu+m}$ and we have

$$\frac{d\overline{P}_j}{\overline{P}_j}\wedge\psi\in\Omega^n(\mathbb{C}^n)$$

by the definition of logarithmic differential form, we can show

$$d\overline{P}_j\wedge\psi\equiv 0\pmod{\overline{P}_j},\quad 1\leq j\leq m. \tag{2.95}$$

For $\mu\geq 0$, when an $(n-1)$-form ψ with coefficients in homogeneous polynomials of degree $\mu+m$ satisfies (2.95), it suffices to show that ψ can be expressed as

$$\psi = \overline{P}_1 \cdots \overline{P}_m \left\{ \psi_0 + \sum_{j=1}^{m} \frac{d\overline{P}_j}{\overline{P}_j} \wedge \psi_j + \cdots + \right. \tag{2.96}$$

$$\left. + \sum_{1 \le j_1 < \cdots < j_{n-1} \le m} \frac{d\overline{P}_{j_1}}{\overline{P}_{j_1}} \wedge \cdots \wedge \frac{d\overline{P}_{j_{n-1}}}{\overline{P}_{j_{n-1}}} \psi_{j_1 \cdots j_{n-1}} \right\},$$

$$\psi_{j_1 \cdots j_\nu} \in \Omega^{n-1-\nu}(\mathbb{C}^n)_\mu.$$

Let us show this by induction on m.

First we show this for $m = 1$. Adjoining $n - 1$ homogeneous polynomials of degree 1 $\overline{P}_2, \ldots, \overline{P}_n$ to $\overline{P}_1$ and regard it as a new coordinate. Then, we remark that the homogeneity is preserved. Simplifying the notation and using

$$*d\overline{P}_i = (-1)^{i-1} d\overline{P}_1 \wedge \cdots \wedge \widehat{d\overline{P}_i} \wedge \cdots \wedge d\overline{P}_n,$$

$$*1 = d\overline{P}_1 \wedge \cdots \wedge d\overline{P}_n,$$

we have the expression:

$$\psi = \sum_{i=1}^{n} a_i(u) * d\overline{P}_i, \quad a_i(u) \in \Omega^0(\mathbb{C}^n)_{\mu-n+2}.$$

By (2.95), we have

$$d\overline{P}_1 \wedge \psi = a_1(u) * 1 \equiv 0 \pmod{\overline{P}_1},$$

which implies that $\overline{P}_1$ divides a_1: $a_1 = \overline{P}_1 \tilde{a}_1$, $\tilde{a}_1 \in \Omega^0(\mathbb{C}^n)_{\mu-n+1}$. Rewriting ψ with this formula, we have

$$\psi = \overline{P}_1 \tilde{a}_1 * d\overline{P}_1 + d\overline{P}_1 \wedge \left(\sum_{i=2}^{n} (-1)^{i-1} a_i(u) d\overline{P}_2 \wedge \cdots \wedge \widehat{d\overline{P}_i} \wedge \cdots \wedge d\overline{P}_n \right).$$

But, as

$$\tilde{a}_1 * d\overline{P}_1 \in \Omega^{n-1}(\mathbb{C}^n)_\mu, \sum_{i=2}^{n} (-1)^{i-1} a_i d\overline{P}_2 \wedge \cdots \wedge \widehat{d\overline{P}_i} \wedge \cdots \wedge d\overline{P}_n \in \Omega^{n-2}(\mathbb{C}^n)_\mu$$

are evident, we could show (2.96) for $m = 1$.

Next, by induction, assuming that the case m is true, we derive (2.96) from (2.95) for $m + 1$. For that, we assume that $\psi \in \Omega^{n-1}(\mathbb{C}^n)_{\mu+m+1}$ satisfies

$$d\overline{P}_j \wedge \psi \equiv 0 \pmod{\overline{P}_j}, \quad 1 \le j \le m+1.$$

ψ can be expressed in the form of (2.96) by induction, but here, we have $\psi_{j_1 \cdots j_\nu} \in \Omega^{n-1-\nu}(\mathbb{C}^n)_{\mu+1}$. Following the proof of Proposition 2.2, we let

$N \in \mathbb{Z}_{\geq 0}$ be the biggest integer satisfying $\psi_{j_1 \cdots j_N} \neq 0$ and prove the assertion by induction on N. Since the proof of Proposition 2.2 applies directly to this case for $N \leq n - 2$, here we omit the detail. For $N = n - 1$, we have $\psi_{j_1 \cdots j_{n-1}} \in \Omega^0(\mathbb{C}^n)_{\mu+1}$ which means that $\psi_{j_1 \cdots j_{n-1}}$ is a homogeneous polynomial of degree $\mu + 1 \ (\geq 1)$. Hence, by using $\overline{P}_{j_1}, \ldots, \overline{P}_{j_{n-1}}, \overline{P}_{m+1}$, we have the expression:

$$\psi_{j_1 \cdots j_{n-1}} = \sum_{k=1}^{n-1} \overline{P}_{j_k} \psi_{j_1 \cdots j_{n-1}; j_k} + \overline{P}_{m+1} \psi_{j_1 \cdots j_{n-1}; m+1},$$

$$\psi_{j_1 \cdots j_{n-1}; \nu} \in \Omega^0(\mathbb{C}^n)_{\mu}.$$

Now, regarding this formula as in (2.69) of Proposition 2.2, by a similar argument, we obtain the result for $N = n - 1$. Thus, the induction has been accomplished.

Here, using the symbol

$$[\overline{p}] \wedge \Omega_{\nu}^{n-1-p} := \sum_{1 \leq j_1 < \cdots < j_p \leq m} \frac{d\overline{P}_{j_1}}{\overline{P}_{j_1}} \wedge \cdots \wedge \frac{d\overline{P}_{j_p}}{\overline{P}_{j_p}} \wedge \Omega^{n-1-p}(\mathbb{C}^n)_{\nu},$$

we can summarize the above results as follows:

$$\Omega^{n-1}(\log \overline{D})_0 = [\overline{n-1}] \wedge \Omega_0^0,$$

$$\Omega^{n-1}(\log \overline{D})_1 = [\overline{n-1}] \wedge \Omega_1^0 + [\overline{n-2}] \wedge \Omega_1^1$$

$$\cdots\cdots\cdots \tag{2.97}$$

$$\Omega^{n-1}(\log \overline{D})_\mu = [\overline{n-1}] \wedge \Omega_\mu^0 + [\overline{n-2}] \wedge \Omega_\mu^1 + \cdots\cdots.$$

Combining Table 9.1 and (2.97), it clearly follows that for $\mu \geq 0$,

$$\sigma_\mu^{n-1} : Gr_\mu^F \Omega^{n-1}(\log D) \longrightarrow \Omega^{n-1}(\log \overline{D})_\mu$$

is an isomorphism, hence, we have $N^{n-1}(\log D)_\mu = 0$. Hence, by (2.93), we have

$$H^n(Gr_\mu^F \Omega^\bullet(\log D), Gr_\mu^F(\nabla_\omega)) = 0, \quad \mu \geq 0.$$

Applying an argument similar to the proof of Theorem 2.7 to this result, we can conclude as follows: any $\varphi \in \Omega^n(\log D)$ can be cohomologous to $\widetilde{\varphi} \in F_{-1}\Omega^n(\log D)$. On the other hand, by the definition of degree, we have

$$F_{-1}\Omega^n(\log D) = \frac{du_1 \wedge \cdots \wedge du_n}{P_1 \cdots P_m} \Omega^0(\mathbb{C}^n)_{\leq m-n-1}.$$

By Lemma 2.22, we have $F_{-1}\Omega^{n-1}(\log D) = 0$ which implies

$$H^n(\Omega^\bullet(\log D), \nabla_\omega) \simeq \frac{du_1 \wedge \cdots \wedge du_n}{P_1 \cdots P_m} \Omega^0(\mathbb{C}^n)_{\leq m-n-1}.$$

As $\dim \Omega^0(\mathbb{C}^n)_{\leq m-n-1} = \binom{m-1}{n}$, the dimension of the above cohomology is $\binom{m-1}{n}$. We summarize all these results in the following theorem.

Theorem 2.11. *For an arrangement of hyperplanes in general position, under the assumption $\alpha_j \notin \mathbb{Z}$ $(1 \leq j \leq m)$, $\sum_{j=1}^{m} \alpha_j \notin \mathbb{Z}$, we have the formulas:*

$$H^p(\Omega^\bullet(*D), \nabla_\omega) = 0 \quad (p \neq n),$$

$$H^n(\Omega^\bullet(*D), \nabla_\omega) \simeq \left\{ \frac{a(u)}{P_1 \cdots P_m} du_1 \wedge \cdots \wedge du_n \mid \deg a \leq m-n-1 \right\},$$

$$\dim H^n(\Omega^\bullet(*D), \nabla_\omega) = \binom{m-1}{n}.$$

Let us provide another expression of $H^n(\Omega^\bullet(*D), \nabla_\omega)$. By Lemma 2.22, we remark that $F_0\Omega^{n-1}(\log D)$ is the vector space with basis $\varphi\langle j_1, \ldots, j_{n-1}\rangle$, $1 \leq j_1 < \cdots < j_{n-1} \leq m$, and $F^0\Omega^n(\log D)$ is the vector space with basis $\varphi\langle j, \ldots, j_n\rangle$, $1 \leq j_1 < \cdots < j_n \leq m$. By the above argument, we can show

$$H^n(\Omega^\bullet(\log D), \nabla_\omega) \simeq F_0\Omega^n(\log D)/\omega \wedge F_0\Omega^{n-1}(\log D).$$

On the other hand, as we have

$$\omega \wedge \varphi\langle j_1, \ldots, j_{n-1}\rangle = \sum_{k=1}^{m} \alpha_k \varphi\langle k, j_1, \ldots, j_{n-1}\rangle,$$

we obtain

$$\varphi\langle m, j_1, \ldots, j_{n-1}\rangle = -\sum_{k=1}^{m-1} \frac{\alpha_k}{\alpha_m} \varphi\langle k, j_1, \ldots, j_{n-1}\rangle$$

in $H^n(\Omega^\bullet(\log D), \nabla_\omega)$. Hence, as a basis of $H^n(\Omega^\bullet(\log D), \nabla_\omega)$, we can choose $\binom{m-1}{n}$ $\varphi\langle j_1, \ldots, j_n\rangle$, $1 \leq j_1 < \cdots < j_n \leq m-1$. Thus, we obtain the following corollary.

Corollary 2.6. *As a basis of $H^n(\Omega^\bullet(*D), \nabla_\omega)$, one can take*

$$\varphi\langle j_1, \ldots, j_n\rangle, \quad 1 \leq j_1 < \cdots < j_n \leq m-1.$$

In [E-S-V], this result was extended to any arrangement of hyperplanes. The contents of this chapter is mainly based on [Ao3] and [Kit-No].

Chapter 3
Arrangement of Hyperplanes and Hypergeometric Functions over Grassmannians

3.1 Classical Hypergeometric Series and Their Generalizations, in Particular, Hypergeometric Series of Type $(n+1, m+1)$

In this section, we introduce a hypergeometric function of several variables with coefficients given by Γ-factors. Under this formulation, we show that the classically known hypergeometric series can be described systematically.

3.1.1 Definition

Let $L = \mathbb{Z}^n$ be a lattice, and $e_1 = (1, 0, \cdots, 0), \cdots, e_n = (0, \cdots, 0, 1)$ be its standard basis. Denoting the dual lattice $\mathrm{Hom}_{\mathbb{Z}}(L, \mathbb{Z})$ of L by $L^{\vee}$, its element $a \in L^{\vee}$ is a $\mathbb{Z}$-valued linear form on L. We use the notation on multi-indices without notice: for

$$x = (x_1, \ldots, x_n) \in \mathbb{C}^n, \quad \nu = (\nu_1, \ldots, \nu_n) \in \mathbb{Z}_{\geq 0}^n,$$

we set

$$x^{\nu} := x_1^{\nu_1} \cdots x_n^{\nu_n}, \quad \nu! = \nu_1! \cdots \nu_n!,$$

$$|\nu| := \sum_{i=1}^{n} \nu_i, \quad \sum_{\nu} \text{ is the sum over all } \nu \in \mathbb{Z}_{\geq 0}^n$$

and

$$\text{for } \alpha, c \in \mathbb{C}, \quad \alpha + c \notin \mathbb{Z}_{\leq 0}, \quad (\alpha; c) := \Gamma(\alpha + c)/\Gamma(c).$$

With these preparations, let us define a hypergeometric series:

Definition 3.1. When a power series

$$F((\alpha_k), x) = \sum_\nu \frac{\prod_{k \in K}(\alpha_k; a_k(\nu))}{\nu!} x^\nu \tag{3.1}$$

defined for $\alpha_k \in \mathbb{C}$ and linear forms $\{a_k \mid k \in K\} \subset L^\vee$ on the lattice L parametrized by a finite number of indices K, satisfies the condition

$$\sum_{k \in K} a_k(e_i) = 1, \qquad 1 \le i \le n, \tag{3.2}$$

this series is called the hypergeometric series associated to $\{a_k \mid k \in K\}$.

The convergence of this series will be shown later in § 3.3. First, we show that classical hypergeometric series can be obtained by taking the lattice L and $\{a_k \mid k \in K\} \subset L^\vee$ appropriately. Let us start from the following simple remark.

Remark 3.1. By the formula of the Γ-function $\Gamma(z)\Gamma(1 - z) = \pi/\sin \pi z$, we have

$$1/(\alpha; n) = (-1)^n (1 - \alpha; -n), \quad n \in \mathbb{Z}_{>0}, \tag{3.3}$$

from which we obtain

$$(\alpha_k; a_k(\nu)) = (-1)^{a_k(\nu)}/(1 - \alpha_k; -a_k(\nu)).$$

Decomposing $K = K_1 \coprod K_2$ and by the above remark, (3.1) can be also expressed as follows:

$$F((\alpha_k), x) = \sum_\nu \frac{\prod_{k \in K_1}(\alpha_k; a_k(\nu))}{\prod_{k \in K_2}(1 - \alpha_k; -a_k(\nu)) \cdot \nu!} \tag{3.4}$$

$$\times \prod_{i=1}^n \left((-1)^{\sum_{k \in K_2} a_k(e_i)} x_i \right)^{\nu_i}.$$

Below, we use an expression suitable for each purpose.

3.1.2 Simple Examples

The power series

$$F(\alpha, (\beta_i), \gamma) = \sum_\nu \frac{(\alpha; a(\nu)) \prod_{i=1}^n (\beta_i; b_i(\nu))}{(\gamma; c(\nu))\nu!} x^\nu$$

defined by $n + 2$ linear forms on $L = \mathbb{Z}^n$

$$a(\nu) = \sum_{i=1}^{n} \nu_i, \quad b_i(\nu) = \nu_i, \quad 1 \leq i \leq n, \quad c(\nu) = \sum_{i=1}^{n} \nu_i$$

satisfies the condition (3.2) and defines a hypergeometric series. This is of classical type, indeed, it corresponds to

$n = 1$ Gauss' hypergeometric series $\qquad F(\alpha, \beta, \gamma; x)$

$n = 2$ Appell's hypergeometric series $\qquad F_1(\alpha, \beta_1, \beta_2, \gamma; x_1, x_2)$

$n \geq 3$ Lauricella's hypergeometric series $F_D(\alpha, (\beta_i), \gamma; x_1, \ldots, x_n)$.

3.1.3 Hypergeometric Series of Type $(n+1, m+1)$

As a direct generalization of the above examples, there is a hypergeometric series studied by several mathematicians in special cases which is defined as follows. Suppose that $n < m$, and consider the power series

$$F((\alpha_i), (\beta_j), \gamma, x) = \sum_{\nu} \frac{\prod_{i=1}^{n}(\alpha_i; a_i(\nu)) \prod_{j=1}^{m-n-1}(\beta_j; b_j(\nu))}{(\gamma; c(\nu))\nu!} x^\nu \qquad (3.5)$$

defined by the lattice L formed by the set $M_{n,m-n-1}(\mathbb{Z})$ of all $n \times (m-n-1)$ matrices with integral coefficients and m linear forms

$$a_i(\nu) = \sum_{j=1}^{m-n-1} \nu_{ij}, \quad 1 \leq i \leq n,$$

$$b_j(\nu) = \sum_{i=1}^{n} \nu_{ij}, \qquad 1 \leq j \leq m-n-1,$$

$$c(\nu) = \sum_{i=1}^{n} \sum_{j=1}^{m-n-1} \nu_{ij}.$$

Here, for $\nu = (\nu_{ij}) \in M_{n,m-n-1}(\mathbb{Z}_{\geq 0})$, $x \in M_{n,m-n-1}(\mathbb{C})$, we used the symbols $x^\nu = \prod x_{ij}^{\nu_{ij}}$, $\nu! = \prod \nu_{ij}!$. This certainly satisfies the condition (3.2) and defines a hypergeometric series. Taking the results obtained in § 3.4 into account, we call it hypergeometric series of type $(n+1, m+1)$. The correspondence with classical types is as follows:

Gauss' hypergeometric series $\qquad\qquad$ type $(2, 4)$,

Appell's hypergeometric series F_1 $\qquad$ type $(2, 5)$,

Lauricella's hypergeometric series F_D type $(2, n+3)$.

3.1.4 Appell–Lauricella Hypergeometric Functions (i)

The power series

$$F(\alpha, (\beta_i), (\gamma_i); x) = \sum_{\nu} \frac{(\alpha; a(\nu)) \prod_{i=1}^{n}(\beta_i; b_i(\nu))}{\prod_{i=1}^{n}(\gamma_i; c_i(\nu)) \cdot \nu!} x^{\nu}$$

defined by $2n + 1$ linear forms on the lattice $L = \mathbb{Z}^n$

$$a(\nu) = \sum_{i=1}^{n} \nu_i, \quad b_i(\nu) = \nu_i, \quad 1 \le i \le n, \quad c_i(\nu) = \nu_i, \quad 1 \le i \le n$$

satisfies the condition (3.2) and defines a hypergeometric series. This is what is called Appell's F_2 for $n = 2$ and Lauricella's F_A for $n \ge 3$.

3.1.5 Appell–Lauricella Hypergeometric Functions (ii)

The hypergeometric series

$$F((\alpha_i), (\beta_i), \gamma; x) = \sum_{\nu} \frac{\prod_{i=1}^{n}(\alpha_i; a_i(\nu)) \prod_{i=1}^{n}(\beta_i; b_i(\nu))}{(\gamma; c(\nu))\nu!} x^{\nu}$$

defined by $2n + 1$-linear forms on the lattice $L = \mathbb{Z}^n$

$$a_i(\nu) = \nu_i, \quad 1 \le i \le n, \quad b_i(\nu) = \nu_i, \quad 1 \le i \le n, \quad c(\nu) = \sum_{i=1}^{n} \nu_i$$

is what is called Appell's F_3 for $n = 2$ and Lauricella's F_B for $n \ge 3$.

3.1.6 Restriction to a Sublattice

Here, to describe a relation between Appell–Lauricella hypergeometric series F_B, F_D and hypergeometric series of type $(n + 1, 2n + 2)$, we begin with the following consideration. Let I be a subset of $\{1, 2, \cdots, n\}$ and take the sublattice $L_I = \sum_{i \in I} \mathbb{Z} e_i$ of the lattice L. Considering the restriction $b_k := a_k|_{L_I}$ of a_k to L_I in the definition of hypergeometric series given in § 3.1.1, $\{b_k \mid k \in K\}$ are linear forms on L_I which apparently satisfy the condition

$$\sum_{k \in K} b_k(e_i) = 1, \quad i \in I.$$

Hence, it defines a hypergeometric series

$$F_I((\alpha_k), x_I) = \sum \frac{\prod_{k \in K}(\alpha_k; b_k(\nu_I))}{\nu_I!} x_I^{\nu_I}.$$

Here, we set, for $I = \{i_1, \cdots, i_t\}$,

$$\nu_I = (\nu_{i_1}, \ldots, \nu_{i_t}) \in L_I, \quad x_I = (x_{i_1}, \ldots, x_{i_t}) \in \mathbb{C}^t.$$

3.1.7 Examples

Let us apply the above consideration. Recall that hypergeometric series F of type $(n+1, 2n+2)$ is defined by $2n+2$ linear forms of the lattice $L = M_n(\mathbb{Z})$

$$a_i(\nu) = \sum_{j=1}^{n} \nu_{ij}, \quad 1 \le i \le n, \quad b_j(\nu) = \sum_{i=1}^{n} \nu_{ij}, \quad 1 \le j \le n, \quad c(\nu) = \sum_{i,j=1}^{n} \nu_{ij}.$$

Here, as a subset of the set of double indices (i, j), $1 \le i, j \le n$, if we take the diagonal component $I = \{(1,1), \cdots, (n,n)\}$, together with § 3.1.5, we obtain Appell–Lauricella hypergeometric series F_B as F_I, and if we take $I = \{(1,1), (1,2), \cdots, (1,n)\}$, together with § 3.1.2, we obtain Gauss–Appell–Lauricella hypergeometric series F_D.

3.1.8 Appell–Lauricella Hypergeometric Functions (iii)

The hypergeometric series

$$F(\alpha, \beta, (\gamma_i); x) = \sum_{\nu} \frac{(\alpha; a(\nu))(\beta; b(\nu))}{\prod_{i=1}^{n}(\gamma_i; c_i(\nu)) \cdot \nu!} x^{\nu}$$

defined by $n+2$ linear forms of the lattice $L = \mathbb{Z}^n$

$$a(\nu) = \sum_{i=1}^{n} \nu_i, \quad b(\nu) = \sum_{i=1}^{n} \nu_i, \quad c_i(\nu) = \nu_i, \quad 1 \le i \le n$$

is what is called Appell's F_4 for $n = 2$ and Laricella's F_C for $n \ge 3$.

3.1.9 Horn's Hypergeometric Functions

Horn found out all the hypergeometric series in two variables with the conditions

$$\sum_{\substack{k\in K \\ a_k(e_i)>0}} a_k(e_i) = 2, \qquad \sum_{\substack{k\in K \\ a_k(e_i)<0}} a_k(e_i) = -1, \quad i = 1, 2$$

that are not the product of hypergeometric series in one variable. There are 14 such series which contain Appell's four hypergeometric series. The result is stated in [Er1], pp.224–225 as Horn's list. Although the linear forms on the lattice L treated above are always of the form $\sum \nu_i$, there are some in Horn's list which are not of this form. For example, the hypergeometric series

$$G_3(\alpha, \beta; x_1, x_2) = \sum_{\nu} \frac{(\alpha; a(\nu))(\beta; b(\nu))}{\nu_1! \nu_2!} x_1^{\nu_1} x_2^{\nu_2}$$

defined by linear forms on $L = \mathbb{Z}^2$

$$a(\nu) = 2\nu_2 - \nu_1, \quad b(\nu) = 2\nu_1 - \nu_2$$

is such a case. Of course, our definition contains all these hypergeometric series. It can be easily checked that the 14 series in Horn's list all satisfy the condition (3.2).

Remark 3.2. According to our definition of hypergeometric series (3.1), the denominator of x^ν is $\nu!$. Hypergeometric functions generalizing further this point are treated in Appendix A.

3.2 Construction of Twisted Cycles (2): For an Arrangement of Hyperplanes in General Positiion

3.2.1 Twisted Homology Group

Let $H_j = \{u \in \mathbb{C}^n \mid P_j(u) = 0\}$, $1 \le j \le m$ be a hyperplane defined by a polynomial of degree 1 with real coefficients

$$P_j(u) = x_{0j} + x_{1j}u_1 + \cdots + x_{nj}u_n, \quad 1 \le j \le m$$

in the complex affine space $\mathbb{C}^n$ with its coordinates $(u_1, \cdots, u_n)$, and consider the affine algebraic variety $M = \mathbb{C}^n \setminus \bigcup_{j=1}^m H_j$. Below, we assume that the arrangement of hyperplanes, obtained by adjoining the hyperplane at infinity H_∞ to H_j, $1 \le j \le m$, is in general position in $\mathbb{P}^n(\mathbb{C})$. That is, we assume that any $(n+1) \times (n+1)$ minor of the real matrix

$$x = \begin{pmatrix} x_{01} & \cdots\cdots & x_{0m} & 1 \\ x_{11} & \cdots\cdots & x_{1m} & 0 \\ \vdots & & \vdots & \vdots \\ x_{n1} & \cdots\cdots & x_{nm} & 0 \end{pmatrix} \in M_{n+1,m+1}(\mathbb{R}) \tag{3.6}$$

never vanishes. Denote the connection form associated to the multi-valued function

$$U(u) = \prod_{j=1}^{m} P_j(u)^{\alpha_j}$$

on M by $\omega = dU/U$ and we use the notation defined in § 2.1 and 2.2 of Chapter 2. First, by Lemma 2.8 and the comparison theorem we have

$$H_p^{lf}(M, \mathcal{L}_\omega) \simeq H^{2n-p}(\mathcal{A}^\bullet(M), \nabla_\omega) \tag{3.7}$$

$$\simeq H^{2n-p}(\Omega^\bullet(*D), \nabla_\omega).$$

Hence, by Lemma 2.9 and Corollary 2.6, we obtain the following lemma:

Lemma 3.1. *Under the assumptions* $\alpha_j \notin \mathbb{Z}$, $1 \le j \le m$ *and* $\sum_{j=1}^m \alpha_j \notin \mathbb{Z}$, *we have:*

1. $H_p^{lf}(M, \mathcal{L}_\omega^\vee) = 0$, $\quad p \neq n$.

2. $H_n^{lf}(M, \mathcal{L}_\omega^\vee) \simeq H^n(M, \mathcal{L}_\omega^\vee) \simeq H_n(M, \mathcal{L}_\omega)^\vee$.

3. $\dim H_n^{lf}(M, \mathcal{L}_\omega^\vee) = \binom{m-1}{n}$.

3.2.2 Bounded Chambers

Denoting $H_j \cap \mathbb{R}^n$ by $H_{j\mathbb{R}}$, the arrangement in $\mathbb{R}^n$ of the real hyperplanes $H_{j\mathbb{R}}, 1 \le j \le m$ decomposes $\mathbb{R}^n$ into several chambers. For two-dimensional cases, it turns out that there are $\binom{m-1}{2}$ bounded chambers (Figure 3.1). Since such a bounded chamber Δ is convex, hence simply connected, we can fix a branch U_Δ of U on Δ and forms $\Delta \otimes U_\Delta$ which defines a twisted homology class of $H_2^{lf}(M, \mathcal{L}_\omega^\vee)$. In fact, we show that these form a basis of $H_2^{lf}(M, \mathcal{L}_\omega^\vee)$. Let us explain this fact for any dimension.

3.2.3 Basis of Locally Finite Homology

Enumerate the bounded chambers of $\mathbb{R}^n \setminus \bigcup_{j=1}^m H_{j\mathbb{R}}$ as $\Delta_\nu, 1 \le \nu \le s$, as we like. Considering the divisor

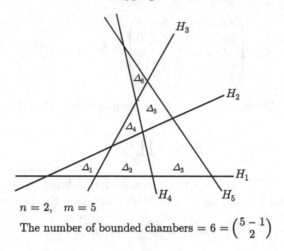

$$n = 2, \quad m = 5$$

The number of bounded chambers $= 6 = \binom{5-1}{2}$

Fig. 3.1

$$\hat{D} := \bigcup_{j=1}^{m} H_j \cup H_\infty$$

of $\mathbb{P}^n(\mathbb{C})$, $\hat{D}$ and $\hat{D} \cup \coprod_\nu \Delta_\nu$ are closed subsets of $\mathbb{P}^n(\mathbb{C})$. We obtain the exact cohomology sequence with coefficients in $\mathcal{L}_\omega$ associated to the triple of spaces $(\mathbb{P}^n(\mathbb{C}), \hat{D} \cup \coprod_\nu \Delta_\nu, \hat{D})$:

$$\longrightarrow H^{p-1}(\hat{D} \cup \coprod \Delta_\nu, \hat{D}) \longrightarrow H^p(\mathbb{P}^n(\mathbb{C}), \hat{D} \cup \coprod_\nu \Delta_\nu) \longrightarrow \qquad (3.8)$$

$$\longrightarrow H^p(\mathbb{P}^n(\mathbb{C}), \hat{D}) \longrightarrow H^p(\hat{D} \cup \coprod_\nu \Delta_\nu, \hat{D}) \longrightarrow \cdots.$$

On the other hand, there are the isomorphisms:

$$H_p^{lf}(\mathbb{P}^n(\mathbb{C}) \setminus \hat{D}, \mathcal{L}_\omega^\vee)$$

$$\simeq H_c^p(\mathbb{P}^n(\mathbb{C}) \setminus \hat{D}, \mathcal{L}_\omega)^\vee \quad \text{(by Lemma 2.9 (2))}$$

$$\simeq H^p(\mathbb{P}^n(\mathbb{C}), \text{a sufficiently small tubular neighborhood of } \hat{D}, \mathcal{L}_\omega)^\vee$$

$$\text{(since } \mathbb{P}^n(\mathbb{C}) \text{ is compact and } \hat{D} \text{ is normal crossing)}$$

$$\simeq H^p(\mathbb{P}^n(\mathbb{C}), \hat{D}, \mathcal{L}_\omega)^\vee$$

($\hat{D}$ is a retract of a sufficiently small tubular neighborhood).

Similarly, we obtain

$$H_p^{lf}(\widehat{D} \cup \coprod_\nu \Delta_\nu \setminus \widehat{D}, \mathcal{L}_\omega^\vee) \simeq H^p(\widehat{D} \cup \coprod_\nu \Delta_\nu, \widehat{D}, \mathcal{L}_\omega)^\vee,$$

$$H_p^{lf}(\mathbb{P}^n(\mathbb{C}) \setminus \left(\widehat{D} \cup \coprod_\nu \Delta_\nu\right), \mathcal{L}_\omega^\vee) \simeq H^p(\mathbb{P}^n(\mathbb{C}), \widehat{D} \cup \coprod_\nu \Delta_\nu, \mathcal{L}_\omega)^\vee.$$

Combining these isomorphisms with the dual of the exact sequence (3.8), noting that $M = \mathbb{P}^n(\mathbb{C}) \setminus \widehat{D}$, we obtain the exact sequence:

$$\longrightarrow H_{n+1}^{lf}(M \setminus \coprod_\nu \Delta_\nu, \mathcal{L}_\omega^\vee) \longrightarrow H_n^{lf}(\coprod_\nu \Delta_\nu, \mathcal{L}_\omega^\vee) \qquad (3.9)$$

$$\longrightarrow H_n^{lf}(M, \mathcal{L}_\omega^\vee) \longrightarrow H_n^{lf}(M \setminus \coprod_\nu \Delta_\nu, \mathcal{L}_\omega^\vee) \longrightarrow H_{n-1}^{lf}(\coprod_\nu \Delta_\nu, \mathcal{L}_\omega^\vee).$$

Since the first isomorphism of (3.7) holds for any C^∞-manifold M, we have

$$H_p^{lf}(M \setminus \coprod_\nu \Delta_\nu, \mathcal{L}_\omega^\vee) \simeq H^{2n-p}(M \setminus \coprod_\nu \Delta_\nu, \mathcal{L}_\omega^\vee).$$

On the other hand, $M \setminus \coprod_\nu \Delta_\nu$ is homotopic to

$$\overline{M} = \mathbb{C}^n \setminus \bigcup_{j=1}^m \{\overline{P}_j = 0\}, \quad \text{where } \overline{P}_j(u) = x_{1j}u_1 + \cdots + x_{nj}u_n.$$

Hence, setting $\overline{\omega} = \sum \alpha_j \frac{d\overline{P}_j}{\overline{P}_j}$, we obtain an isomorphism

$$H^{2n-p}(M \setminus \coprod_\nu \Delta_\nu, \mathcal{L}_\omega^\vee) \simeq H^{2n-p}(\overline{M}, \mathcal{L}_{\overline{\omega}}^\vee). \qquad (3.10)$$

As each $\overline{P}_j$ is homogeneous, by Theorem 2.6 (1), if the condition $\sum_{j=1}^m \alpha_j \notin \mathbb{Z}$ is satisfied, then the twisted cohomologies of the right-hand side of (3.10) all vanish, and in (3.9), we have

$$H_p^{lf}(M \setminus \coprod_\nu \Delta_\nu, \mathcal{L}_\omega^\vee) = 0, \quad p = n+1, n.$$

Hence, we have shown an isomorphism

$$H_n^{lf}(\coprod_\nu \Delta_\nu, \mathcal{L}_\omega^\vee) \simeq H_n^{lf}(M, \mathcal{L}_\omega^\vee). \qquad (3.11)$$

But here, as Δ_ν is convex, it is homeomorphic to $\mathbb{R}^n$, and we can fix a branch U_{Δ_ν} of U on Δ_ν from which we obtain

$$H_n^{lf}(\coprod_\nu \Delta_\nu, \mathcal{L}_\omega^\vee) \simeq \bigoplus_{\nu=1}^s \mathbb{C}[\Delta_\nu \otimes U_{\Delta_\nu}].$$

By Lemma 3.1 (3), (3.11), and the above result, we have shown the following lemma.

Lemma 3.2. *Under the assumption* $\sum_{j=1}^{m} \alpha_j \notin \mathbb{Z}$, *twisted cycles* $\Delta_\nu \otimes U_{\Delta_\nu}$ *defined by bounded chambers* Δ_ν, $1 \le \nu \le \binom{m-1}{n}$, *form a basis of* $H_n^{lf}(M, \mathcal{L}_\omega^\vee)$:

$$H_n^{lf}(M, \mathcal{L}_\omega^\vee) \simeq \bigoplus_{\nu=1}^{\binom{m-1}{n}} \mathbb{C}[\Delta_\nu \otimes U_{\Delta_\nu}].$$

3.2.4 Construction of Twisted Cycles

For the one-dimensional case (in § 2.3), we constructed a twisted cycle $\Delta_j(\omega)$ associated to each interval $\Delta_j = (x_j, x_{j+1})$, and we can construct a twisted cycle $\Delta(\omega) \in H_n(M, \mathcal{L}_\omega^\vee)$ associated to each bounded chamber for an arrangement of hyperplanes in general position. The idea is that since Δ is locally the product of the one-dimensional case, we take the product of part of the twisted cycles of dimension 1, and construct $\Delta(\omega)$ by gluing them following a standard recipe in differential topology.

Since the symbols become complicated for the general case, for simplicity, we assume that a bounded chamber Δ is surrounded by t hyperplanes $H_1, \cdots, H_t$ (Figure 3.2) and for $I \subset \{1, 2, \cdots, t\}$, we use the following notation: $H_I = \bigcap_{i \in I} H_i$, $\Delta_I = \overline{\Delta} \cap H_I$, $T_I(\varepsilon) := \varepsilon$-neighborhood of Δ_I, where ε is sufficiently small.

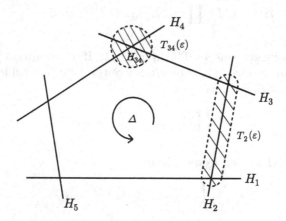

Fig. 3.2

Fix the standard orientation of $\mathbb{R}^n$ defined by the coordinates $(u_1, \cdots, u_n)$ and introduce the induced orientation on Δ. First, we set

$$c_0 := \Delta \setminus \bigcup_I T_I(\varepsilon),$$

and introduce the same orientation as Δ. For any $p \in \partial\Delta$, we can fix the subset $I \subset \{1, \cdots, t\}$ satisfying

$$p \in \Delta_I \setminus \bigcup_{J \supsetneq I} \Delta_J.$$

Then, we can choose a sufficiently small neighborhood $W_I(p)$ of p and local chart $(w_1, \cdots, w_n)$ with the following conditions:

(1) The orientation on $W_I(p) \cap \mathbb{R}^n$ defined by its local charts $(w_1, \cdots, w_n)$ coincides with the standard orientation on $\mathbb{R}^n$.

(2) $W_I(p) = \{w \in \mathbb{C}^n \mid |w_j| \leq 1, \ 1 \leq j \leq n\}$,

$W_I(p) \cap \overline{\Delta} = \{w \in \mathbb{R}^n \mid 0 \leq w_i \leq 1, \ 1 \leq i \leq k, \ |w_j| \leq 1, \ k+1 \leq j \leq n\}$,

$W_I(p) \cap \Delta_I = \{w \in \mathbb{R}^n \mid w_i = 0, \ 1 \leq i \leq k, \ |w_j| \leq 1, \ k+1 \leq j \leq n\}$,

$W_I(p) \cap T_I(\varepsilon) = \{w \in W_I(p) \mid |w_i| < \varepsilon, \ 1 \leq i \leq k\}$.

Now, let $S^1_\varepsilon(0)$ be the circle with the center at 0 with radius ε having the positive orientation and with the fixed starting point ε. We consider the n-chain in $W_I(p)$

$$c_I(p) := \frac{1}{d_{i_1}} S^1_\varepsilon(0) \times \cdots \times \frac{1}{d_{i_k}} S^1_\varepsilon(0) \times [-1, 1]^{n-k}.$$

Here, we assume that $\{w_k = 0\}$ corresponds to the hyperplane H_{i_k}, setting $d_{i_k} = \exp(2\pi\sqrt{-1}\alpha_{i_k}) - 1$, and introduce the product of the orientations on $c_I(p)$ (Figure 3.3).

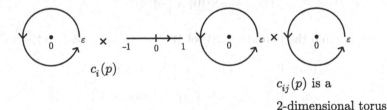

$c_i(p)$

$c_{ij}(p)$ is a
2-dimensional torus

Fig. 3.3

By using the uniqueness theorem of tubular neighborhood ([Mat], p.21), gluing $c_I(p)$ constructed on each point of $\Delta_I \setminus \bigcup_{J \supsetneq I} T_J(\varepsilon)$, we can construct a twisted chain c_I. From Figure 3.4, the reader may understand what can be obtained by constructing "earthen pipes" around Δ_I is this c_I. Now, by setting

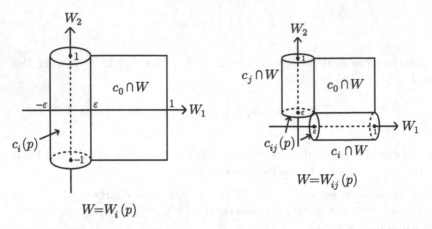

Fig. 3.4 Conceptual diagram

$$\Delta(\omega) = c_0 + \sum_{|I|=1}^{n} \sum_{I} c_I,$$

we obtain the following lemma:

Lemma 3.3. $\partial_\omega \Delta(\omega) = 0.$

Proof. By the construction of each $c_I(p)$, if we can verify $\partial_\omega \Delta(\omega) = 0$ mod $\partial W_I(p)$, as the tubular neighborhoods glue together, we can conclude $\partial_\omega \Delta(\omega) = 0$. Let us show this by induction on $|I|$: for $|I| = 1$, assuming $I = \{1\}$ and setting $W := W_1(p)$ for simplicity (cf. Figure 3.4), we have

$$c_0 \cap W = [\varepsilon, 1] \times [-1, 1]^{n-1},$$

$$c_1 \cap W = \frac{1}{d_1} S_\varepsilon^1(0) \times [-1, 1]^{n-1}.$$

As we have, modulo the boundary ∂W of $W = \{w \in \mathbb{C}^n \mid |w_i| \leq 1\}$,

$$\partial_\omega(c_0 \cap W) \equiv -\langle \varepsilon \rangle \times [-1, 1]^{n-1} \;(\text{mod } \partial W),$$

$$\partial_\omega(c_1 \cap W) \equiv \langle \varepsilon \rangle \times [-1, 1]^{n-1} \;(\text{mod } \partial W),$$

this implies $\partial_\omega(\Delta(\omega) \cap W) \equiv 0 \;(\text{mod } \partial W)$. For $|I| = 2$, similarly, setting $I = \{1, 2\}$, $W = W_{12}(p)$ (cf. Figure 3.4), we obtain

$$c_0 \cap W = [\varepsilon, 1] \times [\varepsilon, 1] \times [-1, 1]^{n-2}, \qquad (3.12)$$

$$c_1 \cap W = \frac{1}{d_1} S^1_\varepsilon(0) \times [\varepsilon, 1] \times [-1, 1]^{n-2},$$

$$c_2 \cap W = [\varepsilon, 1] \times \frac{1}{d_2} S^1_\varepsilon(0) \times [-1, 1]^{n-2},$$

$$c_{12} \cap W = \frac{1}{d_1} S^1_\varepsilon(0) \times \frac{1}{d_2} S^1_\varepsilon(0) \times [-1, 1]^{n-2}.$$

Here, let us say a few words on $c_2 \cap W$: by the definition of $c_2(p)$, expressing, by using the local charts $(w_2, w_1, \cdots, w_n)$, as

$$c_2 \cap W = \frac{1}{d_2} S^1_\varepsilon(0) \times [-1, 1]^{n-1}, \qquad (3.13)$$

$(w_2, w_1, \cdots, w_n)$ defines an opposite orientation to that on $\mathbb{R}^n$ defined by $(w_1, w_2, \cdots, w_n)$, hence, the orientation on $c_{2\cap}W$ has to be opposite to the product of the orientations in the right-hand side of (3.13). Hence, the formulas (3.12) hold with the orientation. Similarly as before, setting $I = [-1, 1]$, we see that

$$\partial_\omega(c_0 \cap W) \equiv -\langle \varepsilon \rangle \times [\varepsilon, 1] \times I^{n-2} + [\varepsilon, 1] \times \langle \varepsilon \rangle \times I^{n-2} \, (\mathrm{mod} \ \partial W),$$

$$\partial_\omega(c_1 \cap W) \equiv \langle \varepsilon \rangle \times [\varepsilon, 1] \times I^{n-2} + \frac{1}{d_1} S^1_\varepsilon(0) \times \langle \varepsilon \rangle \times I^{n-2} \, (\mathrm{mod} \ \partial W),$$

$$\partial_\omega(c_2 \cap W) \equiv -\langle \varepsilon \rangle \times \frac{1}{d_2} S^1_\varepsilon(0) \times I^{n-2} - [\varepsilon, 1] \times \langle \varepsilon \rangle \times I^{n-2} \, (\mathrm{mod} \ \partial W),$$

$$\partial_\omega(c_{12} \cap W) \equiv \langle \varepsilon \rangle \times \frac{1}{d_2} S^1_\varepsilon(0) \times I^{n-2} - \frac{1}{d_1} S^1_\varepsilon(0) \times \langle \varepsilon \rangle \times I^{n-2} \, (\mathrm{mod} \ \partial W),$$

from which we obtain $\partial_\omega(\Delta(\omega) \cap W) \equiv 0 \, (\mathrm{mod} \ \partial W)$. Repeating this argument for each $W_I(p)$, we obtain $\partial_\omega \Delta(\omega) = 0$.

3.2.5 Regularization of Integrals

We have constructed in § 3.2.4 the twisted cycle $\Delta_\nu(\omega)$ for a bounded chamber Δ_ν, $1 \le \nu \le \binom{m-1}{n}$. By construction, as $\Delta_\nu(\omega)$ is $\varepsilon > 0$ away from $D = \cup \{P_j = 0\}$, the integral $\int_{\Delta_\nu(\omega)} \prod P_j^{\alpha_j} du \cdots du_n$ makes sense for any $\alpha_j \notin \mathbb{Z}$ and for $\alpha_j \in (\mathbb{C} \setminus \mathbb{Z}) \cap \{\Re \alpha_j > -1\}$ and we have

$$\int_{\Delta_\nu(\omega)} \prod P_j^{\alpha_j} du_1 \cdots du_n = \int_{\Delta_\nu} \prod P_j^{\alpha_j} du_1 \cdots du_n. \qquad (3.14)$$

In this sense, the left-hand side of (3.14) provides an analytic continuation of the right-hand side of (3.14), which diverges when $\Re \alpha_j \leq -1$, with respect to the parameters $\alpha_1, \cdots, \alpha_m$, and we have concretely constructed the so-called Hadamard finite part of a divergent integral. For a chamber Δ_ν, the operation to construct the twisted cycle $\Delta_\nu(\omega)$ is called a regularization. Then, since, by Lemma 3.2, under the assumption $\sum_{j=1}^{m} \alpha_j \notin \mathbb{Z}$, Δ_ν, $1 \leq \nu \leq \binom{m-1}{n}$ form a basis of $H_n^{lf}(M, \mathcal{L}_\omega^\vee)$, the map

$$\mathrm{reg} : H_n^{lf}(M, \mathcal{L}_\omega^\vee) \longrightarrow H_n(M, \mathcal{L}_\omega^\vee)$$

$$[\Delta_\nu] \longrightarrow [\Delta_\nu(\omega)]$$

defines a homomorphism from $H_n^{lf}(M, \mathcal{L}_\omega^\vee)$ to $H_n(M, \mathcal{L}_\omega^\vee)$. On the other hand, there also is a natural homomorphism

$$\kappa : H_n(M, \mathcal{L}_\omega^\vee) \longrightarrow H_n^{lf}(M, \mathcal{L}_\omega^\vee).$$

By construction of a twisted cycle, the support of $\kappa \cdot \mathrm{reg}(\Delta_\nu) - \Delta_\nu$ is contained in any ε-neighborhood of $\partial \Delta_\nu$ which implies that for any $\varphi \in \mathcal{A}_c^n(M)$, we have

$$\int_{\kappa \cdot \mathrm{reg}(\Delta_\nu) - \Delta_\nu} U \cdot \varphi = 0.$$

By Lemma 2.9 (2), this means that the formula

$$\kappa \cdot \mathrm{reg}(\Delta_\nu) - \Delta_\nu \backsim 0$$

in $H_n^{lf}(M, \mathcal{L}_\omega^\vee)$ which implies $\kappa \cdot \mathrm{reg} = 1$. As $\dim H_n(M, \mathcal{L}_\omega^\vee) = \dim H_n^{lf}(M, \mathcal{L}_\omega^\vee)$, reg becomes an isomorphism. Summarizing this, we obtain the following theorem.

Theorem 3.1. *Under the assumptions $\alpha_j \notin \mathbb{Z}$, $1 \leq j \leq m$ and $\sum_{j=1}^{m} \alpha_j \notin \mathbb{Z}$, $\binom{m-1}{n}$ bounded chambers Δ_ν, $1 \leq \nu \leq \binom{m-1}{n}$ form a basis of $H_n^{lf}(M, \mathcal{L}_\omega^\vee)$, and the regularization map*

$$\mathrm{reg} : H_n^{lf}(M, \mathcal{L}_\omega^\vee) \overset{\sim}{\longrightarrow} H_n(M, \mathcal{L}_\omega^\vee)$$

$$[\Delta_\nu] \longrightarrow [\Delta_\nu(\omega)]$$

becomes an isomorphism. Therefore, $\Delta_\nu(\omega)$, $1 \leq \nu \leq \binom{m-1}{n}$ provide a basis of $H_n(M, \mathcal{L}_\omega^\vee)$.

For $J = \{j_1, \cdots, j_n\}$, we abbreviate as $\varphi \langle J \rangle := \frac{dP_{j_1}}{P_{j_1}} \wedge \cdots \wedge \frac{dP_{j_n}}{P_{j_n}}$. By the above theorem, together with Lemma 2.9 (1), Theorem 2.5 and Corollary 2.6, we obtain the following corollary:

Corollary 3.1. *Under the same assumptions, we have*

$$\det\left(\int_{\Delta_\nu(\omega)} U \cdot \varphi\langle J\rangle\right) \neq 0.$$

Here, $1 \leq \nu \leq \binom{m-1}{n}$ and J runs over all $J \subset \{1, \cdots, m-1\}$, $|J| = n$.

3.3 Kummer's Method for Integral Representations and Its Modernization via the Twisted de Rham Theory: Integral Representations of Hypergeometric Series of Type $(n+1, m+1)$

3.3.1 Kummer's Method

To express an infinite sum series as an integral, Kummer [Ku] used an excellent method described as follows. That is, suppose that, for an infinite series $\sum_{k=0}^{\infty} a_k b_k$, we could find nice functions satisfying:

(1)   a continuous function $U(u)$ on a bounded closed domain Δ of $\mathbb{R}^n$ and a series of functions $\varphi_k(u)$, $k \in \mathbb{Z}_{>0}$;

(2) $\int_\Delta U(u)\varphi_k(u)du = b_k \int_\Delta U(u)\varphi_0(u)du, \quad k \in \mathbb{Z}_{>0}$

       and $\int_\Delta U(u)\varphi_0(u)du \neq 0$;

(3)   the sum $\sum_{k=1}^{\infty} a_k\varphi_k(u)$ converges uniformly to a continuous function $\Phi(u)$ on Δ.

Then, we may have

$$\sum_{k=1}^{\infty} a_k b_k = \sum_{k=1}^{\infty} a_k \int_\Delta U(u)\varphi_k(u)du \Big/ \int_\Delta U(u)\varphi_0(u)du \qquad (3.15)$$

$$= \int_\Delta U(u)\Phi(u)du \Big/ \int_\Delta U(u)\varphi_0(u)du,$$

namely, the convergence of the series $\sum_{k=1}^{\infty} a_k b_k$ and an integral representation of the sum are obtained at the same time. Below, we call this Kummer's method.

3.3.2 One-Dimensional Case

By the above Kummer's method, we would like to show that a general hypergeometric series defined in § 3.1 converges and possesses an integral representation. First, let us show this in the most simple one-dimensional case.

As was shown in Theorem 2.11, for $M = \mathbb{C} \setminus \{x_1, \cdots, x_m\}$, $U = \prod_{j=1}^{m} (u - x_j)^{\alpha_j}$, $x_j \in \mathbb{R}$, under the assumptions $\alpha_j \notin \mathbb{Z}(1 \leq j \leq m)$, $\sum \alpha_j \notin \mathbb{Z}$, we have $\dim H^1(\Omega^\bullet(*D), \nabla_\omega) = m - 1$. Hence, only in the case when $m = 2$, we have $\dim H^1(\Omega^\bullet(*D), \nabla_\omega) = 1$, and in this case, for any $k \in \mathbb{Z}_{\geq 0}$, there exists a constant $b_k \in \mathbb{C}$ such that $u^k du \backsim b_k du$ in $H^1(\Omega(*D), \nabla_\omega)$. Hence, using the regularization $\text{reg}(\Delta) = \Delta(\omega)$ of the interval $\Delta = (x_1, x_2)$, we have

$$\int_{\Delta(\omega)} U \cdot u^k du = b_k \int_{\Delta(\omega)} U du, \quad k \in \mathbb{Z}_{\geq 0}$$

which corresponds to the case $\varphi_k(u) = u^k$ in the notation of § 3.3.1. Hence, if the power series $\sum_{k=0}^{\infty} a_k u^k$ converges uniformly to $\Phi(u)$ on $\Delta(\omega)$, the sum $\sum_{k=0}^{\infty} a_k b_k$ converges, and at the same time, we have $\sum_{k=0}^{\infty} a_k b_k = \int_{\Delta(\omega)} U \Phi du / \int_{\Delta(\omega)} U du$. In particular, taking the configuration of two points as $x_1 = 0$, $x_2 = 1$ and reducing the exponents α_1, α_2 by 1, Euler's beta function

$$\int_{\Delta(\omega)} u^{\alpha_1 - 1}(1 - u)^{\alpha_2 - 1} du = \Gamma(\alpha_1)\Gamma(\alpha_2)/\Gamma(\alpha_1 + \alpha_2)$$

appears as the integral $\int_{\Delta(\omega)} U du$. Hence, by an identity of the Γ-function, we obtain

$$\int_{\Delta(\omega)} U \cdot u^k du = \frac{(\alpha_1; k)}{(\alpha_1 + \alpha_2; k)} \int_{\Delta(\omega)} U du. \tag{3.16}$$

3.3.3 Higher-Dimensional Case

Using the notation defined in § 3.2, let us consider an n-dimensional case. As $\dim H^n(\Omega^\bullet(*D), \nabla_\omega) = \binom{m-1}{n}$, only in the case $m - 1 = n$, the twisted cohomology $H^n(\Omega^\bullet(*D), \nabla_\omega)$ becomes one-dimensional and an argument similar to § 3.3.2 applies. Normalizing an arrangement of $n + 1$ hyperplanes as

$$H_i = \{u_i = 0\}, \quad 1 \leq i \leq n, \quad H_{n+1} = \{1 - u_1 - \cdots - u_n = 0\},$$

set

$$U(u) = \prod_{i=1}^{n} u_i^{\alpha_i - 1} \cdot (1 - \sum_{i=1}^{n} u_i)^{\alpha_{n+1} - 1}, \quad \omega = dU/U$$

$$\Delta^n = \{u \in \mathbb{R}^n | u_i > 0, \ 1 \leq i \leq n, \ 1 - \sum u_i > 0\}. \tag{3.17}$$

First, as a generalization of Euler's beta function, recall the Dirichlet integral ([W-W] p.258):

$$\int_{\Delta^n(\omega)} U \, du = \prod_{i=1}^{n+1} \Gamma(\alpha_i)/\Gamma(\alpha_1 + \cdots + \alpha_{n+1}). \tag{3.18}$$

Using a multi-index $\nu = (\nu_1, \cdots, \nu_n) \in \mathbb{Z}^n$, by (3.18), we obtain

$$\frac{(\alpha_1; \nu_1)\cdots(\alpha_n; \nu_n)}{\left(\sum_{i=1}^{n+1}\alpha_i; |\nu|\right)} = \frac{\Gamma\left(\sum_{i=1}^{n+1}\alpha_i\right)}{\prod_{i=1}^{n+1}\Gamma(\alpha_i)} \int_{\Delta^n(\omega)} U \cdot u^\nu \, du.$$

Summarizing, we obtain:

Lemma 3.4. *Let $\Delta^n(\omega)$ be the regularization of an n-simplex Δ^n in $\mathbb{R}^n$, and suppose that the exponents of the multi-valued function U satisfy $\alpha_i \notin \mathbb{Z}$, $1 \le i \le n+1$, $\sum_{i=1}^{n+1}\alpha_i \notin \mathbb{Z}$. Then, for $\nu \in \mathbb{Z}^n$, we have*

$$\frac{(\alpha_1; \nu_1)\cdots(\alpha_n; \nu_n)}{\left(\sum_{i=1}^{n+1}\alpha_i; |\nu|\right)} = \frac{\Gamma\left(\sum_{i=1}^{n+1}\alpha_i\right)}{\prod_{i=1}^{n+1}\Gamma(\alpha_i)} \int_{\Delta^n(\omega)} U \cdot u^\nu \, du. \tag{3.19}$$

3.3.4 Elementary Integral Representations

To treat the hypergeometric series introduced in § 3.1, we rewrite (3.19) a little bit by using (3.3). By $1/\left(\sum_{i=1}^{n+1}\alpha_i; |\nu|\right) = (-1)^{|\nu|}\left(1 - \sum_{i=1}^{n+1}\alpha_i; -|\nu|\right)$, we have

$$(\alpha_1; \nu_1)\cdots(\alpha_n; \nu_n)(\beta; -|\nu|) = c\int_{\Delta^n(\omega)} U \cdot (-u)^\nu \, du, \tag{3.20}$$

$$\text{where} \quad \beta = 1 - \sum_{i=1}^{n+1}\alpha_i, \quad c = \Gamma\left(\sum_{i=1}^{n+1}\alpha_i\right)/\prod_{i=1}^{n+1}\Gamma(\alpha_i).$$

Introducing a redundant parameter $\gamma \notin \mathbb{Z}$ to the hypergeometric series (3.1), by using (3.3), we rewrite it as

$$F((\alpha_k), x) = \sum_\nu \prod_{k \in K} (\alpha_k; a_k(\nu)) \frac{(\gamma; |\nu|)}{(\gamma; |\nu|)} \frac{x^\nu}{\nu!}$$

$$= \sum_\nu \prod_{k \in K} (\alpha_k; a_k(\nu)) \cdot (1 - \gamma; -|\nu|) \cdot \frac{(\gamma; |\nu|)(-x)^\nu}{\nu!}.$$

Below, for simplicity, we set $K = \{1, \cdots, m\}$ and define the linear form a_{m+2} on $L = \mathbb{Z}^n$ and the exponent α_{m+2} by

$$a_{m+2}(\nu) := -\nu_1 - \cdots - \nu_n, \quad \alpha_{m+2} := 1 - \gamma, \quad K^* = K \cup \{m+2\}.$$

By the condition (3.2), we have

$$\sum_{k \in K^*} a_k(e_i) = 0, \qquad 1 \le i \le n$$

from which we deduce

$$\sum_{k \in K^*} a_k(\nu) = 0$$

for any $\nu \in \mathbb{Z}^n$. If we transform (3.20) by

$$n \longrightarrow m, \quad \alpha_i \longrightarrow \alpha_k, \quad \nu_i \longrightarrow a_k(\nu), \quad \beta \longrightarrow \alpha_{m+2}, \quad -|\nu| \longrightarrow a_{m+2}(\nu),$$

by simple computation, we obtain

$$\prod_{k \in K^*} (\alpha_k; a_k(\nu)) = c \int_{\Delta^m(\omega)} U \cdot \prod_{k \in K} (-u_k)^{a_k(\nu)} d^m u, \qquad (3.21)$$

where

$$\begin{cases} U = \prod_{k \in K} u_k^{\alpha_k - 1} \cdot \left(1 - \sum_{k \in K} u_k\right)^{-\sum\limits_{k \in K^*} \alpha_k}, \quad d^m u = du_1 \cdots du_m, \\ c = \Gamma(\gamma)/\Gamma\left(\gamma - \sum_{k \in K} \alpha_k\right) \cdot \prod_{k \in K} \Gamma(\alpha_k). \end{cases} \qquad (3.22)$$

Hence, by Kummer's method, we obtain

$$F((\alpha_k), x) = c \sum_\nu \frac{(\gamma; |\nu|)(-x)^\nu}{\nu!} \int_{\Delta^m(\omega)} U \cdot \prod_{k \in K} (-u_k)^{a_k(\nu)} d^m u$$

$$= c \sum_\nu \frac{(\gamma; |\nu|)}{\nu!} \int_{\Delta^m(\omega)} U \cdot \prod_{i=1}^n \left(-x_i \prod_{k \in K} (-u_k)^{a_k(e_i)}\right)^{\nu_i} d^m u.$$

On the other hand, by construction, $\Delta^m(\omega)$ is compact, its distance with each axe is no less than ε, and $\varepsilon \le |u_i| \le 1$, $1 \le i \le n$ on it. Hence, for a sufficiently small $|x|$, $\left|-x_i \prod_{k \in K} (-u_k)^{a_k(e_i)}\right|$ becomes sufficiently small, and the multi-series

$$\sum_\nu \frac{(\gamma; |\nu|)}{\nu!} \prod_{i=1}^n \left(-x_i \prod_{k \in K} (-u_k)^{a_k(e_i)}\right)^{\nu_i}$$

converges on $\Delta^m(\omega)$ uniformly to

$$\left(1 + \sum_{i=1}^{n} x_i \prod_{k \in K} (-u_k)^{a_k(e_i)}\right)^{-\gamma}.$$

Hence, the power series (3.1) converges for sufficiently small $|x|$, it possesses an integral representation

$$F((\alpha_k), x) = c \int_{\Delta^m(\omega)} U \cdot \left(1 + \sum_{i=1}^{n} x_i \prod_{k \in K} (-u_k)^{a_k(e_i)}\right)^{-\gamma} d^m u,$$

where c, U etc. are defined in (3.22). Thus, we have shown the following theorem.

Theorem 3.2. *The hypergeometric series*

$$F((\alpha_k), x) = \sum_{\nu} \prod_{k \in K} (\alpha_k; a_k(\nu)) \cdot \frac{x^\nu}{\nu!}$$

converges in sufficiently small neighborhood of the origin. If we add a redundant parameter $\gamma \notin \mathbb{Z}$, *under the conditions* $\alpha_k \notin \mathbb{Z}$, $k \in K$, $\gamma - \sum_{k \in K} \alpha_k \notin \mathbb{Z}$, *we have the following elementary integral representation:*

$$F((\alpha_k), x) = c \int_{\Delta^m(\omega)} \prod_{k \in K} u_k^{\alpha_k - 1} \cdot \left(1 - \sum_{k \in K} u_k\right)^{\gamma - \sum_{k \in K} \alpha_k - 1}.$$

$$\cdot \left(1 + \sum_{i=1}^{n} x_i \cdot \prod_{k \in K} (-u_k)^{a_k(e_i)}\right)^{-\gamma} d^m u,$$

$$c = \Gamma(\gamma)/\Gamma(\gamma - \sum_{k \in K} \alpha_k) \cdot \prod_{k \in K} \Gamma(\alpha_k), \quad m = |K|.$$

3.3.5 Hypergeometric Function of Type $(3, 6)$

By the form of a hypergeometric function, namely, the way we take linear forms a_k, $k \in K$ on the lattice $L = \mathbb{Z}^n$, it might be possible that one does not need to use the redundant parameter γ to obtain an integral representation. Here, we show its idea with an important example. The hypergeometric series of type $(3, 6)$

$$F((\alpha_i), (\beta_j), \gamma; x)$$

$$= \sum_{\nu} \frac{(\alpha_1; a_1(\nu))(\alpha_2; a_2(\nu))(\beta_1; b_1(\nu))(\beta_2; b_2(\nu))}{(\gamma; c(\nu))\nu!} x^\nu$$

is defined by setting $n = 2$, $m = 5$ in § 3.1.3 and taking five linear forms on the lattice $L = M_2(\mathbb{Z})$:

$$
\begin{aligned}
a_1(\nu) &= \nu_{11} + \nu_{12}, & b_1(\nu) &= \nu_{11} + \nu_{21}, \\
a_2(\nu) &= \nu_{21} + \nu_{22}, & b_1(\nu) &= \nu_{12} + \nu_{22}, \\
c(\nu) &= \nu_{11} + \nu_{12} + \nu_{21} + \nu_{22}. &
\end{aligned}
$$

By $a_1(\nu) + a_2(\nu) = c(\nu)$, replacing (3.19) by

$$
n \longrightarrow 2, \quad \alpha_i \longrightarrow \alpha_i, \quad \nu_i \longrightarrow a_i(\nu), \quad \sum \alpha_i \longrightarrow \gamma,
$$

we obtain

$$
\frac{(\alpha_1; a_1(\nu))(\alpha_2; a_2(\nu))}{(\gamma; c(\nu))} = c \int_{\Delta^2(\omega)} U \cdot u_1^{a_1(\nu)} u_2^{a_2(\nu)} d^2 u,
$$

$$
U = u_1^{\alpha_1 - 1} u_2^{\alpha_2 - 1} (1 - u_1 - u_2)^{\gamma - \alpha_1 - \alpha_2 - 1},
$$

$$
c = \Gamma(\gamma)/\Gamma(\alpha_1)\Gamma(\alpha_2)\Gamma(\gamma - \alpha_1 - \alpha_2).
$$

By simple computation using Kummer's method, we obtain

$$
F((\alpha_i), (\beta_j), \gamma; x)
$$

$$
= c \sum_\nu \int_{\Delta^2(\omega)} U \cdot \sum_{\nu_{11}, \nu_{21}} \frac{(\beta_1; \nu_{11} + \nu_{21})}{\nu_{11}! \, \nu_{21}!} (x_{11} u_1)^{\nu_{11}} (x_{21} u_2)^{\nu_{21}}
$$

$$
\cdot \sum_{\nu_{12}, \nu_{22}} \frac{(\beta_2; \nu_{12} + \nu_{22})}{\nu_{12}! \, \nu_{22}!} (x_{12} u_1)^{\nu_{12}} (x_{12} u_2)^{\nu_{22}} d^2 u.
$$

When $|x|$ is sufficiently small, by the multinomial theorem, the above two series converge to $(1 - x_{11} u_1 - x_{21} u_2)^{-\beta_1}$ and $(1 - x_{12} u_1 - x_{22} u_2)^{-\beta_2}$ uniformly on $\Delta^2(\omega)$, respectively. Consequently, we obtain

$$
F((\alpha_i), (\beta_j), \gamma; x)
$$

$$
= c \int_{\Delta^2(\omega)} U \cdot (1 - x_{11} u_1 - x_{21} u_2)^{-\beta_1} (1 - x_{12} u_1 - x_{22} u_2)^{-\beta_2} d^2 u.
$$

By $b_1(\nu) + b_2(\nu) = c(\nu)$, almost the same computation yields the second integral representation of $F((\alpha_i), (\beta_j), \gamma; x)$:

$$F((\alpha_i), (\beta_j), \gamma; x) \tag{3.23}$$

$$= c' \int_{\Delta^2(\omega)} u_1^{\beta_1 - 1} u_2^{\beta_2 - 1} (1 - u_1 - u_2)^{\gamma - \beta_1 - \beta_2 - 1}$$

$$\cdot (1 - x_{11} u_1 - x_{12} u_2)^{-\alpha_1} (1 - x_{21} u_1 - x_{22} u_2)^{-\alpha_2} d^2 u,$$
$$c' = \Gamma(\gamma) / \Gamma(\beta_1) \Gamma(\beta_2) \Gamma(\gamma - \beta_1 - \beta_2).$$

This can be regarded as a duality of $E(3, 6, \alpha)$ explained later in § 3.3.7.

3.3.6 Hypergeometric Functions of Type $(n + 1, m + 1)$

We can clearly apply the method used for hypergeometric series of type $(3, 6)$ to a general hypergeometric series (3.5) of type $(n + 1, m + 1)$. We use the notation defined in § 3.1.3. By $\sum_{i=1}^{n} a_i(\nu) = c(\nu)$, we obtain the first integral representation, and by $\sum_{j=1}^{m-n-1} b_j(\nu) = c(\nu)$, we obtain the second integral representation. The results are stated as follows:

Theorem 3.3. *In the hypergeometric series of type $(n + 1, m + 1)$*

$$F((\alpha_i), (\beta_j), \gamma; x) = \sum_{\nu} \frac{\prod_{i=1}^{n}(\alpha_i; a_i(\nu)) \prod_{j=1}^{m-n-1}(\beta_j; b_j(\nu))}{(\gamma; c(\nu))\nu!} x^\nu,$$

suppose that the parameters satisfy the conditions:

$$\alpha_i \notin \mathbb{Z}, \quad 1 \le i \le n, \quad \beta_j \notin \mathbb{Z}, \quad 1 \le j \le m - n - 1,$$

$$\gamma - \sum_{i=1}^{n} \alpha_i \notin \mathbb{Z}, \quad \gamma - \sum_{j=1}^{m-n-1} \beta_j \notin \mathbb{Z}.$$

Then, the above series admits two integral representations:

$$F((\alpha_i), (\beta_j), \gamma; x) \tag{3.24}$$

$$= c_1 \int_{\Delta^n(\omega_1)} U_1(u) \cdot \prod_{j=1}^{m-n-1} (1 - x_{1j} u_1 - \cdots - x_{nj} u_n)^{-\beta_j} d^n u,$$

$$U_1(u) = \prod_{i=1}^{n} u_i^{\alpha_i - 1} \cdot (1 - u_1 - \cdots - u_n)^{\gamma - \sum \alpha_i - 1},$$

$$c_1 = \Gamma(\gamma) / \Gamma \left(\gamma - \sum_{i=1}^{n} \alpha_i \right) \cdot \prod_{i=1}^{n} \Gamma(\alpha_i),$$

$$F((\alpha_i), (\beta_j), \gamma; x) \tag{3.25}$$

$$= c_2 \int_{\Delta^{m-n-1}(\omega_2)} U_2(u)$$

$$\cdot \prod_{j=1}^{n} (1 - x_{i1} u_1 - \cdots - x_{i,m-n-1} u_{m-n-1})^{-\alpha_i} d^{m-n-1} u,$$

$$U_2(u) = \prod_{j=1}^{m-n-1} u_j^{\beta_j - 1} \cdot (1 - u_1 - \cdots - u_{m-n-1})^{\gamma - \sum \beta_j - 1},$$

$$c_2 = \Gamma(\gamma) / \Gamma \left(\gamma - \sum_{j=1}^{m-n-1} \beta_j \right) \cdot \prod_{j=1}^{m-n-1} \Gamma(\beta_j).$$

3.3.7 Horn's Cases

We can apply a similar method to find an integral representation for 14 hypergeometric series in Horn's list ([Er1], pp.224–225). Here, we treat two such examples.

Example 3.1. Appell's F_4 is introduced as a power series in § 3.1.8 which has the form:

$$F_4(\alpha, \beta, \gamma_1, \gamma_2; x_1, x_2) = \sum_{\nu_1, \nu_2 = 0}^{\infty} \frac{(\alpha; \nu_1 + \nu_2)(\beta; \nu_1 + \nu_2)}{(\gamma_1; \nu_1)(\gamma_2; \nu_2)\nu_1! \nu_2!} x_1^{\nu_1} x_2^{\nu_2}.$$

By formula (3.3) and Lemma 3.4, we have

$$\frac{(\alpha; \nu_1 + \nu_2)}{(\gamma_1; \nu_1)(\gamma_2; \nu_2)} = \frac{(1 - \gamma_1; -\nu_1)(1 - \gamma_2; -\nu_2)}{(1 - \alpha; -\nu_1 - \nu_2)}$$

$$= c \iint_{\Delta^2(\omega)} u_1^{-\gamma_1} u_2^{-\gamma_2} (1 - u_1 - u_2)^{\gamma_1 + \gamma_2 - \alpha - 2} u_1^{-\nu_1} u_2^{-\nu_2} du_1 du_2,$$

$$c = \frac{\Gamma(1 - \alpha)}{\Gamma(1 - \gamma_1)\Gamma(1 - \gamma_2)\Gamma(\gamma_1 + \gamma_2 - \alpha - 1)}.$$

With Kummer's method, we obtain

$$F_4(\alpha, \beta, \gamma_1, \gamma_2; x_1, x_2)$$

$$= c \iint_{\Delta^2(\omega)} u_1^{-\gamma_1} u_2^{-\gamma_2} (1 - u_1 - u_2)^{\gamma_1 + \gamma_2 - \alpha - 2} \left(1 - \frac{x_1}{u_1} - \frac{x_2}{u_2} \right)^{-\beta} du_1 \wedge du_2$$

by simple computation. One can also obtain an integral representation of F_C in almost the same way as above.

Example 3.2. In Horn's list, G_3 is introduced as a power series in § 3.1.9 which has the form:

$$G_3(\alpha, \beta; x_1, x_2) = \sum_{\nu_1, \nu_2=0}^{\infty} \frac{(\alpha; 2\nu_2 - \nu_1)(\beta; 2\nu_1 - \nu_2)}{\nu_1! \, \nu_2!} x_1^{\nu_1} x_2^{\nu_2}.$$

Here, introducing the redundant parameter κ, we first rewrite the above series as

$$G_3 = \sum_{\nu_1, \nu_2=0}^{\infty} \frac{(\alpha; 2\nu_2 - \nu_1)(\beta; 2\nu_1 - \nu_2)}{(\kappa; \nu_1 + \nu_2)} \cdot \frac{(\kappa; \nu_1 + \nu_2)}{\nu_1! \, \nu_2!} x_1^{\nu_1} x_2^{\nu_2}.$$

Rewriting Lemma 3.4 by

$$n \longrightarrow 2, \quad \alpha_1 \longrightarrow \alpha, \quad \alpha_2 \longrightarrow \beta, \quad \nu_1 \longrightarrow 2\nu_2 - \nu_1,$$

$$\nu_2 \longrightarrow 2\nu_1 - \nu_2, \quad \alpha_1 + \alpha_2 + \alpha_3 \longrightarrow \kappa,$$

we obtain

$$\frac{(\alpha; 2\nu_2 - \nu_1)(\beta; 2\nu_1 - \nu_2)}{(\kappa; \nu_1 + \nu_2)}$$

$$= c \iint_{\Delta^2(\omega)} u_1^{\alpha-1} u_2^{\beta-1} (1 - u_1 - u_2)^{\kappa-\alpha-\beta-1} u_1^{2\nu_2-\nu_1} u_2^{2\nu_1-\nu_2} \, du_1 \wedge du_2,$$

$$c = \Gamma(\kappa)/\Gamma(\alpha)\Gamma(\beta)\Gamma(\kappa - \alpha - \beta).$$

Applying also Kummer's method, we obtain an integral representation

$$G_3 = c \iint_{\Delta^2(\omega)} u_1^{\alpha-1} u_2^{\beta-1} (1 - u_1 - u_2)^{\kappa-\alpha-\beta-1}$$

$$\cdot \sum_{\nu_1, \nu_2=0}^{\infty} \frac{(\kappa; \nu_1 + \nu_2)}{\nu_1! \, \nu_2!} \left(\frac{u_2^{\,2}}{u_1} x_1 \right)^{\nu_1} \left(\frac{u_1^{\,2}}{u_2} x_2 \right)^{\nu_2} du_1 \wedge du_2$$

$$= c \iint_{\Delta^2(\omega)} u_1^{\alpha-1} u_2^{\beta-1} (1 - u_1 - u_2)^{\kappa-\alpha-\beta-1}$$

$$\cdot \left(1 - \frac{u_2^{\,2}}{u_1} x_1 - \frac{u_1^{\,2}}{u_2} x_2 \right)^{-\kappa} du_1 \wedge du_2.$$

Remark 3.3. The results obtained by systematic computations for Horn's list can be found in a table in [Kit1], pp.56–58. From another viewpoint, [Dw-Lo] also obtain an integral representation, but the domain of the integral is not calculated. Here, we can again find the effectiveness of the twisted cycles.

3.4 System of Hypergeometric Differential Equations $E(n+1, m+1; \alpha)$

3.4.1 Hypergeometric Integral of Type $(n+1, m+1; \alpha)$

Let us start from the integral representation of the hypergeometric series $F((\alpha_i), (\beta_j), \gamma; x)$ of type $(n+1, m+1)$ obtained in § 3.3.6:

$$F((\alpha_i), (\beta_j), \gamma; x) \tag{3.26}$$

$$= \sum_{\nu} \frac{\prod_{i=1}^{n}(\alpha_i; \nu_{i1} + \cdots + \nu_{i,m-n-1}) \prod_{j=1}^{m-n-1}(\beta_j; \nu_{1j} + \cdots + \nu_{nj})}{(\gamma; \sum_{i,j} \nu_{ij}) \nu!} x^{\nu}$$

$$= \frac{\Gamma(\gamma)}{\Gamma(\gamma - \sum_{i=1}^{n} \alpha_i) \prod_{i=1}^{n} \Gamma(\alpha_i)} \int_{\Delta^n(\omega)} \prod_{i=1}^{n} u_i^{\alpha_i - 1} \cdot \left(1 - \sum_{i=1}^{n} u_i\right)^{\gamma - \sum_{i=1}^{n} \alpha_i - 1}$$

$$\cdot \prod_{j=1}^{m-n-1} \left(1 - \sum_{i=1}^{n} x_{ij} u_i\right)^{-\beta_j} du_1 \cdots du_n.$$

Now, we associate a hyperplane $a_0 + a_1 u_1 + \cdots + a_n u_n = 0$ in $\mathbb{C}^n$ to a column vector $^t(a_0, a_1, \cdots, a_n)$. Arranging the hyperplanes appearing in the above integral and the hyperplane at infinity, we obtain the $(n+1) \times (m+1)$ matrix

$$\begin{pmatrix} 1 & 0 & \cdots & \cdots & \cdots & 0 & 1 & 1 & \cdots \cdots & 1 \\ 0 & 1 & 0 & \cdots & \cdots & 0 & -1 & -x_{11} & \cdots \cdots & -x_{1,m-n-1} \\ 0 & 0 & 1 & \ddots & & \vdots & \vdots & \vdots & & \vdots \\ \vdots & \vdots & \ddots & \ddots & \ddots & \vdots & \vdots & \vdots & & \vdots \\ \vdots & \vdots & & \ddots & 1 & 0 & \vdots & \vdots & & \vdots \\ 0 & 0 & \cdots & \cdots & 0 & 1 & -1 & -x_{n1} & \cdots \cdots & -x_{n,m-n-1} \end{pmatrix}. \tag{3.27}$$

We can regard the above matrix as a matrix defining an arrangement of $m+1$ hyperplanes in $\mathbb{P}^n$. For an integral representation, changing the defining equation of a hyperplane corresponds to a multiplication by a scalar, and it has no essential effect. This corresponds to the multiplication of an $(m+1) \times (m+1)$ diagonal matrix from the right to the matrix (3.27). Changing the homogeneous coordinates of $\mathbb{P}^n$ corresponds to the multiplication of a matrix of GL_{n+1} from the left to the matrix (3.27). Hence, the most general form of a matrix (3.27) with bigger symmetry is

$$x = \begin{pmatrix} x_{00} & \cdots\cdots & x_{0m} \\ \vdots & & \vdots \\ x_{n0} & \cdots\cdots & x_{nm} \end{pmatrix} \in M_{n+1,\,m+1}(\mathbb{C}), \tag{3.28}$$

where, each column vector $\neq 0$.

This can be regarded as defining an arrangement of $m+1$ hyperplanes in $\mathbb{P}^n$. Below, we consider mainly the open subset of $M_{n+1,\,m+1}(\mathbb{C})$ defined by

$$X := \{x \in M_{n+1,\,m+1}(\mathbb{C}) | \text{any } (n+1) \times (n+1) \text{ minor of } x \text{ is not zero}\}.$$

On X, $GL_{n+1}(\mathbb{C})$ acts from the left and $H_{m+1} = \{\text{diag}(h_0, \cdots, h_m) | h_j \in \mathbb{C}^*\}$ from the right. To derive an integral representation associated to an arrangement of hyperplanes (3.28) which corresponds to the integral (3.26), we use homogeneous coordinates of $\mathbb{P}^n$. First, we take an n-form on the affine space $\mathbb{C}^{n+1}$ with coordinates $(v_0, \cdots, v_n)$

$$\tau = \sum_{i=0}^{n} (-1)^i v_i dv_0 \wedge \cdots \wedge \widehat{dv_i} \wedge \cdots \wedge dv_n, \tag{3.29}$$

and consider the n-form on $\mathbb{C}^{n+1}$

$$\prod_{j=0}^{m} (x_{0j} v_0 + \cdots + x_{nj} v_n)^{\alpha_j} \cdot \tau.$$

As this form has the weight $\sum_{j=0}^{m} \alpha_j + n + 1$, if it satisfies the condition

$$\sum_{j=0}^{m} \alpha_j + n + 1 = 0, \tag{3.30}$$

this defines a multi-valued n-form on $M_x := \mathbb{P}^n \setminus \bigcup_{j=0}^{m} \{\sum_{i=0}^{n} x_{ij} v_i = 0\}$. From now, we provide a formulation to regard an integral of this n-form on an appropriate twisted cycle as a function of x.

$$\mathcal{M} := \bigcup_{x \in X} M_x$$

$$= \mathbb{P}^n \times X - \bigcup_{j-0}^{m} \left\{ (v, x) \in \mathbb{P}^n \times X | \sum_{i=0}^{n} x_{ij} v_i = 0 \right\}$$

is an $n + (n+1)(m+1)$-dimensional complex manifold. Denoting the holomorphic map on $\mathcal{M}$ induced by the projection $\mathbb{P}^n \times X \longrightarrow X$ by

$$\pi : \mathcal{M} \longrightarrow X,$$

$(\mathcal{M}, \pi, X)$ is a complex analytic family with the space of parameters X whose fiber is the complement of an arrangement of hyperplanes in $\mathbb{P}^n$. Below, we use inhomogeneous coordinates $u_1 = v_1/v_0, \cdots, u_n = v_n/v_0$ of $\mathbb{P}^n$, and assume $^t(x_{00}, \cdots, x_{n0}) = {}^t(1, 0, \cdots, 0)$ for simplicity. Then, $U(u; x) = \prod_{j=0}^{m}(x_{0j} + x_{1j}u_1 + \cdots + x_{nj}u_n)^{\alpha_j}$ is a multi-valued holomorphic function on $\mathcal{M}$ and $\omega(u; x) = dU(u; x)/U(u; x)$ is a single-valued holomorphic 1-form on $\mathcal{M}$ which defines the covariant differential $\nabla_{\omega(u;x)} := d + \omega(u; x) \wedge$ on $\mathcal{M}$. As in § 2.1.5, the solutions of $\nabla_{\omega(u;x)}h = 0$, $h \in \mathcal{O}_\mathcal{M}$ defines a local system $\mathcal{L}_{\omega(u;x)}$ of rank 1 on $\mathcal{M}$, and the restriction of $\mathcal{L}_{\omega(u;x)}$ to M_x clearly coincides with the local system $\mathcal{L}_{\omega_x}$ defined by the covariant differential on M_x associated to $\omega_x := \omega(u; x)|_{M_x}$.

As the arrangement of hyperplanes $\mathbb{P}^n$ defined by x is in general position (cf. § 2.9.1), varying x a little bit in X, the M_x's are isotopic to each other. Hence, the function

$$x \longmapsto H_n(M_x, \mathcal{L}_{\omega_x}^\vee)$$

becomes locally constant, and

$$\mathcal{H}_n^\vee := \bigcup_{x \in X} H_n(M_x, \mathcal{L}_{\omega_x}^\vee)$$

forms a flat vector bundle on X with fiber $\mathbb{C}^{\binom{m-1}{n}}$ whose transition matrices are constant, or in other words, a local system of rank $\binom{m-1}{n}$. Considering an integral

$$F(x; \alpha) = \int_{\sigma_x} U(u; x) du_1 \wedge \cdots \wedge du_n \qquad (3.31)$$

for a local section σ_x (x varies in an open neighborhood W) of $\mathcal{H}_n^\vee$, this becomes a holomorphic function of x. We call this integral the hypergeometric integral of type $(n+1, m+1)$. Take any continuous curve γ in X, and choose a local section $\mathcal{H}_n^\vee$ along this path. Continuing $U(u; x)$ analytically along γ at the same time, we obtain the analytic continuation of (3.31) along γ. Repeating this for all γ, we see that $F(x; \alpha)$ can be analytically continued to the whole X.

Now, to derive a differential equation satisfied by the function $F(x; \alpha)$ in x, let us first study the induced action of $GL_{n+1}(\mathbb{C})$ and H_{m+1} on $F(x; \alpha)$.

3.4.2 Differential Equation $E(n+1, m+1; \alpha)$

(i) The action of H_{m+1}

As (3.31) is homogeneous with respect to $x_{0j}, \cdots, x_{nj}$ of degree α_j,

$$F((\lambda_j x_{ij}); \alpha) = \prod_{j=0}^{m} \lambda_j^{\alpha_j} \cdot F((x_{ij}); \alpha). \tag{3.32}$$

(ii) The action of $GL_{n+1}(\mathbb{C})$

For $v = (v_0, \cdots, v_n)$, $g \in GL_{n+1}(\mathbb{C})$, under the coordinate transform $v' = vg^{-1}$, by using $x_j = {}^t(x_{0j}, \cdots, x_{nj})$, the hyperplane $vx_j = 0$ is transformed to $v'gx_j = 0$, hence, the matrix defining an arrangement (3.28) is transformed to gx. Denoting (3.29) corresponding to v' by τ', we have

$$\tau' = (\det g)^{-1} \tau, \tag{3.33}$$

by simple computation. See Figure 3.5.

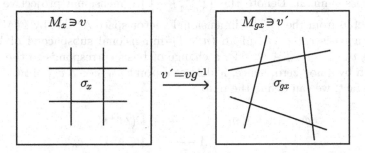

Fig. 3.5

Now, by definition, we have

$$F(gx; \alpha) = \int_{\sigma_{gx}} \prod_{j=0}^{m} (v'gx_j)^{\alpha_j} \tau'.$$

Since $GL_{n+1}(\mathbb{C})$ is connected, the unit matrix 1_{n+1} and g is joined by a continuous curve $\gamma : [0,1] \longrightarrow GL_{n+1}(\mathbb{C})$. Then, $M_{\gamma(t)x}$ continuously varies from M_x to M_{gx} without changing its isotopic type, and locally, it becomes the product of $M_{\gamma(t)x}$ and a small interval containing t. Hence, setting

$$M_\gamma = \bigcup_{t \in [0,1]} M_{\gamma(t)x},$$

σ_x and σ_{gx} are homologous in $H_n(M_\gamma, \mathcal{L}_{\omega(u;x)}|M_\gamma)$ from which we conclude

$$\int_{\sigma_{gx}} \prod_{j=0}^{m} (v'gx_j)^{\alpha_j} \tau' = \int_{\sigma_x} \prod_{j=0}^{m} (vx_j)^{\alpha_j} (\det g)^{-1} \tau.$$

Thus, we obtain the $GL_{n+1}(\mathbb{C})$-invariance

$$F(gx; \alpha) = (\det g)^{-1}F(x; \alpha). \qquad (3.34)$$

Remark 3.4. We denote the subgroup of $GL_{n+1}(\mathbb{C})$ of the matrices with determinant 1 by $SL_{n+1}(\mathbb{C})$. By (3.34), we have

$$F(gx; \alpha) = F(x; \alpha), \quad g \in SL_{n+1}(\mathbb{C}).$$

Hence, we can regard $F(x; \alpha)$ as a function on the quotient space $SL_{n+1}(\mathbb{C})\backslash X$ of X by $SL_{n+1}(\mathbb{C})$.

We briefly explain this below.

Denote the set of complex $(n+1)$-dimensional subspaces of the $(m+1)$-dimensional complex vector space $V = \mathbb{C}^{m+1}$ by $G(n+1, m+1)$ and call it a Grassmannian. Denote the $\left(\binom{m+1}{n+1} - 1\right)$-dimensional projective space constructed from the $\binom{m+1}{n+1}$-dimensional vector space $\wedge^{n+1}V$ by $\mathbb{P}(\wedge^{n+1}V)$. Taking a basis $v_0, \cdots, v_n$ of an $(n+1)$-dimensional subspace Λ of V and forming $v_0 \wedge \cdots \wedge v_n \in \wedge^{n+1}V$, a change of basis corresponds to the multiplication by a non-zero, hence, it defines a point $[v_0 \wedge \cdots \wedge v_n]$ of $\mathbb{P}(\wedge^{n+1}V)$. By this fact, we can define the map

$$\Phi : G(n+1, m+1) \longrightarrow \mathbb{P}(\wedge^{n+1}V)$$
$$\Lambda \longmapsto [v_0 \wedge \cdots \wedge v_n].$$

It is known that Φ is an embedding and its image is an $(n+1)(m-n)$-dimensional non-singular projective algebraic variety. Hence, via this Φ, $G(n+1, m+1)$ can be regarded as an algebraic variety. Fixing a basis $e_0, \cdots, e_m$ of V, a basis of Λ defines a matrix $\widetilde{\Lambda}$ by the formula

$$\begin{pmatrix} v_0 \\ \vdots \\ v_n \end{pmatrix} = \widetilde{\Lambda} \begin{pmatrix} e_0 \\ e_1 \\ \vdots \\ e_m \end{pmatrix}, \quad \widetilde{\Lambda} \in M_{n+1,\,m+1}(\mathbb{C}), \quad \text{rank } \widetilde{\Lambda} = n+1.$$

A change of basis of Λ corresponds to the multiplication of an element of $GL_{n+1}(\mathbb{C})$ from the left, and $G(n+1, m+1)$ can be expressed as the quotient space:

$$G(n+1, m+1) = GL_{n+1}(\mathbb{C})\backslash Z,$$
$$Z = \{z \in M_{n+1,m+1}(\mathbb{C}) | \text{rank } z = n+1\}.$$

Let us choose a basis of $\wedge^{n+1}V$ by $\{e_{j_0} \wedge \cdots \wedge e_{j_n} \mid 0 \leq j_0 < \cdots < j_n \leq m\}$ and arrange them, for example, by the lexicographic order. Then, expanding $v_0 \wedge \cdots \wedge v_n$ with respect to this basis, the coefficient of $e_{j_0} \wedge \cdots \wedge e_{j_n}$ is equal to the minor $\widetilde{\Lambda}\begin{pmatrix} 0 & 1 & \cdots & n \\ j_0 & j_1 & \cdots & j_n \end{pmatrix}$ of the matrix $\widetilde{\Lambda}$ obtained by picking up

the j_0th, $\cdots$, j_nth column vectors. In this sense, $\binom{m+1}{n+1}$

$$\widetilde{\Lambda}\begin{pmatrix} 0 & 1 & \cdots & n \\ j_0 & j_1 & \cdots & j_n \end{pmatrix}, \quad 0 \le j_0 < \cdots < j_n \le m$$

is called the Plücker coordinates of Λ. As the image of Φ is an algebraic variety, there are some relations among the Plücker coordinates, among which, as one of the simplest examples, the following Plücker relation is known: for $2n+2$ indices $i_1, \cdots, i_n, j_0, j_1, \cdots, j_{n+1}$,

$$\sum_{k=0}^{n+1} (-1)^k \widetilde{\Lambda}\begin{pmatrix} 0 & 1 & \cdots & n \\ i_1 & \cdots & i_n & j_k \end{pmatrix} \widetilde{\Lambda}\begin{pmatrix} 0 & 1 & \ldots\ldots & n \\ j_0 & \cdots & \widehat{j_k} & \cdots & j_{n+1} \end{pmatrix} = 0.$$

Now, by $X \subset Z$, we have

$$GL_{n+1}(\mathbb{C}) \backslash X \subset G(n+1, m+1).$$

There also is a natural map

$$p : SL_{n+1}(\mathbb{C})\backslash Z \longrightarrow GL_{n+1}(\mathbb{C})\backslash Z = G(n+1, m+1).$$
$$SL_{n+1}(\mathbb{C}) \cdot x \longmapsto GL_{n+1}(\mathbb{C}) \cdot x$$

Gelfand [Ge] formulated this map p geometrically as follows. First, he defined a natural $(n+1)$-dimensional vector bundle (tautological bundle) on $G(n+1, m+1)$, as a sub-bundle of the trivial vector bundle $G(n+1, m+1) \times V$, by

$$S = \cup\, (\Lambda \text{ as a point of } G(n+1, m+1)) \times (\Lambda \text{ as a subspace of } V).$$

Taking the $(n+1)$th exterior product of the dual bundle S^* to S, it follows that

$$\wedge^{n+1} S^* = SL_{n+1}(\mathbb{C})\backslash Z.$$

Hence, the hypergeometric function $F(x; \alpha)$ can be regarded as a function defined on the line bundle $\wedge^{n+1} S^*$ over the Grassmannian $G(n+1, m+1)$. In this sense, a hypergeometric function of type $(n+1, m+1)$ is said to be a hypergeometric function on a Grassmannian.

Next, let us calculate the infinitesimal transformations of (i), (ii). By (3.32), we obviously have the differential equations:

$$\sum_{i=0}^{n} x_{ij} \frac{\partial F}{\partial x_{ij}} = \alpha_j F, \quad 0 \le j \le m. \tag{3.35}$$

To compute the infinitesimal transformations of the action of $GL_{n+1}(\mathbb{C})$, rewrite (3.34) with the one-parameter subgroup $g = \exp(tA)$ generated by $A = (a_{pq}) \in M_{n+1}(\mathbb{C})$:

$$F(\exp(tA) \cdot x; \alpha) = \exp(-t\mathrm{Tr}(A))F(x; \alpha).$$

Differentiate the both sides of this formula by t, and setting $t = 0$, we have

$$\sum_{i,j} \frac{\partial F}{\partial x_{ij}} \cdot \sum_p a_{ip}x_{pj} = -\sum_p a_{pp} \cdot F,$$

and since a_{pq} can be choosen arbitrarily, we finally obtain

$$\sum_{j=0}^m x_{ij}\frac{\partial F}{\partial x_{pj}} = -\delta_{ip}F, \quad 0 \le i,\, p \le n. \tag{3.36}$$

The differential equations (3.35), (3.36) are the infinitesimal version of the action of H_{m+1} and $GL_{n+1}(\mathbb{C})$. Next, we derive a second-order differential equation which can be obtained immediately from the form of the integral (3.31): for $0 \le j,\, q \le m,\, j \ne q$, we have

$$\frac{\partial^2 F}{\partial x_{ij}\partial x_{pq}} = \int_\sigma \alpha_j\alpha_q \cdot \frac{u_iu_p}{P_jP_q}U\,du_1 \wedge \cdots \wedge du_n$$

$$= \frac{\partial^2 F}{\partial x_{iq}\partial x_{pj}},$$

where $P_j = \sum_{i=0}^n x_{ij}u_i$. As this formula becomes trivial for $j = q$, we finally obtain

$$\frac{\partial^2 F}{\partial x_{ij}\partial x_{pq}} = \frac{\partial^2 F}{\partial x_{iq}\partial x_{pj}}, \quad 0 \le i,\, p \le n, \quad 0 \le j,\, q \le m.$$

The above three sets of differential equations satisfied by the hypergeometric integral (3.31) is denoted by $E(n+1, m+1; \alpha)$ and is called the system of hypergeometric differential equations of type $(n+1, m+1; \alpha)$:

$E(n+1, m+1; \alpha)$

$$\begin{cases} (1) \displaystyle\sum_{i=0}^n x_{ij}\frac{\partial F}{\partial x_{ij}} = \alpha_j F, \quad 0 \le j \le m, \\[2ex] (2) \displaystyle\sum_{j=0}^m x_{ij}\frac{\partial F}{\partial x_{pj}} = -\delta_{ip}F, \quad 0 \le i,\, p \le n, \\[2ex] (3) \dfrac{\partial^2 F}{\partial x_{ij}\partial x_{pq}} = \dfrac{\partial^2 F}{\partial x_{iq}\partial x_{pj}}, \quad 0 \le i,\, p \le n, \quad 0 \le j,\, q \le m. \end{cases} \tag{3.37}$$

Remark 3.5. Below, we introduce a system of differential equations which is equivalent to the above system and which kills the action of the groups $GL_{n+1}(\mathbb{C})$ and H_{m+1}. When there is no risk to be confused, we will denote it by the same symbol $E(n+1, m+1; \alpha)$. The above system of equations was introduced in [Ge-Ge] under a general setting, so it is also called the Gelfand system. Obviously, the system killing the action of $GL_{n+1}(\mathbb{C})$ and H_{m+1} corresponds to the classical equations:

$$E(2, 4; \alpha) \quad \cdots\cdots \text{ Gauss' differential equation,}$$
$$E(2, 5; \alpha) \quad \cdots\cdots \text{ Appell's differential equation } F_1,$$
$$E(2, 3+l; \alpha) \cdots\cdots \text{ Lauricella's differential equation } F_D.$$

It should be clear to the reader why the above symbol (introduced in [M-S-Y]) is convenient.

3.4.3 Equivalent System

Let us compute the system of equations killing the action of $GL_{n+1}(\mathbb{C})$. We decompose X into the blocks of the form $x = (x_1, x_2) \in X$, $x_1 \in GL_{n+1}(\mathbb{C})$. As $\det x_1 \neq 0$, we have

$$x = x_1(1_{n+1}, x'), \quad x' = x_1^{-1} x_2 \in M_{n+1, m-n}(\mathbb{C}),$$

hence by (3.34), we obtain

$$F(x; \alpha) = (\det x_1)^{-1} F((1_{n+1}, x'); \alpha). \tag{3.38}$$

So, we define the $(n+1)(m-n)$-dimensional closed sub-manifold X' of X by

$$X' = \left\{ x = (1_{n+1}, x') = \begin{pmatrix} 1 & & x_{0,n+1} & \cdots & x_{0,m} \\ & \ddots & \vdots & & \vdots \\ & 1 & x_{n,n+1} & \cdots & x_{n,m} \end{pmatrix} \in X \right\}.$$

Let us derive the system of equations satisfied by $F(x; \alpha)$ restricted to X':

$$F(x'; \alpha) = \int_\sigma u_1^{\alpha_1} \cdots u_n^{\alpha_n} \tag{3.39}$$

$$\cdot \prod_{j=n+1}^{m} (x_{0j} + x_{1j}u_1 + \cdots + x_{nj}u_n)^{\alpha_j} \, du_1 \wedge \cdots \wedge du_n.$$

For the column vectors of the matrix $(1_{n+1}, x')$, as in § 3.4.2, we obtain

$$\sum_{i=0}^{n} x_{ij} \frac{\partial F}{\partial x_{ij}} = \alpha_j F, \quad n+1 \le j \le m. \tag{3.40}$$

Next, to obtain the equations associated to the row vectors that correspond to (2) of $E(n+1, m+1; \alpha)$, let us recall some notation:

$$U = u_1^{\alpha_1} \cdots u_n^{\alpha_n} \prod_{j=n+1}^{m} (x_{0j} + x_{1j}u_1 + \cdots + x_{nj}u_n)^{\alpha_j} = \prod_{j=1}^{m} P_j^{\alpha_j},$$

$$\omega = \frac{dU}{U} = \sum_{i=1}^{n} \alpha_i \frac{du_i}{u_i} + \sum_{j=n+1}^{m} \alpha_j \frac{dP_j}{P_j}.$$

Although it seems to be unnatural for the reader, we calculate as follows:

$$\nabla_\omega ((-1)^{i-1} u_i du_1 \wedge \cdots \wedge \widehat{du_i} \wedge \cdots \wedge du_n)$$

$$= \left\{ (\alpha_i + 1) + \sum_{j=n+1}^{m} \frac{\alpha_j x_{ij} u_i}{P_j} \right\} du_1 \wedge \cdots \wedge du_n.$$

Passing to an integral, we have

$$\sum_{j=n+1}^{m} \int_\sigma U \cdot \frac{\alpha_j x_{ij} u_i}{P_j} du_1 \wedge \cdots \wedge du_n = -(1 + \alpha_i) F. \tag{3.41}$$

On the other hand, the left-hand side of (3.41) is equal to

$$\sum_{j=n+1}^{m} x_{ij} \frac{\partial F(x'; \alpha)}{\partial x_{ij}}$$

by simple computation, and we obtain the following differential equations:

$$\sum_{j=n+1}^{m} x_{ij} \frac{\partial F(x'; \alpha)}{\partial x_{ij}} = -(1 + \alpha_i) F, \quad 0 \le i \le n. \tag{3.42}$$

The differential equations which correspond to (3) of $E(n+1, m+1; \alpha)$ can be derived in the same way. Finally, a system of differential equations satisfied by $F(x'; \alpha)$ is given as follows:

$E'(n+1, m+1; \alpha)$

$$
\begin{cases}
(1) \quad \displaystyle\sum_{i=0}^{n} x_{ij} \frac{\partial F}{\partial x_{ij}} = \alpha_j F, & n+1 \le j \le m, \\[2mm]
(2) \quad \displaystyle\sum_{j=n+1}^{m} x_{ij} \frac{\partial F}{\partial x_{ij}} = -(1+\alpha_i)F, & 0 \le i \le n, \\[2mm]
(3) \quad \dfrac{\partial^2 F}{\partial x_{ij} \partial x_{pq}} = \dfrac{\partial^2 F}{\partial x_{iq} \partial x_{pj}}, & 0 \le i, \ p \le n, \quad n+1 \le j, \ q \le m.
\end{cases}
\tag{3.43}
$$

Remark 3.6. For an equivalence of the systems of equations (3.37) and (3.43), see [Hor], Appendix A.

Remark 3.7. (1), (2) of the above system of equations can be considered as representing the homogeneity of x' with respect to the action of $(\mathbb{C}^*)^{m+1}$ corresponding to a scalar multiplication to column and row vectors. In [G-G-Z], generalizing such a situation, when an algebraic torus is acting on a vector space V, they derived Lie group theoretically a system of differential equations which corresponds to (3.43) by using a character of the torus. Notice that the formulation in § 3.1 corresponds to killing the action of an algebraic torus. See Appendix A.

3.5 Integral Solutions of $E(n+1, m+1; \alpha)$ and Wronskian

The integrals (3.31) satisfy the system of hypergeometric differential equations $E(n+1, m+1; \alpha)$, but when the arrangement of hyperplanes in $\mathbb{P}^n$ defined by the integrand of (3.31) is in general position, it can be shown that they exhaust the solutions of $E(n+1, m+1; \alpha)$. We show this here and in § 3.6. Also in this section, taking appropriate twisted cycles, we show that there are $\binom{m-1}{n}$ independent solutions of the form (3.31).

3.5.1 Hypergeometric Integrals as a Basis

We denote the $(n+1) \times (n+1)$ minor of the $(n+1) \times (m+1)$ matrix

$$
x = \begin{pmatrix} x_{00} & \cdots\cdots & x_{0m} \\ \vdots & & \vdots \\ x_{n0} & \cdots\cdots & x_{nm} \end{pmatrix}
$$

obtained by extracting the j_0th, j_1th, $\cdots$, j_nth column vectors by $x(j_0, j_1, \cdots, j_n)$. Assume that the arrangement of hyperplanes in $\mathbb{P}^n$ defined by the above matrix x is in general position; i.e.,

$$x(j_0, j_1, \ldots, j_n) \neq 0, \quad 0 \le j_0 < j_1 < \cdots < j_n \le m, \qquad (3.44)$$

and that x is a real matrix. Then, the real hyperplane in $\mathbb{P}^n(\mathbb{R})$

$$H_{j\mathbb{R}} := \{u \in P^n(\mathbb{R}) | \sum_{i=0}^{n} x_{ij} u_i = 0\}, \quad 0 \le j \le m$$

decomposes $\mathbb{P}^n(\mathbb{R})$ into some open connected components. Below, we call each component a chamber. By $x(01 \cdots n) \neq 0$, we apply the reduction of § 3.4.3 (killing the action of $GL_{n+1}(\mathbb{C})$), and consider the arrangement of hyperplanes defined by

$$x' = \begin{pmatrix} 1 & & x_{0,n+1} & \cdots\cdots & x_{0,m} \\ & \ddots & \vdots & & \vdots \\ & & 1 & x_{n,n+1} & \cdots\cdots & x_{n,m} \end{pmatrix} \in X' \cap M_{n+1,m+1}(\mathbb{R}).$$

Recall some notation:

$$P_i(u) = u_i, \quad 1 \le i \le n, \quad P_j(u) = x_{0j} + \sum_{i=1}^{n} x_{ij} u_i, \quad n+1 \le j \le m,$$

$$U(u) = \prod_{j=1}^{m} P_j^{\alpha_j}, \quad \omega = dU/U, \quad M = \mathbb{C}^n \setminus \bigcup_{j=1}^{m} H_j.$$

Here, we always assume the conditions:

$$\alpha_j \notin \mathbb{Z}, \quad 1 \le j \le m, \quad \sum_{j=1}^{m} \alpha_j \notin \mathbb{Z}.$$

A basis of n-dimensional twisted homology group $H_n(M, \mathcal{L}_\omega^\vee)$ associated to the multi-valued function U was determined in § 3.2. There are $r := \binom{m-1}{n}$ bounded chambers in $\mathbb{R}^n \setminus \bigcup_{j=1}^{m} H_{j\mathbb{R}}$, so we enumerate them as Δ_ν, $1 \le \nu \le r$. By Lemma 3.2, Δ_ν, $1 \le \nu \le r$ form a basis of $H_n^{lf}(M, \mathcal{L}_\omega^\vee)$, and by Theorem 3.1, $\Delta_\nu(\omega) = \mathrm{reg}\Delta_\nu$, $1 \le \nu \le r$ form a basis of $H_n(M, \mathcal{L}_\omega^\vee)$. On the other hand, by Corollary 2.6, we see that

$$H^n(M, \mathcal{L}_\omega) \simeq H^n(\Omega(\log D), \nabla_\omega)$$

$$\simeq \{\{\varphi\langle j_1, \ldots, j_n\rangle \mid 1 \leq j_1 < \cdots < j_n \leq m - 1\}\},$$

$$\text{where} \quad \varphi\langle j_1, \ldots, j_n\rangle = \frac{dP_{j1}}{P_{j1}} \wedge \cdots \wedge \frac{dP_{jn}}{P_{jn}}.$$

To make the following computations smooth, instead of the integral (3.39), we start with the hypergeometric integral:

$$\widehat{\varphi}(x'; \alpha') = \int_\sigma U \cdot \varphi\langle 1, 2, \ldots, n\rangle. \tag{3.45}$$

Here, we remark

$$\alpha' = (\alpha_0, \alpha_1 - 1, \ldots, \alpha_n - 1, \alpha_{n+1}, \ldots, \alpha_m). \tag{3.46}$$

Hence, the system of hypergeometric differential equations satisfied by (3.45) becomes $E'(n+1, m+1; \alpha')$ (cf. (3.43)). Replacing σ in (3.45) by independent $\Delta_\nu(\omega)$'s, we consider

$$\widehat{\varphi}_\nu(x'; \alpha') = \int_{\Delta_\nu(\omega)} U \cdot \varphi\langle 12 \cdots n\rangle, \quad 1 \leq \nu \leq r. \tag{3.47}$$

The aim is to show that r $\widehat{\varphi}_\nu$ become independent solutions of $E'(n+1, m+1; \alpha')$, but since, for a general case, the notation becomes complicated, we start by discussing some important examples.

3.5.2 Gauss' Equation $E'(2, 4; \alpha')$

$$x' = \begin{pmatrix} 1 & 0 & x_{02} & x_{03} \\ 0 & 1 & x_{12} & x_{13} \end{pmatrix},$$

$$P_1 = u, \quad P_2 = x_{02} + x_{12}u, \quad P_3 = x_{03} + x_{13}u,$$

for $\Delta_1(\omega), \Delta_2(\omega)$, see Figure 2.5 of § 2.3.1.

We set

$$\widehat{\varphi}_\nu\langle j\rangle := \int_{\Delta_\nu(\omega)} U \cdot \varphi\langle i\rangle, \quad 1 \leq j, \nu \leq 2.$$

By simple computation, we have

$$\frac{\partial \widehat{\varphi}_\nu\langle 1\rangle}{\partial x_{12}} = \frac{\alpha_2}{x_{12}} \int_{\Delta_\nu(\omega)} U \cdot \varphi\langle 2\rangle = \frac{\alpha_2}{x_{12}} \widehat{\varphi}_\nu\langle 2\rangle. \tag{3.48}$$

Taking the Wronskian of solutions of $E'(2, 4; \alpha')$ as

$$W = \det \begin{pmatrix} \widehat{\varphi}_1\langle 1 \rangle & \widehat{\varphi}_2\langle 1 \rangle \\ \dfrac{\partial \widehat{\varphi}_1\langle 1 \rangle}{\partial x_{12}} & \dfrac{\partial \widehat{\varphi}_2\langle 1 \rangle}{\partial x_{12}} \end{pmatrix},$$

by (3.48) we obtain

$$W = \frac{\alpha_2}{x_{12}} \det(\widehat{\varphi}_\nu\langle j \rangle).$$

By Corollary 3.1, we have $\det(\widehat{\varphi}_\nu\langle j \rangle) \neq 0$, i.e., $W \neq 0$ and $\widehat{\varphi}_1\langle 1 \rangle$, $\widehat{\varphi}_2\langle 1 \rangle$ become independent solutions of $E'(2.4; \alpha')$.

Remark 3.8. Since all 2×2 minor of x' are not zero, the condition $x_{12} \neq 0$ follows from the assumption "being in general position".

3.5.3 Appell–Lauricella Hypergeometric Differential Equation $E'(2, m + 1; \alpha')$

$$x' = \begin{pmatrix} 1 & 0 & x_{02} & x_{03} & \cdots & x_{0m} \\ 0 & 1 & x_{12} & x_{13} & \cdots & x_{1m} \end{pmatrix} \in M_{2,m+1}(\mathbb{R}).$$

To simplify the notation, we assume $-\frac{x_{01}}{x_{11}} = 0 < -\frac{x_{02}}{x_{12}} < \cdots < -\frac{x_{0m}}{x_{1m}}$, and denote the regularization of the open interval $\Delta_\nu = \left(-\frac{x_{0\nu}}{x_{1\nu}}, -\frac{x_{0,\nu+1}}{x_{1,\nu+1}} \right)$, $\nu = 1, \cdots, m-1$ by $\Delta_\nu(\omega)$. As in the previous case, we set

$$P_1(u) = u, \quad P_j(u) = x_{0j} + x_{1j}u, \quad 2 \leq j \leq m$$

etc., and

$$\widehat{\varphi}_\nu\langle j \rangle = \int_{\Delta_\nu(\omega)} U \cdot \varphi\langle j \rangle, \quad 1 \leq j, \ \nu \leq m - 1.$$

For $m - 1$ solutions of $E'(2, m + 1; \alpha')$

$$\widehat{\varphi}_\nu\langle 1 \rangle = \int_{\Delta_\nu(\omega)} U \cdot \varphi\langle 1 \rangle,$$

by simple computation, we obtain

$$\frac{\partial \widehat{\varphi}_\nu\langle 1 \rangle}{\partial x_{1j}} = \frac{\alpha_j}{x_{1j}} \widehat{\varphi}_\nu\langle j \rangle, \quad 2 \leq j \leq m - 1, \ 1 \leq \nu \leq m - 1. \qquad (3.49)$$

Here, using the abbreviated symbols $\partial_{1j} = \frac{\partial}{\partial x_{1j}}$ and taking

$$W := \det \begin{pmatrix} \widehat{\varphi}_1\langle 1\rangle & \cdots\cdots & \widehat{\varphi}_{m-1}\langle 1\rangle \\ \partial_{12}\widehat{\varphi}_1\langle 1\rangle & \cdots\cdots & \partial_{12}\widehat{\varphi}_{m-1}\langle 1\rangle \\ \vdots & & \vdots \\ \partial_{1,m-1}\widehat{\varphi}_1\langle 1\rangle & \cdots\cdots & \partial_{1,m-1}\widehat{\varphi}_{m-1}\langle 1\rangle \end{pmatrix}$$

as Wronskian of $\widehat{\varphi}_1\langle 1\rangle, \cdots, \widehat{\varphi}_{m-1}\langle 1\rangle$, by (3.49), we obtain

$$W = \prod_{j=2}^{m-1} \frac{\alpha_j}{x_{1j}} \cdot \det(\widehat{\varphi}_\nu\langle j\rangle),$$

and by the same reasoning as in § 3.5.2, since we have $\det(\varphi_\nu\langle j\rangle) \neq 0$, $x_{1j} \neq 0$, $2 \leq j \leq m$, i.e., $W \neq 0$, and $m-1$ integral solutions $\widehat{\varphi}_\nu\langle j\rangle$ of $E'(2, m+1; \alpha')$ become independent.

3.5.4 Equation $E'(3.6; \alpha')$

$$x' = \begin{pmatrix} 1 & & x_{03} & x_{04} & x_{05} \\ & 1 & x_{13} & x_{14} & x_{15} \\ & & 1 & x_{23} & x_{24} & x_{25} \end{pmatrix} \in M_{3,6}(\mathbb{R}),$$

$$P_1 = u_1, \quad P_2 = u_2, \quad P_j = x_{0j} + x_{1j}u_1 + x_{2j}u_2, \quad j = 3, 4, 5.$$

As in Figure 3.1 in § 3.2, five lines decompose $\mathbb{R}^2$ into several chambers among which six are bounded. Take

$$\varphi\langle 1, 2\rangle, \quad \varphi\langle 1, 3\rangle, \quad \varphi\langle 1, 4\rangle, \quad \varphi\langle 2, 3\rangle, \quad \varphi\langle 2, 4\rangle, \quad \varphi\langle 3, 4\rangle$$

as a basis of $H^2(\Omega^\bullet(\log D), \nabla_\omega)$. As in the previous case, we set

$$\widehat{\varphi}_\nu\langle j_1, j_2\rangle := \iint_{\Delta_\nu(\omega)} U \cdot \varphi\langle j_1, j_2\rangle, \quad 1 \leq j_1 \leq j_2 \leq 4.$$

Notice that the following fundamental relations can be obtained by simple computation:

$$\partial_{13}\widehat{\varphi}_\nu\langle 12\rangle = -\frac{\alpha_3}{x_{13}}\widehat{\varphi}_\nu\langle 23\rangle, \quad \partial_{23}\widehat{\varphi}_\nu\langle 12\rangle = -\frac{\alpha_3}{x_{23}}\widehat{\varphi}_\nu\langle 13\rangle, \qquad (3.50)$$

$$\partial_{14}\widehat{\varphi}_\nu\langle 12\rangle = -\frac{\alpha_4}{x_{14}}\widehat{\varphi}_\nu\langle 24\rangle, \quad \partial_{24}\widehat{\varphi}_\nu\langle 12\rangle = -\frac{\alpha_4}{x_{24}}\widehat{\varphi}_\nu\langle 14\rangle,$$

$$\partial_{13}\partial_{24}\widehat{\varphi}_\nu\langle 12\rangle = \frac{\alpha_3\alpha_4}{x\binom{1\ 2}{3\ 4}}\widehat{\varphi}_\nu\langle 34\rangle.$$

Setting

$$W = \det \begin{pmatrix} \widehat{\varphi}_1\langle 12 \rangle & \cdots\cdots & \widehat{\varphi}_6\langle 12 \rangle \\ \partial_{13}\widehat{\varphi}_1\langle 12 \rangle & \cdots\cdots & \partial_{13}\widehat{\varphi}_6\langle 12 \rangle \\ \vdots & & \vdots \\ \partial_{24}\widehat{\varphi}_1\langle 12 \rangle & \cdots\cdots & \partial_{24}\widehat{\varphi}_6\langle 12 \rangle \\ \partial_{13}\partial_{24}\widehat{\varphi}_1\langle 12 \rangle & \cdots\cdots & \partial_{13}\partial_{24}\widehat{\varphi}_6\langle 12 \rangle \end{pmatrix}$$

as Wronskian of the six integral solutions $\widehat{\varphi}_\nu\langle 12 \rangle$, $1 \le \nu \le 6$, by (3.50), we obtain

$$W = \frac{(\alpha_3\alpha_4)^3}{\displaystyle\prod_{\substack{1\le i\le 2 \\ 3\le j\le 4}} x_{ij} \cdot x\binom{1\ 2}{3\ 4}} \det(\widehat{\varphi}_\nu\langle j_1, j_2 \rangle).$$

On the other hand, by Corollary 3.1, we have $\det(\varphi_\nu\langle j_1, j_2 \rangle) \ne 0$. As the arrangement of lines in $\mathbb{P}^2$ defined by x' is in general position, we have $x_{ij} \ne 0$, $x\binom{1\ 2}{3\ 4} \ne 0$ which implies $W \ne 0$. Hence, the six integral solutions $\widehat{\varphi}_\nu\langle 12 \rangle$, $1 \le \nu \le 6$ are linearly independent.

3.5.5 Equation $E'(4, 8; \alpha')$

$$x = \begin{pmatrix} 1 & & & & x_{04}\ x_{05}\ x_{06}\ x_{07} \\ & 1 & & & x_{14}\ x_{15}\ x_{16}\ x_{17} \\ & & 1 & & x_{24}\ x_{25}\ x_{26}\ x_{27} \\ & & & 1 & x_{34}\ x_{35}\ x_{36}\ x_{37} \end{pmatrix} \in M_{4,8}(\mathbb{R}),$$

$$P_i = u_1, \quad 1 \le i \le 3, \quad P_j = x_{0j} + \sum_{i=1}^3 x_{ij}u_i, \quad 4 \le j \le 7.$$

Seven hyperplanes in $\mathbb{R}^3$ determine 20 bounded chambers Δ_ν, $1 \le \nu \le 20$. Moreover, taking

$$\varphi\langle j_1 j_2 j_3 \rangle, \quad 1 \le j_1 < j_2 < j_3 \le 6$$

as a basis of $H^3(\Omega^\bullet(\log D), \nabla_\omega)$, setting

$$\widehat{\varphi}_\nu\langle j_1 j_2 j_3 \rangle := \iiint_{\Delta_\nu(\omega)} U \cdot \varphi\langle j_1 j_2 j_3 \rangle$$

as in the previous case, we obtain the fundamental relations:

$$\partial_{ij}\widehat{\varphi}_\nu\langle 123 \rangle = \frac{\mathrm{sgn}\begin{pmatrix} 1 & 2 & 3 \\ i & l_1 & l_2 \end{pmatrix}\alpha_j}{x_{ij}}\widehat{\varphi}_\nu\langle l_1 l_2 j \rangle, \tag{3.51}$$

$$1 \le i \le 3, \quad 4 \le j \le 6$$

where we set $\{1,2,3\}\backslash\{i\} = \{l_1, l_2\}$,

$$\partial_{i_1 j_1}\partial_{i_2 j_2}\widehat{\varphi}_\nu\langle 123 \rangle = \frac{\mathrm{sgn}\begin{pmatrix} 1 & 2 & 3 \\ i_1 & i_2 & l \end{pmatrix}\alpha_j}{x\begin{pmatrix} i_1 & i_2 \\ j_1 & j_2 \end{pmatrix}}\widehat{\varphi}_\nu\langle l, j_1, j_2 \rangle, \tag{3.52}$$

$$1 \le i_1 < i_2 \le 3, \quad 4 \le j_1 < j_2 \le 6$$

where we set $\{1,2,3\}\backslash\{i_1, i_2\} = \{l\}$, and

$$\partial_{14}\partial_{25}\partial_{36}\widehat{\varphi}_\nu\langle 123 \rangle = \frac{\alpha_4\alpha_5\alpha_6}{x\begin{pmatrix} 1 & 2 & 3 \\ 4 & 5 & 6 \end{pmatrix}}\widehat{\varphi}_\nu\langle 456 \rangle. \tag{3.53}$$

We let the Wronskian of the 20 integral solutions $\widehat{\varphi}_\nu\langle 123 \rangle$, $1 \le \nu \le 20$ be the determinant of the matrix obtained by arranging $\widehat{\varphi}_\nu\langle 123 \rangle$, the left-hand sides of (3.51), (3.52), (3.53), respectively, in the lexicographic order:

$$W = \det \begin{pmatrix} \widehat{\varphi}_1\langle 123 \rangle & \cdots\cdots & \widehat{\varphi}_{20}\langle 123 \rangle \\ \partial_{14}\widehat{\varphi}_1\langle 123 \rangle & \cdots\cdots & \partial_{14}\widehat{\varphi}_{20}\langle 123 \rangle \\ \vdots & & \vdots \\ \partial_{14}\partial_{25}\widehat{\varphi}_1\langle 123 \rangle & \cdots\cdots & \partial_{14}\partial_{25}\widehat{\varphi}_{20}\langle 123 \rangle \\ \vdots & & \vdots \\ \partial_{14}\partial_{25}\partial_{36}\widehat{\varphi}_1\langle 123 \rangle & \cdots\cdots & \partial_{14}\partial_{25}\partial_{36}\widehat{\varphi}_{20}\langle 123 \rangle \end{pmatrix}$$

By a simple computation with (3.51)—(3.53), we obtain

$$W = \frac{(\alpha_4\alpha_5\alpha_6)^{10}}{\prod x_{ij} \cdot \prod x\begin{pmatrix} i_1 & i_2 \\ j_1 & j_2 \end{pmatrix} \cdot x\begin{pmatrix} 1 & 2 & 3 \\ 4 & 5 & 6 \end{pmatrix}} \det(\widehat{\varphi}_\nu\langle j_1 j_2 j_3 \rangle),$$

where the products are taken over $1 \le i \le 3$, $4 \le j \le 6$ and $1 \le i_1 < i_2 \le 3$, $4 \le j_1 < j_2 \le 6$. Hence, as before, we have $W \ne 0$, i.e., the 20 integral solutions $\widehat{\varphi}_\nu\langle 123 \rangle$ of $E'(4, 8; \alpha')$ become independent.

3.5.6 General Cases

From the above observations, choosing appropriate partial derivatives of the integral solution $\widehat{\varphi}_\nu\langle 12\cdots n\rangle$ may correspond to replacing the logarithmic n-form $\varphi\langle 12\cdots n\rangle$ with another $\varphi\langle j_1,\cdots,j_n\rangle$ in the basis $\{\varphi\langle j_1,\cdots,j_n\rangle \mid 1 \leq j < \cdots < j_n \leq m-1\}$ of the twisted cohomology $H^n(\Omega^\bullet(\log D),\nabla_\omega)$. Below, we show that this is still valid in general. Assume that the arrangement of hyperplanes in $\mathbb{P}^n$ defined by the real matrix

$$x = \begin{pmatrix} 1 & & & x_{0,n+1} & \cdots & x_{0,m} \\ & 1 & & \vdots & & \vdots \\ & & \ddots & \vdots & & \vdots \\ & & & 1 & x_{n,n+1} & \cdots & x_{n,m} \end{pmatrix} \in M_{n+1,m+1}(\mathbb{R}) \qquad (3.54)$$

is in general position. By this condition, it follows that

$$x \begin{pmatrix} i_1 \cdots i_p \\ k_1 \cdots p_p \end{pmatrix} \neq 0, \quad 0 \leq i_1 < \cdots < i_p \leq n, \qquad (3.55)$$

$$n + 1 \leq k_1 < \cdots < k_p \leq m.$$

Below, we use the notation introduced in § 3.5.6. We consider $r := \binom{m-1}{n}$ integral solutions $\widehat{\varphi}_\nu\langle 12\cdots n\rangle$ of $E'(n+1, m+1; \alpha')$. To proceed with our computations effectively, we also use the following notation: for

$$I = \{i_1,\ldots,i_p\}, \quad 1 \leq i_1 < \cdots < i_p \leq n,$$

$$K = \{k_1,\ldots,k_p\}, \quad n+1 \leq k_1 < \cdots < k_p \leq m-1,$$

we set

$$L := \{1, 2, \ldots, n\} \backslash I = \{l_1,\ldots,l_{n-p}\}, \quad 1 \leq l_1 < \cdots < l_{n-p} \leq n$$

and

$$\mathrm{sgn}\begin{pmatrix} 1\ 2\ \cdots\ n \\ L\ I \end{pmatrix} := \mathrm{sgn}\begin{pmatrix} 1 & 2 & \cdots & & \cdots & n \\ l_1 & \cdots & l_{n-p} & i_1 & \cdots & i_p \end{pmatrix},$$

$$x\begin{pmatrix} I \\ K \end{pmatrix} := x\begin{pmatrix} i_1 & \cdots & i_p \\ k_1 & \cdots & k_p \end{pmatrix}, \quad \varphi\langle LK\rangle = \varphi\langle l_1,\ldots,l_{n-p}, k_1,\ldots,k_p\rangle,$$

for short. We have

$$\partial_{i_1 k_1} \cdots \partial_{i_p k_p} \widehat{\varphi}_\nu \langle 12 \cdots n \rangle \qquad (3.56)$$

$$= \prod_{k \in K} \alpha_k \int_{\Delta_\nu(\omega)} \frac{U \prod_{i \in I} u_i}{\prod_{k \in K} P_k} \varphi \langle 12 \cdots n \rangle$$

$$= \prod_{k \in K} \alpha_k \int_{\Delta_\nu(\omega)} \frac{U}{\prod_{l \in L} u_l \prod_{k \in K} P_k} du_1 \wedge \cdots \wedge du_n.$$

On the other hand, by simple computation, we also have

$$du_{l_1} \wedge \cdots \wedge du_{l_{n-p}} \wedge dP_{k_1} \wedge \cdots \wedge dP_{k_p}$$

$$= x \begin{pmatrix} i_1 & \cdots & i_p \\ k_1 & \cdots & k_p \end{pmatrix} du_{l_1} \wedge \cdots \wedge du_{l_{n-p}} \wedge du_{i_1} \wedge \cdots \wedge du_{i_p}$$

$$= \operatorname{sgn} \begin{pmatrix} 1\ 2\ \cdots\ n \\ L\ \ I \end{pmatrix} x \begin{pmatrix} I \\ K \end{pmatrix} du_1 \wedge \cdots \wedge du_n.$$

By this together with (3.56), we obtain the fundamental relation:

$$\partial_{i_1 k_1} \cdots \partial_{i_p k_p} \widehat{\varphi}_\nu \langle 12 \cdots n \rangle \qquad (3.57)$$

$$= \frac{\operatorname{sgn} \begin{pmatrix} 1\ 2\ \cdots\ n \\ L\ \ I \end{pmatrix}}{x \begin{pmatrix} I \\ K \end{pmatrix}} \cdot \prod_{k \in K} \alpha_k \cdot \widehat{\varphi}_\nu \langle LK \rangle.$$

$\{\varphi\langle LK \rangle\}$ is a subset of the basis $\{\varphi\langle j_1, \cdots, j_n \rangle \mid 1 \leq j_1 < \cdots < j_n \leq m-1\}$ of $H^n(\Omega^\bullet(\log D), \nabla_\omega)$ and the cardinality of $\{\varphi\langle LK \rangle\}$ is given by

$$\sum_{p=0}^{n} \binom{n}{n-p} \binom{m-n-1}{p}.$$

Let us compute this sum.

Lemma 3.5. $\displaystyle \sum_{p=0}^{n} \binom{n}{n-p} \binom{m-n-1}{p} = \binom{m-1}{n}.$

Proof. Let us prove this by using generating series: compare the coefficients of x^n in the formula obtained by multiplying each side of

$$(1+x)^n = \sum_{p=0}^{n} \binom{n}{p} x^p,$$

$$(1+x)^{m-n-1} = \sum_{p=0}^{m-n-1} \binom{m-n-1}{p} x^p.$$

By this lemma, $\{\varphi\langle LK\rangle\}$ becomes a basis of $H^n(\Omega^\bullet(\log D), \nabla_\omega)$, as we have computed before; we adapt the Wronskian as the determinant of the matrix obtained by arranging $\widehat{\varphi}_\nu\langle 12\cdots n\rangle$ and all of the partial derivatives appearing in (3.57), for example, in the lexicographic order. Denoting this determinant symbolically as

$$W = \det\left(\partial_{i_1 k_1} \cdots \partial_{i_p k_p} \widehat{\varphi}_\nu\langle 12\cdots n\rangle\right),$$

from (3.57), we obtain

$$W = \prod_{I,K}\left(\frac{\mathrm{sgn}\binom{12\cdots n}{LI}}{x\binom{I}{K}}\right)\prod_{k\in K}\alpha_k \cdot \det(\widehat{\varphi}_\nu\langle LK\rangle). \qquad (3.58)$$

We have $\det(\varphi_\nu\langle LK\rangle) \neq 0$ by Corollary 3.1, hence (3.55) implies $W \neq 0$. Thus, the $\binom{m-1}{n}$ integral solutions $\widehat{\varphi}_\nu\langle 12\cdots n\rangle$ of $E'(n+1, m+1; \alpha')$ are linearly independent. Therefore, we obtain the following theorem.

Theorem 3.4. *The arrangement of hyperplanes in general position defined by the real matrix (3.54) determines $r = \binom{m-1}{n}$ bounded chambers Δ_ν, $1 \leq \nu \leq r$ in $\mathbb{R}^n$. We assume $\alpha_j \notin \mathbb{Z}$, $1 \leq j \leq m$, $\sum_{j=1}^m \alpha_j \notin \mathbb{Z}$ on exponents. Then, r integrals*

$$\widehat{\varphi}_\nu\langle 12\cdots n\rangle = \int_{\Delta_\nu(\omega)} U \cdot \varphi\langle 12\cdots n\rangle$$

become solutions of $E'(n+1, m+1; \alpha_0, \alpha_1 - 1, \cdots, \alpha_n - 1, \alpha_{n+1}, \cdots, \alpha_m)$ $= E'(n+1, m+1; \alpha')$. Moreover, the Wronskian obtained by arranging $\widehat{\varphi}_\nu$ $\langle 12\cdots n\rangle$ and all of the partial derivatives of (3.57) satisfies

$$W = \det(\partial_{i_1 k_1} \cdots \partial_{i_p k_p} \widehat{\varphi}_\nu\langle 12\cdots n\rangle) \neq 0,$$

hence the above r solutions are linearly independent.

3.5.7 Wronskian

To rewrite the Wronskian in a simple form including the Γ-function, let us first rewrite the right-hand side of (3.58). By $W = \det(\partial_{i_1 k_1} \cdots \partial_{i_p k_p} \widehat{\varphi}_\nu\langle 12\cdots n\rangle)$ and the choice of $K = \{k_1, \cdots, k_p\}$, the right-hand side of (3.58) must be symmetric with respect to $\alpha_{n+1}, \cdots, \alpha_{m-1}$, hence, we obtain

$$\prod_{I,K}\left(\prod_{k\in K}\alpha_k\right) = (\alpha_{n+1}\cdots\alpha_{m-1})^s, \qquad s \in \mathbb{Z}_{>0}. \qquad (3.59)$$

On the other hand, the number of the factors appearing in the product of the left-hand side is

$$\sum_{p=0}^{n} p \binom{n}{p} \binom{m-n-1}{p}$$

by $|K| = p$. We compute this sum.

Lemma 3.6.

$$\sum_{p=0}^{n} p \binom{n}{p} \binom{m-n-1}{p} = (m-n-1) \binom{m-2}{n-1}.$$

Proof. Let us prove this by using generating series. Taking the binomial expansion of $(1+x)^{m-n-1}$ and differentiating it, we obtain

$$(m-n-1)(1+x)^{m-n-2} = \sum_{q=0}^{m-n-1} q \binom{m-n-1}{q} x^{q-1}.$$

Multiplying each side by

$$(1+x)^n = \sum_{p=0}^{n} \binom{n}{p} x^p,$$

we obtain

$$(m-n-1)(1+x)^{m-2} = \sum_{p,q} q \binom{n}{p} \binom{m-n-1}{q} x^{p+q-1}.$$

It is enough to compare the coefficients of x^{n-1} in both sides.

By Lemma 3.6, s in (3.59) becomes $\binom{m-2}{n-1}$, and (3.58) can be rewritten as follows:

$$W = (\alpha_{n+1} \cdots \alpha_{m-1})^{\binom{m-2}{n-1}} \tag{3.60}$$

$$\cdot \prod_{I,K} \frac{\operatorname{sgn} \begin{pmatrix} 1 \ 2 \ \cdots \ n \\ L \ \ I \end{pmatrix}}{x \binom{I}{K}} \cdot \det(\widehat{\varphi}_\nu \langle LK \rangle).$$

3.5.8 Varchenko's Formula

Varchenko, in [V1], [V2], obtained a concrete form of the determinant $\det(\widehat{\varphi}_\nu \langle LK \rangle)$ of hypergeometric integrals. Here, using this result, we further rewrite our Wronskian. First, we state Varchenko's result. As $\binom{m-1}{n}$ bounded chambers Δ_ν are simply connected, for each P_j, $1 \le j \le m$, fixing its argument on each chamber Δ_ν, we fix a branch of the multi-valued function $P_j^{\alpha_j}$ on Δ_ν. The branch of $U = \prod_{j=1}^m P_j^{\alpha_j}$ on Δ_ν is well-determined, hence so is the twisted cycle $\Delta_\nu(\omega)$. We denote the point of the closure $\overline{\Delta}_\nu$ of Δ_ν that attains the maximum of the absolute value $|P_j^{\alpha_j}(u)|$, of the branch $P_j^{\alpha_j}$ in the above sense, by $a_{\nu,j} \in \overline{\Delta}_\nu$ and the value of the branch $P_j^{\alpha_j}$ at this point by $c(P_j^{\alpha_j}, \Delta_\nu)$: $c(P_j^{\alpha_j}, \Delta_\nu) := P_j^{\alpha_j}(a_{\nu,j})$. Then, we have the following lemma:

Lemma 3.7 ([V1],[V2]).

$$\det\left(\int_{\Delta_\nu(\omega)} U \cdot \varphi\langle LK\rangle\right) = \pm \frac{1}{(\alpha_1 \cdots \alpha_{m-1})^{\binom{m-2}{n-1}}} \cdot B \cdot \prod_{j=1}^m \prod_{\nu=1}^{\binom{m-1}{n}} c(P_j^{\alpha_j}, \Delta_\nu),$$

where we set

$$B = \left(\prod_{j=1}^m \Gamma(\alpha_j + 1)/\Gamma\left(\sum_{j=1}^m \alpha_j + 1\right)\right)^{\binom{m-2}{n-1}}.$$

Example 3.3. For $n = 1$, let us compute $c(P_j^{\alpha_j}, \Delta_\nu)$ in Lemma 3.7 and give an explicit expression. Let $x_1 < \cdots < x_m$ be m real points in $\mathbb{C}$. Setting $P_j = u - x_j$, we have $U(u) = \prod_{j=1}^m (u - x_j)^{\alpha_j}$. Here, to facilitate the computation of $c(P_j^{\alpha_j}, \Delta_\nu)$, we assume: $\Re\alpha_j > 0$, $1 \le j \le m$. We fix $\arg(u - x_j)$ as in (2.44) (cf. Figure 2.5) and use the twisted cycles $\Delta_\nu(\omega)$ constructed there. When u is real, by

$$|(u - x_j)^{\alpha_j}| = \exp\{(\Re\alpha_j) \log|u - x_j|\}$$

$$= |u - x_j|^{\Re\alpha_j},$$

this is an increasing function of $|u - x_j|$ by the assumption $\Re\alpha_j > 0$. Hence, we obtain the formula:

$$c(P_j^{\alpha_j}, \Delta_\nu) = \begin{cases} (x_{\nu+1} - x_j)^{\alpha_j}, & \nu \ge j, \\ \exp(-\pi\sqrt{-1}\alpha_j) \cdot (x_j - x_\nu)^{\alpha_j}, & \nu < j. \end{cases}$$

From this, we easily obtain

$$\prod_{j=1}^m \prod_{\nu=1}^{m-1} c(P_j^{\alpha_j}, \Delta_\nu) = \exp\left(-\sqrt{-1}\sum_{j=1}^m (j-1)\pi\alpha_j\right) \cdot \prod_{k \neq j} |x_j - x_k|^{\alpha_j}.$$

Hence, the Varchenko formula takes the following form:

$$\det \begin{pmatrix} \int_{\Delta_1(\omega)} U \cdot \varphi\langle 1 \rangle & \cdots\cdots & \int_{\Delta_1(\omega)} U \cdot \varphi\langle m-1 \rangle \\ \vdots & & \vdots \\ \int_{\Delta_{m-1}(\omega)} U \cdot \varphi\langle 1 \rangle & \cdots\cdots & \int_{\Delta_{m-1}(\omega)} U \cdot \varphi\langle m-1 \rangle \end{pmatrix}$$

$$= \frac{1}{\alpha_1 \cdots \alpha_{m-1}} \cdot \exp\left(-\sqrt{-1} \sum_{j=1}^{m} (j-1)\pi\alpha_j\right) \prod_{k \neq j} |x_j - x_k|^{\alpha_j}$$

$$\times \prod_{j=1}^{m} \Gamma(\alpha_j + 1) / \Gamma\left(1 + \sum_{j=1}^{m} \alpha_j\right),$$

where $\varphi\langle j \rangle = du/(u - x_j)$.

By Lemma 3.7 and (3.60), rewriting the Wronskian as the product of Γ-factors and the "critical-values" of $P_j^{\alpha_j}$ on Δ_ν, we obtain the following theorem.

Theorem 3.5. *The Wronskian W can be written as follows:*

$$W = \frac{\pm 1}{(\alpha_1 \cdots \alpha_n)^{\binom{m-2}{n-1}} \prod_{I,K} x\binom{I}{K}} \cdot B \cdot \prod_{j=1}^{m} \prod_{\nu=1}^{\binom{m-1}{n}} c\left(P_j^{\alpha_j}, \Delta_\nu\right).$$

Here, B is the one given in Lemma 3.7, the multi-indicies I, K are taken over $I = \{i_1, \cdots, i_p\}$, $1 \leq i_1 < \cdots < i_p \leq n$, $K = \{k_1, \cdots, k_p\}$, $n+1 \leq k_1 < \cdots < k_p \leq m-1$ and p from 0 to n.

Remark 3.9. This determinant formula is extended to any arrangement of hyperplanes (cf. [Do-Te]).

3.5.9 Intersection Number (ii)

By Lemma 2.9 (3), the non-degenerate bilinear form

$$H^n(M, \mathcal{L}_\omega) \times H_c^n(M, \mathcal{L}_\omega^\vee) \longrightarrow \mathbb{C} \tag{3.61}$$

can be regarded as defining the intersection number of twisted cocycles. In § 2.3.3, we have computed the intersection number of twisted cycles explicitly for $n = 1$, in [Cho-Ma], it was shown that a similar computation can be carried out for twisted cocycles. We leave the detail to the above paper, and here, we briefly state its essence.

Below, making no exception to the infinity, we take $m+1$ points $x_0, x_1, \cdots,$ x_m on $\mathbb{P}^1$, and set $M := \mathbb{P}^1 \setminus \{x_0, \cdots, x_m\}$,

$$\omega := \sum_{j=0}^{m} \alpha_j \frac{du}{u - x_j}, \quad \sum_{j=0}^{m} \alpha_j = 0, \quad \nabla_\omega := d + \omega \wedge,$$

$$\nabla_\omega^\vee := d - \omega \wedge, \quad j : M \longrightarrow \mathbb{P}^1 \quad \text{(inclusion)}.$$

To compute the cohomology with compact suport $H_c^1(M, \mathcal{L}_\omega^\vee)$, we denote the sheaf on $\mathbb{P}^1$, obtained by extending the sheaf $\mathcal{L}_\omega^\vee$ on M to the sheaf on $\mathbb{P}^1$ by 0 at each point of $\mathbb{P}^1 \setminus M$ (extension by zero), by $j_!\mathcal{L}_\omega^\vee$. First, we have

$$H_c^1(M, \mathcal{L}_\omega^\vee) \simeq H^1(\mathbb{P}^1, j_!\mathcal{L}_\omega^\vee).$$

Let D be the divisor of $\mathbb{P}^1$ determined by the $m + 1$ points $x_0, \cdots, x_m$, and denote the sheaf of logarithmic p-forms that may have logarithmic poles only along D by $\Omega_{\mathbb{P}^1}^p(\log D)$, and the sub-sheaf of $\Omega_{\mathbb{P}^1}^p(\log D)$ of p-forms which have poles along D of, at least, order 1 by $\Omega_{\mathbb{P}^1}^p(\log D)(-D)$. Then, we have the exact sequence of sheaves:

$$0 \longrightarrow j_!\mathcal{L}_\omega^\vee \longrightarrow \Omega_{\mathbb{P}^1}^0(\log D)(-D) \xrightarrow{\nabla_\omega^\vee} \Omega_{\mathbb{P}^1}^1(\log D)(-D) \longrightarrow 0.$$
$$\wr|\qquad\qquad\qquad\qquad \wr|$$
$$\mathcal{O}_{\mathbb{P}^1}(-D) \qquad\qquad\qquad \Omega_{\mathbb{P}^1}^1$$

As in § 2.4.6, we obtain

$$H^1(\mathbb{P}^1, j_!\mathcal{L}_\omega^\vee) \simeq \mathbb{H}^1(\mathbb{P}^1, (\Omega_{\mathbb{P}^1}^\bullet(\log D)(-D), \nabla_\omega^\vee)).$$

To rewrite the right-hand side of this formula, we consider the following exact sequence of complexes of sheaves:

$$0 \longrightarrow (\Omega^\bullet(\log D)(-D), \nabla_\omega^\vee) \xrightarrow{\iota} (\Omega^\bullet(\log D), \nabla_\omega^\vee) \longrightarrow$$

$$\longrightarrow \left(\coprod_{j=0}^{m} \mathbb{C}_{x_j} \xrightarrow{\times \mathrm{res}} \coprod_{j=0}^{m} \mathbb{C}_{x_j} \right) \longrightarrow 0,$$

namely,

$$
\begin{array}{ccccc}
0 & & 0 & & 0 \\
\downarrow & & \downarrow & & \downarrow \\
0 \longrightarrow \mathcal{O}_{\mathbb{P}^1}(-D) \longrightarrow & \mathcal{O}_{\mathbb{P}^1} & \longrightarrow & \coprod_{j=0}^{m} \mathbb{C}_{x_j} & \longrightarrow 0 \\
\downarrow \nabla_\omega^\vee & & \downarrow \nabla_\omega^\vee & & \downarrow \times \mathrm{res} \\
0 \longrightarrow \Omega_{\mathbb{P}^1}^1 & \longrightarrow & \Omega_{\mathbb{P}^1}^1(\log D) \xrightarrow{\mathrm{Res}} & \coprod_{j=0}^{m} \mathbb{C}_{x_j} & \longrightarrow 0. \\
\downarrow & & \downarrow & & \downarrow \\
0 & & 0 & & 0
\end{array}
$$

Here, we set $\times \mathrm{res} : (c_0, \ldots, c_m) \longmapsto (-\alpha_0 c_0, \ldots, -\alpha_m c_m)$ and Res signifies taking the residue.

Under the assumptions $\alpha_j \neq 0$, the complex in the right-hand side becomes exact, hence, we obtain an isomorphism of hypercohomologies:

$$\iota : \mathbb{H}^1(\mathbb{P}^1, (\varOmega^\bullet(\log D)(-D), \nabla_\omega^\vee)) \simeq \mathbb{H}^1(\mathbb{P}^1, (\varOmega^\bullet(\log D), \nabla_\omega^\vee)). \quad (3.62)$$

By the fact that the spectral sequence of hypercohomologies, discussed in § 2.4.4, $''E_1^{p,q}$ degenerates (the former has $''E_1^{p,q} = 0$ for $q = 1$, and the latter has $''E_1^{p,q} = 0$ for $q = 0$), (3.62) can be rewritten as follows:

$$\iota : \ker\{-\omega : H^1(\mathbb{P}^1, \mathcal{O}_{\mathbb{P}^1}(-D)) \longrightarrow H^1(\mathbb{P}^1, \varOmega^1_{\mathbb{P}^1})\} \quad (3.63)$$
$$\xrightarrow{\sim} \varGamma(\mathbb{P}^1, \varOmega^1(\log D))/\mathbb{C} \cdot (-\omega).$$

With the aid of the isomorphism (3.63), we define the intersection number of logarithmic differential forms that may have poles along D as follows: for $\varphi, \psi \in \varGamma(\mathbb{P}^1, \varOmega^1_{\mathbb{P}^1}(\log D))$, setting $\varphi^+ = \varphi \bmod \mathbb{C} \cdot \omega$, $\psi^- = \psi \bmod \mathbb{C} \cdot (-\omega)$, we have $\iota^{-1}(\psi^-) \in H^1(\mathbb{P}^1, \mathcal{O}_{\mathbb{P}^1}(-D))$. Hence, $\varphi^+ \wedge \iota^{-1}(\psi^-) \in H^1(\mathbb{P}^1, \varOmega^1_{\mathbb{P}^1})$ does not depend on the choice of the representatives. We fix the Serre duality

$$\varGamma(\mathbb{P}^1, \varOmega^1_{\mathbb{P}^1}(\log D)) \times H^1(\mathbb{P}^1, \mathcal{O}_{\mathbb{P}^1}(-D)) \longrightarrow H^1(\mathbb{P}^1, \varOmega^1_{\mathbb{P}^1}),$$

the Dolbeault isomorphism, and an isomorphism via an integral over $\mathbb{P}^1$

$$H^1(\mathbb{P}^1, \varOmega^1_{\mathbb{P}^1}) \longrightarrow \mathbb{C}$$
$$\frac{C^\infty(1,1) - \text{forms}}{\bar\partial\{C^\infty(0,1) - \text{forms}\}} \ni [\tau] \longmapsto \iint_{\mathbb{P}^1} \tau.$$

By this, we obtain the following non-degenerate bilinear form:

$$I : \varGamma(\mathbb{P}^1, \varOmega^1_{\mathbb{P}^1}(\log D)) \times H^1(\mathbb{P}^1, \mathcal{O}_{\mathbb{P}^1}(-D)) \longrightarrow \mathbb{C}.$$

Definition 3.2. We define the intersection number $I_c(\varphi^+, \psi^-)$ of φ^+, ψ^- by

$$I_c(\varphi^+, \psi^-) := I(\varphi, \iota^{-1}(\psi^-)).$$

This value does not depend on the choice of representatives and it is non-degenerate.

By an argument relating the Serre duality and the residue calculus ([Fo], § 17), we can compute the intersection numbers explicitly.

Theorem 3.6 ([Cho-Ma]). *For logarithmic 1-forms*

$$\omega_{ij} = \frac{du}{u - x_i} - \frac{du}{u - x_j} \in \varGamma(\mathbb{P}^1, \varOmega^1(\log D)), \quad 0 \leq i \neq j \leq m,$$

we have

$$I_c(\omega_{pq}^+, \omega_{ij}^-) = 2\pi\sqrt{-1}\left\{\frac{1}{\alpha_i}(\delta_{ip} - \delta_{iq}) - \frac{1}{\alpha_j}(\delta_{jp} - \delta_{jq})\right\}.$$

Here, δ_{ip} is the Kronecker delta.

3.5.10 Twisted Riemann's Period Relations and Quadratic Relations of Hypergeometric Functions

We denote a basis of the twisted homologies $H_1(M, \mathcal{L}_\omega^\vee)$ and $H_1(M, \mathcal{L}_\omega)$ by $\{\gamma_1^+, \cdots, \gamma_{m-1}^+\}, \{\delta_1^-, \cdots, \delta_{m-1}^-\}$, respectively, and by using the isomorphism $\mathrm{reg} : H_1^{lf}(M, \mathcal{L}_\omega) \xrightarrow{\sim} H_1(M, \mathcal{L}_\omega)$ stated in Theorem 3.1, we define the intersection number $\gamma_i^+ \cdot \delta_j^-$ of γ_i^+ and δ_j^- as the intersection number of γ_i^+ and $\mathrm{reg}^{-1}(\delta_j^-)$. We assume that $\xi_1, \cdots, \xi_{m-1}, \eta_1, \cdots, \eta_{m-1} \in \Gamma(\mathbb{P}^1, \Omega^1(\log D))$ defines a basis $\{\xi_1^+, \cdots, \xi_{m-1}^+\}, \{\eta_1^-, \cdots \eta_{m-1}^-\}$ of $\Gamma(\mathbb{P}^1, \Omega^1(\log D))/\mathbb{C} \cdot \omega$, $\Gamma(\mathbb{P}^1, \Omega^1(\log D))/\mathbb{C} \cdot (-\omega)$, respectively. Then, setting $U(u) = \prod_{j=0}^m (u - x_j)^{\alpha_j}$, we obtain two twisted period matrices:

$$\prod_+ := \begin{pmatrix} \int_{\gamma_1^+} U \cdot \xi_1 & \cdots\cdots & \int_{\gamma_{m-1}^+} U \cdot \xi_1 \\ \vdots & & \vdots \\ \int_{\gamma_1^+} U \cdot \xi_{m-1} & \cdots\cdots & \int_{\gamma_{m-1}^+} U \cdot \xi_{m-1} \end{pmatrix},$$

$$\prod_- := \begin{pmatrix} \int_{\delta_1^-} U^{-1} \cdot \eta_1 & \cdots\cdots & \int_{\delta_{m-1}^-} U^{-1} \cdot \eta_1 \\ \vdots & & \vdots \\ \int_{\delta_1^-} U^{-1} \cdot \eta_{m-1} & \cdots\cdots & \int_{\delta_{m-1}^-} U^{-1} \cdot \eta_{m-1} \end{pmatrix}.$$

By defining the intersection matrices of cycles and cocycles as

$$A = \begin{pmatrix} \gamma_1^+ \cdot \delta_1^- & \cdots\cdots & \gamma_1^+ \cdot \delta_{m-1}^- \\ \vdots & & \vdots \\ \gamma_{m-1}^+ \cdot \delta_1^- & \cdots\cdots & \gamma_{m-1}^+ \cdot \delta_{m-1}^- \end{pmatrix},$$

$$B = \begin{pmatrix} I_c(\xi_1^+, \eta_1^-) & \cdots\cdots & I_c(\xi_1^+, \eta_{m-1}^-) \\ \vdots & & \vdots \\ I_c(\xi_{m-1}^+, \eta_1^-) & \cdots\cdots & I_c(\xi_{m-1}^+, \eta_{m-1}^-) \end{pmatrix},$$

we obtain the following theorem.

Theorem 3.7 (Twisted Riemann's Period Relations).

$$(1)\ \prod_+ {}^t A^{-1} \prod_- = B,$$
$$(2)\ {}^t \prod_- B^{-1} \prod_+ = A.$$

For the proof, see [Cho-Ma]. By the above theorem, we obtain new types of quadratic relations of hypergeometric functions. Let us explain this by two examples.

Example 3.4. The beta function

$$B(\alpha, \beta) = \int_0^1 u^{\alpha-1}(1-u)^{\beta-1} du$$

$$= \int_0^1 u^{\alpha}(1-u)^{\beta} \left(\frac{du}{u-1} - \frac{du}{u} \right)$$

corresponds to the case $m = 2$, $x_0 = \infty$, $x_1 = 0$, $x_2 = 1$, and by the results obtained in § 2.3.3, setting $\gamma^+ = \mathrm{reg}(0,1) \in H_1(M, \mathcal{L}_\omega^\vee)$, $\delta^- = \mathrm{reg}(0,1) \in H_1(M, \mathcal{L}_\omega)$, $c_1 = \exp(2\pi i \alpha)$, $c_2 = \exp(2\pi i \beta)$, we have

$$\gamma^+ \cdot \delta^- = -\frac{c_1 c_2 - 1}{(c_1 - 1)(c_2 - 1)}.$$

Denote the classes defined by $\frac{du}{u-1} - \frac{du}{u}$ as $\xi^+ \in \Gamma(\mathbb{P}^1, \Omega^1(\log D))/\mathbb{C}\cdot\omega$ and $\eta^- \in \Gamma(\mathbb{P}^1, \Omega^1(\log D))/\mathbb{C}\cdot(-\omega)$, Theorem 3.6 implies

$$I_c(\xi^+, \eta^-) = 2\pi\sqrt{-1}\left(\frac{1}{\alpha} + \frac{1}{\beta} \right).$$

Hence, the twisted period relation becomes

$$\frac{1}{2\pi\sqrt{-1}} B(-\alpha, -\beta)\left(\frac{1}{\alpha} + \frac{1}{\beta} \right) B(\alpha, \beta) = -\frac{c_1 c_2 - 1}{(c_1 - 1)(c_2 - 1)}.$$

This can be proved directly by using $B(\alpha, \beta) = \Gamma(\alpha)\Gamma(\beta)/\Gamma(\alpha+\beta)$, but this was the fact that motivated the research effectuated in [Cho-Ma].

Example 3.5. (Quadratic relations of Lauricella's F_D)
The hypergeometric series of type $(2, m+3)$ introduced in § 3.1.2 has the following integral representation by Theorem 3.3:

$$F_D(\alpha, \beta_1, \ldots, \beta_m, \gamma; z_1, \ldots, z_m) \tag{3.64}$$

$$= \frac{\Gamma(\gamma)}{\Gamma(\alpha)\Gamma(\gamma-\alpha)} \int_0^1 u^{\alpha-1}(1-u)^{\gamma-\alpha-1} \prod_{j=1}^m (1-z_j u)^{-\beta_j} du.$$

Here, we set $x_0 = \infty$, $x_1 = 0$, $x_2 = 1 = \frac{1}{z_0}$, $x_3 = \frac{1}{z_1}, \ldots, x_{m+2} = \frac{1}{z_m}$,

$$\xi_j = \frac{du}{u} - \frac{du}{u - x_{j+1}} = \frac{du}{u(1 - z_{j-1}u)},$$

$$\eta_j = \frac{du}{1 - z_{j-1}u} = -\frac{1}{z_{j-1}}\frac{du}{u - x_{j+1}}, \quad 1 \le j \le m+1,$$

$$U(u) = u^\alpha(1 - u)^{\gamma - \alpha}\prod(1 - z_j u)^{-\beta_j}.$$

Notice that η_j has the residue $\frac{1}{z_{j-1}}$ at $x_0 = \infty$. Hence, by Theorem 3.6, we have

$$I_c(\xi_j^+, \eta_k^-) = 0 \quad (j \ne k),$$

$$I_c(\xi_j^+, \eta_k^-) = \begin{cases} \dfrac{2\pi\sqrt{-1}}{\gamma - \alpha} & (j = 1) \\[2mm] -\dfrac{2\pi\sqrt{-1}}{\beta_{j-1}z_{j-1}} & (2 \le j \le m+1), \end{cases}$$

which implies

$$B^{-1} = \frac{1}{2\pi\sqrt{-1}}\mathrm{diag}[\gamma - \alpha, -\beta_1 z_1, \ldots, -\beta_m z_m].$$

As in Example 3.4, taking $\mathrm{reg}(0,1)$ as γ_1^+, δ_1^-, by comparing the $(1,1)$ component of the both sides of Theorem 3.7, (2), we obtain

$$\left(\int_0^1 U^{-1}\cdot\eta_1, \ldots, \int_0^1 U^{-1}\cdot\eta_{m+1}\right)B^{-1^t}\left(\int_0^1 U\cdot\xi_1, \ldots, \int_0^1 U\cdot\xi_{m+1}\right)$$

$$= \left(-\frac{e^{2\pi i\gamma} - 1}{(e^{2\pi i\alpha} - 1)(e^{2\pi i(\gamma - \alpha)} - 1)}\right).$$

Rewriting the above relation to a relation among F_D by using (3.64) and $\Gamma(1 - s)\Gamma(s) = \pi/\sin\pi s$, we obtain the quadratic relation.

Setting $\beta = (\beta_1, \ldots, \beta_m)$, $e_j = (0, \ldots, \overset{j}{1}, 0\cdots 0)$, we have

$$F_D(\alpha, \beta, \gamma; z)F_D(1 - \alpha, -\beta, 1 - \gamma; z) - 1$$

$$= \frac{\gamma - \alpha}{\gamma(\gamma - 1)}\sum_{j=1}^m \beta_j z_j F_D(\alpha, \beta + e_j, \gamma; z)F_D(1 - \alpha, -\beta + e_j, 2 - \gamma; z).$$

For further references about uniformization or quadratic relation, see [De-Mo], [Ha-Y], [Matu], [Yos3], [Yos4].

3.6 Determination of the Rank of $E(n+1, m+1; \alpha)$

The maximal number of linearly independent solutions of the system of hypergeometric differential equations $E(n+1, m+1; \alpha)$ is called the rank of this system of equations and is denoted by rank $E(n+1, m+1; \alpha)$. We showed in § 3.5 that this system of equations admits $\binom{m-1}{n}$ linearly independent integral solutions. Next, if we can show only from the form of this system of differential equations that the number of linearly independent solutions is at most $\binom{m-1}{n}$, we can conclude rank $E(n+1, m+1; \alpha) = \binom{m-1}{n}$. To facilitate the computations, let us consider $E'(n+1, m+1; \alpha')$ obtained by killing the GL_{n+1}-action on $E(n+1, m+1; \alpha)$. Here, we recall that α' is defined in (3.46). In this section, we use the notation $\vartheta_{ij} = x_{ij}\partial_{ij}$ without notice.

3.6.1 Equation $E'(n+1, m+1; \alpha')$

The equation $E'(n+1, m+1; \alpha')$ is given by

$$
E'(n+1, m+1; \alpha')
$$

$$
\begin{cases}
(1) & \displaystyle\sum_{i=0}^{n} x_{ij}\frac{\partial F}{\partial x_{ij}} = \alpha_j F, \qquad n+1 \le j \le m, \\[2ex]
(2) & \displaystyle\sum_{j=n+1}^{m} x_{0j}\frac{\partial F}{\partial x_{0j}} = -(1+\alpha_0)F, \\[2ex]
& \displaystyle\sum_{j=n+1}^{m} x_{ij}\frac{\partial F}{\partial x_{ij}} = -\alpha_i F, \quad 1 \le i \le n, \\[2ex]
(3) & \displaystyle\frac{\partial^2 F}{\partial x_{ij}\partial x_{pq}} = \frac{\partial^2 F}{\partial x_{iq}\partial x_{pj}}, \quad 0 \le i, p \le n, \quad n+1 \le j, q \le m,
\end{cases}
\tag{3.65}
$$

$$
\text{where } \sum_{j=0}^{m} \alpha_j + 1 = 0.
$$

By the form of the Wronskian calculated in § 3.5, letting a general solution of (3.65) be F, one may expect that the partial derivatives of F besides

$$
F, \partial_{i_1 k_1}\cdots\partial_{i_p k_p}F,
\tag{3.66}
$$

$$
\text{where } 1 \le i_1 < \cdots < i_p \le n, \quad n+1 \le k_1 < \cdots < k_p \le m-1
$$

can be expressed as linear combinations of the members in (3.66) over the rational function field $\mathbb{C}(x) = \mathbb{C}(x_{ij} \mid 0 \le i \le n, n+1 \le j \le m)$. Once this is proved, it apparently follows that the rank of $E(n+1, m+1; \alpha)$ is at most $\binom{m-1}{n}$. As in the case of Wronskian, we start from some important examples.

3.6.2 Equation $E'(2,4;\alpha')$

For easy memorization, we write the following table:

Table 6.1

$$\begin{pmatrix} x_{02} & x_{03} \\ x_{12} & x_{13} \end{pmatrix} \quad \begin{matrix} -(\alpha_0 + 1) \\ -\alpha_1, \end{matrix} \qquad \alpha_0 + \alpha_1 + \alpha_2 + \alpha_3 + 1 = 0$$

$$\alpha_2 \quad \alpha_3$$

The meaning of this table is as follows:

1. From each column vector,

$$(\vartheta_{02} + \vartheta_{12})F = \alpha_2 F, \quad (\vartheta_{03} + \vartheta_{13})F = \alpha_3 F. \tag{3.67}$$

2. From each row vector,

$$(\vartheta_{02} + \vartheta_{03})F = -(\alpha_0 + 1)F, \quad (\vartheta_{12} + \vartheta_{13})F = -\alpha_1 F. \tag{3.68}$$

3. From the 2×2 minor,

$$(\partial_{02}\partial_{13} - \partial_{03}\partial_{12})F = 0. \tag{3.69}$$

By simple computation, we have

$$\vartheta_{02}F = -\vartheta_{12}F + \alpha_2 F,$$

$$\vartheta_{03}F = \vartheta_{12}F + (\alpha_1 + \alpha_3)F = \vartheta_{12}F - (\alpha_0 + \alpha_2 + 1)F, \tag{3.70}$$

$$\vartheta_{13}F = -\vartheta_{12}F - \alpha_1 F.$$

In this case, as (3.66) consists of $F, \partial_{12}F$, it is sufficient to calculate the partial derivatives of $\partial_{12}F$. As $\vartheta_{02}, \vartheta_{03}, \vartheta_{12}, \vartheta_{13}$ are commutative, it makes sense to consider the determinant in the commutative ring $\mathbb{C}[\vartheta_{02}, \vartheta_{03}, \vartheta_{12}, \vartheta_{13}]$. Writing (3.67) in the matrix form, we have

$$\begin{pmatrix} \vartheta_{02} & \vartheta_{12} \\ \vartheta_{03} & \vartheta_{13} \end{pmatrix} \begin{pmatrix} F \\ F \end{pmatrix} = \begin{pmatrix} \alpha_2 F \\ \alpha_3 F \end{pmatrix},$$

and multiplying by the cofactor matrix

$$\begin{pmatrix} \vartheta_{13} & -\vartheta_{12} \\ -\vartheta_{03} & \vartheta_{02} \end{pmatrix}$$

of the matrix in the left-hand side from the left, we obtain

$$\det(\vartheta_{ij}) \cdot \begin{pmatrix} F \\ F \end{pmatrix} = \begin{pmatrix} \alpha_2 \vartheta_{13} F - \alpha_3 \vartheta_{12} F \\ -\alpha_2 \vartheta_{03} F + \alpha_3 \vartheta_{02} F \end{pmatrix}.$$

On the other hand,

$$\det(\vartheta_{ij})F = (\vartheta_{02}\vartheta_{13} - \vartheta_{03}\vartheta_{12})F$$
$$= (x_{02}x_{13}\partial_{02}\partial_{13} - x_{03}x_{12}\partial_{03}\partial_{12})F$$

together with (3.69) implies

$$\det(\vartheta_{ij})F = (x_{02}x_{13} - x_{03}x_{12})\partial_{03}\partial_{12}F$$
$$= x \begin{pmatrix} 0 & 1 \\ 2 & 3 \end{pmatrix} \partial_{03}\partial_{12}F.$$

Hence, we obtain

$$\partial_{03}\partial_{12}F = \frac{1}{x \begin{pmatrix} 0 & 1 \\ 2 & 3 \end{pmatrix}} \{\alpha_2 \vartheta_{13} F - \alpha_3 \vartheta_{12} F\},$$

and by (3.70), the right-hand side can be rewritten as a linear combination of F and $\partial_{12}F$ over $\mathbb{C}(x)$. For simplicity, setting

$$S := \mathbb{C}(x)F + \mathbb{C}(x)\partial_{12}F,$$

the above fact can be expressed as $\vartheta_{03}\vartheta_{12}F \in S$. Hence, by the second formula of (3.70), we have $\vartheta_{12}^2 F + (\alpha_1 + \alpha_3)\vartheta_{12}F \in S$ which implies $\vartheta_{12}^2 F \in S$. Next, by taking $\vartheta_{12}^2 = x_{12}\partial_{12}^2 + x_{12}\partial_{12}$ into account, we have $\partial_{12}^2 F \in S$. Again, by (3.70), we obtain $\vartheta_{02}\vartheta_{12}F$, $\vartheta_{13}\vartheta_{12}F \in S$. Hence, by simple computation, we obtain $\partial_{ij}\partial_{12}F \in S$. Thus, we have rank $E'(2,4;\alpha') \leq 2$. On the other hand, since $E'(2,4;\alpha')$ admits two linearly independent integral solutions, we showed rigorously that the rank of the system of equations $E'(2,4;\alpha')$ is 2.

3.6.3 Equation $E'(2, m+1; \alpha')$

As in § 3.6.2, let us start from

Table 6.2

$$\vartheta = \begin{pmatrix} \vartheta_{02} & \vartheta_{03} & \cdots & \vartheta_{0m} \\ \vartheta_{12} & \vartheta_{13} & \cdots & \vartheta_{1m} \end{pmatrix} \begin{matrix} -(\alpha_0 + 1) \\ -\alpha_1, \end{matrix} \qquad \sum_{j=0}^{m} \alpha_j + 1 = 0.$$
$$\quad \alpha_2 \;\; \alpha_3 \qquad\quad \alpha_m$$

From this table, we obtain

$$
\begin{cases}
(\vartheta_{02} + \vartheta_{12})F = \alpha_2 F, \\
\quad\vdots \qquad\qquad \vdots \\
(\vartheta_{0m} + \vartheta_{1m})F = \alpha_m F.
\end{cases}
\tag{3.71}
$$

$$
\begin{cases}
\displaystyle\sum_{j=2}^{m} \vartheta_{0j}F = -(\alpha_0 + 1)F, \\
\displaystyle\sum_{j=2}^{m} \vartheta_{1j}F = -\alpha_1 F.
\end{cases}
\tag{3.72}
$$

$$
(\partial_{0j}\partial_{1k} - \partial_{0k}\partial_{1j})F = 0, \quad 2 \le j,\ k \le m.
\tag{3.73}
$$

By (3.71) and (3.72), we have

$$
\begin{cases}
\displaystyle\vartheta_{02}F = \sum_{j=3}^{m} \vartheta_{1j}F - \Big(1 + \alpha_0 + \sum_{j=3}^{m} \alpha_j\Big)F \\
\qquad = \displaystyle\sum_{j=3}^{m} \vartheta_{1j}F + (\alpha_1 + \alpha_2)F, \\
\vartheta_{0j}F = -\vartheta_{1j}F + \alpha_j F, \quad 3 \le j \le m, \\
\displaystyle\vartheta_{12}F = -\sum_{j=3}^{m} \vartheta_{1j}F - \alpha_1 F.
\end{cases}
\tag{3.74}
$$

Moreover, picking up two formulas from (3.71), we obtain the formula in a way similar to § 3.6.2:

$$
\det\begin{pmatrix} \vartheta_{0j} & \vartheta_{0k} \\ \vartheta_{1j} & \vartheta_{1k} \end{pmatrix}\begin{pmatrix} F \\ F \end{pmatrix} = \begin{pmatrix} \alpha_j\vartheta_{1k}F - \alpha_k\vartheta_{1j}F \\ -\alpha_j\vartheta_{0k}F + \alpha_k\vartheta_{0j}F \end{pmatrix}.
$$

Hence, by (3.73), the above formula becomes

$$
x\begin{pmatrix} 0 & 1 \\ j & k \end{pmatrix}\partial_{0j}\partial_{1k}F = \alpha_j\vartheta_{1k}F - \alpha_k\vartheta_{1j}F.
\tag{3.75}
$$

Setting $\mathbb{C}(x) = \mathbb{C}(x_{02}, \cdots, x_{1m})$ and using the notation

$$
S = \mathbb{C}(x) + \sum_{j=3}^{m} \mathbb{C}(x)\partial_{1j}F,
$$

(3.74) and (3.75) are expressed as

$$
\partial_{ij}F \in S, \quad \partial_{0j}\partial_{1k}F \in S \quad (j \ne k).
\tag{3.76}
$$

For simplicity, we consider $\partial_{13}F$. By (3.76), we have $\partial_{02}\partial_{13}F, \partial_{04}\partial_{13}F, \ldots,$ $\partial_{0m}\partial_{13}F \in S$, and by (3.71), we also have $\partial_{12}\partial_{13}F, \partial_{14}\partial_{13}F, \cdots, \partial_{1m}\partial_{13}F \in S$. By the last result and (3.72), we have

$$\partial_{03}\partial_{13}F, \qquad \partial_{13}^2 F \in S.$$

Similarly for $\partial_{1k}F$, $4 \leq k \leq m$, we can show

$$\partial_{ij}\partial_{1k}F \in S, \quad i = 0, 1, \quad 2 \leq j \leq m,$$

and we see that any partial derivatives of F can be expressed as a linear combination of $F, \partial_{13}F, \cdots, \partial_{1m}F$ over $\mathbb{C}(x)$, which implies

$$\operatorname{rank} E'(2, m+2; \alpha') \leq m - 1.$$

Thus, from this together with the result obtained in § 3.5.3, we can conclude

$$\operatorname{rank} E'(2, m+2; \alpha') = m - 1.$$

3.6.4 Equation $E'(3, 6; \alpha')$

As we saw in the above two examples, if we do not set up the notation cleverly, the description becomes cumbersome. Diverting the notation for minor, we denote a minor of the matrix with coefficients in the commutative ring $\mathbb{C}[\vartheta_{ij} \mid 0 \leq i \leq 2, 3 \leq j \leq 5]$

$$\vartheta = \begin{pmatrix} \vartheta_{03} & \vartheta_{04} & \vartheta_{05} \\ \vartheta_{13} & \vartheta_{14} & \vartheta_{15} \\ \vartheta_{23} & \vartheta_{24} & \vartheta_{25} \end{pmatrix}$$

by

$$\vartheta \begin{pmatrix} 0 & 1 \\ 3 & 4 \end{pmatrix} = \det \begin{pmatrix} \vartheta_{03} & \vartheta_{04} \\ \vartheta_{13} & \vartheta_{14} \end{pmatrix}$$

etc. As the results obtained in § 3.5.4 suggest, we set

$$S = \mathbb{C}(x)F + \sum_{\substack{1 \leq i \leq 2 \\ 4 \leq j \leq 5}} \mathbb{C}(x)\partial_{ij}F + \mathbb{C}(x)\partial_{14}\partial_{25}F.$$

First, we start from the following table:

Table 6.3

$$\vartheta = \begin{pmatrix} \vartheta_{03} & \vartheta_{04} & \vartheta_{05} \\ \vartheta_{13} & \vartheta_{14} & \vartheta_{15} \\ \vartheta_{23} & \vartheta_{24} & \vartheta_{25} \end{pmatrix} \begin{matrix} -(\alpha_0+1) \\ -\alpha_1 \\ -\alpha_2 \end{matrix} \qquad \sum_{j=0}^{5} \alpha_j + 1 = 0.$$

$$\alpha_3 \quad \alpha_4 \quad \alpha_5$$

$$\begin{cases} (\vartheta_{03} + \vartheta_{04} + \vartheta_{05})F = -(\alpha_0+1)F, \\ (\vartheta_{13} + \vartheta_{14} + \vartheta_{15})F = -\alpha_1 F, \\ (\vartheta_{23} + \vartheta_{24} + \vartheta_{25})F = -\alpha_2 F, \end{cases} \tag{3.77}$$

$$\sum_{i=0}^{2} \vartheta_{ij}F = \alpha_j F, \quad 3 \le j \le 5, \tag{3.78}$$

$$\partial_{ij}\partial_{pq}F = \partial_{iq}\partial_{pj}F, \quad 0 \le i,\, p \le 2, \quad 3 \le j,\, q \le 5. \tag{3.79}$$

By (3.77) and (3.78), we have $\partial_{ij}F \in S$. Next, for the 2×2 minor $\vartheta\begin{pmatrix} i_1 & i_2 \\ j_1 & j_2 \end{pmatrix}$ of ϑ, let us show

$$\vartheta \begin{pmatrix} i_1 & i_2 \\ j_1 & j_2 \end{pmatrix} F \in S. \tag{3.80}$$

First, one can show

$$\vartheta \begin{pmatrix} 1 & 2 \\ 4 & 5 \end{pmatrix} F = x \begin{pmatrix} 1 & 2 \\ 4 & 5 \end{pmatrix} \partial_{14}\partial_{25} F \in S.$$

It is sufficient to show that the other minors are expressed as $\mathbb{C}$-linear combinations of $\vartheta\begin{pmatrix} 1 & 2 \\ 4 & 5 \end{pmatrix}F$ and $\vartheta_{ij}F$. By (3.78), we have

$$\vartheta \begin{pmatrix} 0 & i \\ j & k \end{pmatrix} F = -\vartheta \begin{pmatrix} i' & i \\ j & k \end{pmatrix} F + \alpha_j \vartheta_{ik}F - \alpha_k \vartheta_{ij}F$$

$$\text{where} \quad \{0, i, i'\} = \{0, 1, 2\}.$$

This formula means that as an operator acting on F, the action of the 2×2 minor of the matrix ϑ containing the first row can be expressed as a $\mathbb{C}$-linear combination of 2×2 minors, not containing the first row, and ϑ_{ij}'s. Similarly, the action of a 2×2 minor of ϑ containing the first column can be expressed as a $\mathbb{C}$-linear combination of 2×2 minors, not containing the first column, and ϑ_{ij}'s. Hence, (3.80) is proved. Now, writing (3.77) in matrix form

$$\vartheta \begin{pmatrix} F \\ F \\ F \end{pmatrix} = \begin{pmatrix} -(\alpha_0 + 1)F \\ -\alpha_1 F \\ -\alpha_2 F \end{pmatrix},$$

and multiplying the above formula by the cofactor matrix Θ of ϑ from the left, we have

$$\det \vartheta \begin{pmatrix} F \\ F \\ F \end{pmatrix} = \Theta \begin{pmatrix} -(\alpha_0 + 1)F \\ -\alpha_1 F \\ -\alpha_2 F \end{pmatrix}. \tag{3.81}$$

Since we have

$$\det \vartheta = x \begin{pmatrix} 0 & 1 & 2 \\ 3 & 4 & 5 \end{pmatrix} \partial_{03} \partial_{14} \partial_{25} F,$$

from (3.80), it follows that each component of the right-hand side of (3.81) belongs to S. Hence, we showed $\partial_{03} \partial_{14} \partial_{25} F \in S$. By (3.80), the above formula and (3.79), we have

$$\partial_{i_1 j_1} \partial_{i_2 j_2} F \in S, \quad 0 \le i_1 < i_2 \le 2, \quad 3 \le j_1 < j_2 \le 5, \tag{3.82}$$

$$\partial_{0 j_0} \partial_{1 j_1} \partial_{2 j_2} F \in S, \quad \{j_0, j_1, j_2\} = \{3, 4, 5\}.$$

For simplicity we consider $\partial_{14} F$. The second-order partial derivatives that are not contained in (3.82) can be obtained by applying $\vartheta_{i,j}$, which appears either in the second column or the second row, to $\partial_{14} F$, i.e., they are given by $\partial_{13} \partial_{14} F$, $\partial_{14}^2 F$, $\partial_{15} \partial_{14} F$, $\partial_{04} \partial_{14} F$, $\partial_{24} \partial_{14} F$. For example, we have

$$\partial_{13} \partial_{14} F = \frac{1}{x_{13} x_{14}} \vartheta_{13} \vartheta_{14} F$$

$$= \frac{1}{x_{13} x_{14}} \{-\vartheta_{03} - \vartheta_{23} + \alpha_3\} \vartheta_{14} F \in S \quad \text{(by (3.78))}.$$

Similarly, we can show $\partial_{04} \partial_{14} F$, $\partial_{15} \partial_{14} F$, $\partial_{24} \partial_{14} F \in S$. On the other hand, adding three formulas in (3.78), we have

$$\vartheta_{14} F = - \sum_{(i,j) \ne (1,4)} \vartheta_{ij} F + (\alpha_3 + \alpha_4 + \alpha_5) F,$$

applying ∂_{14} to this formula from the left, we obtain

$$\partial_{14}^2 F = \frac{1}{x_{14}} \left\{ - \sum_{(i,j) \ne (1,4)} x_{ij} \partial_{ij} \partial_{14} F + (\alpha_3 + \alpha_4 + \alpha_5 - 1) \partial_{14} F \right\}.$$

The right-hand side can be shown to be in S as before. A similar argument shows $\partial_{1j} \partial_{1i} F \in S$ etc. Finally, we should show $\partial_{ij} \partial_{14} \partial_{25} F \in S$. For simplicity, let us show $\partial_{04} \partial_{14} \partial_{25} F \in S$. The other cases can be similarly proved.

First, considering the 2×2 minor $\vartheta\left(\begin{smallmatrix} 1 & 2 \\ 4 & 5 \end{smallmatrix}\right)$ of the matrix ϑ obtained by removing the first row and the second column, by the same argument as before, we have

$$\vartheta\begin{pmatrix} 1 & 2 \\ 4 & 5 \end{pmatrix} F = -\vartheta\begin{pmatrix} 1 & 2 \\ 3 & 5 \end{pmatrix} F - \alpha_1 \vartheta_{25} F + \alpha_2 \vartheta_{15} F$$

which implies

$$\partial_{14}\partial_{25} F = -\frac{1}{x\begin{pmatrix} 1 & 2 \\ 4 & 5 \end{pmatrix}} \left\{ x\begin{pmatrix} 1 & 2 \\ 3 & 5 \end{pmatrix} \partial_{13}\partial_{25} F + \alpha_1 \vartheta_{25} F - \alpha_2 \vartheta_{15} F \right\}.$$

Differentiating this formula by ∂_{04}, and using the second formula of (3.82), we obtain $\partial_{04}\partial_{14}\partial_{25} F \in S$. Thus, any partial derivative of F can be expressed as a linear combination of F, $\partial_{14} F$, $\partial_{15} F$, $\partial_{24} F$, $\partial_{25} F$, $\partial_{14}\partial_{25} F$ over $\mathbb{C}(x)$. In particular, by the result obtained in § 3.5.4, we conclude rank $E'(3, 6; \alpha') = 6$.

This argument can be extended to general cases. In § 3.6.5, we verify it.

3.6.5 Equation $E'(n + 1, m + 1; \alpha')$

We start from the following table:

$$\vartheta = \begin{pmatrix} \vartheta_{0,n+1} & \cdots & \cdots & \vartheta_{0,m} \\ \vdots & & & \vdots \\ \vdots & & & \vdots \\ \vartheta_{n,n+1} & \cdots & \cdots & \vartheta_{n,m} \end{pmatrix} \quad \begin{matrix} -\beta_0 = -(\alpha_0 + 1) \\ -\beta_1 = -\alpha_1 \\ \\ -\beta_n = -\alpha_n \end{matrix} \qquad \sum_{j=0}^{m} \alpha_j + 1 = 0.$$
$$\qquad\qquad \alpha_{n+1} \quad \cdots\cdots \quad \alpha_m$$

From the above table, let us write the differential equations:

$$\sum_{j=n+1}^{m} \vartheta_{ij} F = -\beta_i F, \quad 0 \le i \le n, \tag{3.83}$$

$$\sum_{i=0}^{n} \vartheta_{ij} F = \alpha_j F, \quad n+1 \le j \le m, \tag{3.84}$$

$$(\partial_{ij}\partial_{pq} - \partial_{iq}\partial_{pj}) F = 0, \quad 0 \le i, p \le n, \quad n+1 \le j, q \le m. \tag{3.85}$$

Since the column vectors and the row vectors in ϑ play the same role, for simplicity, we assume $n + 1 \le m - n$, and following the arguments developed in § 3.6.4, let us discuss this in two steps. First, we define the notation as follows:

$$I = \{i_1, \cdots, i_p\}, \quad 0 \le i_1 < \cdots < i_p \le n,$$

$$K = \{k_1, \cdots, k_p\}, \quad n+1 \le k_1 < \cdots < k_p \le m,$$

$$L = \{0, 1, \cdots, n\} \backslash I = \{l_0, l_1, \cdots, l_{n-p}\}, \quad 0 \le l_0 < \cdots < l_{n-p} \le n,$$

$$H = \{n+1, \cdots, m\} \backslash K = \{h_1, \cdots, h_{m-n-p}\},$$

$$n+1 \le h_1 < \cdots < h_{m-n-p} \le m,$$

$$\vartheta \binom{I}{K} = \det \begin{pmatrix} \vartheta_{i_1 k_1} & \cdots\cdots & \vartheta_{i_1 k_p} \\ \vdots & & \vdots \\ \vartheta_{i_p k_1} & \cdots\cdots & \vartheta_{i_p k_p} \end{pmatrix} \in \mathbb{C}[\vartheta_{ij}].$$

We show the following proposition.

Proposition 3.1. *Let F be a general solution of (3.83), (3.84), (3.85) and $\mathbb{C}(x)$ be the field of the rational functions of x_{ij}, $0 \le i \le n$, $n+1 \le j \le m$.*

(1) $S = \mathbb{C}(x)F + \sum\limits_{p=1}^{n} \sum\limits_{\substack{I,K \\ |I|=|K|=p}} \mathbb{C}(x)\vartheta\binom{I}{K}F$ *is stable under ∂_{ij}, $0 \le i \le n$,*

$n+1 \le j \le m$.

(2) *Taking all $\vartheta\binom{I}{K}F$, $|I| = |K| = p$, $p \le n$, $I \subset \{1, \cdots, n\}$, $K \subset \{n+2, \cdots, m\}$, corresponding to the minors of the matrix obtained by removing the first column and row from the matrix ϑ*

$$\begin{pmatrix} \vartheta_{1,n+2} & \cdots\cdots & \vartheta_{1m} \\ \vdots & & \vdots \\ \vartheta_{n,n+2} & \cdots\cdots & \vartheta_{nm} \end{pmatrix},$$

and F itself, any element of S in (1) can be expressed as a $\mathbb{C}(x)$-linear combination of them.

For the proof of this proposition, let us prepare some lemmata.

Lemma 3.8. *For $|K| = n+1$, $K = \{k_0, \ldots, k_n\}$, we have:*

(1) $\vartheta \begin{pmatrix} 0 & 1 & \cdots & n \\ & K & \end{pmatrix} F \in S.$

(2) $\vartheta \begin{pmatrix} 0 & 1 & \cdots & n \\ & K & \end{pmatrix} F = x \begin{pmatrix} 0 & 1 & \cdots & n \\ & K & \end{pmatrix} \partial_{0k_0} \cdots \partial_{nk_n} F$

hence, $\partial_{0k_0} \cdots \partial_{nk_n} F \in S$.

(3) *For $p \le n$,*

$$\vartheta \begin{pmatrix} i_1 & \cdots & i_p \\ k_1 & \cdots & k_p \end{pmatrix} F = x \begin{pmatrix} i_1 & \cdots & i_p \\ k_1 & \cdots & k_p \end{pmatrix} \partial_{i_1 k_1} \cdots \partial_{i_p k_p} F$$

$$\text{hence,} \quad \partial_{i_1 k_1} \cdots \partial_{i_p k_p} F \in S.$$

Proof. Picking up the column vectors in (3.84) corresponding to K, it takes the following matrix form

$$\begin{pmatrix} \vartheta_{0k_0} & \cdots & \vartheta_{nk_0} \\ \vdots & & \vdots \\ \vartheta_{0k_n} & \cdots & \vartheta_{nk_n} \end{pmatrix} \begin{pmatrix} F \\ \vdots \\ F \end{pmatrix} = \begin{pmatrix} \alpha_{k_0} F \\ \vdots \\ \alpha_{k_n} F \end{pmatrix}.$$

Multiplying by the cofactor matrix of the matrix in the left-hand side from the left, we obtain (1). Next, we show (2). By the definition of determinant, we have

$$\vartheta \begin{pmatrix} 0 & 1 & \cdots & n \\ & & K & \end{pmatrix} F = \sum_{\sigma \in \mathfrak{S}_{n+1}} \text{sgn} \begin{pmatrix} 0 & 1 & \cdots & n \\ k_{\sigma(0)} & \cdots & & k_{\sigma(n)} \end{pmatrix} \vartheta_{0k_{\sigma(0)}} \cdots \vartheta_{nk_{\sigma(n)}}$$

$$= \sum_{\sigma \in \mathfrak{S}_{n+1}} \text{sgn} \begin{pmatrix} 0 & 1 & \cdots & n \\ k_{\sigma(0)} & \cdots & & k_{\sigma(n)} \end{pmatrix} x_{0k_{\sigma(0)}} \cdots x_{nk_{\sigma(n)}} \cdot \partial_{0k_{\sigma(0)}} \cdots \partial_{nk_{\sigma(n)}} F,$$

but since (3.85) implies

$$\partial_{0k_{\sigma(0)}} \cdots \partial_{nk_{\sigma(n)}} F = \partial_{0k_0} \cdots \partial_{nk_n} F,$$

the above formula implies the first half of (2). This together with (1) implies the latter half. (3) can be proved similarly.

Next, to compute $\partial_{\iota\kappa} \vartheta \binom{I}{K} F$, $|I| = |K| \leq n$, we prepare two lemmata.

Lemma 3.9 (Operations with repect to rows).

$$\vartheta \begin{pmatrix} i_1 & i_2 & \cdots & i_p \\ k_1 & k_2 & \cdots & k_p \end{pmatrix} F = -\sum_{l \in L} \vartheta \begin{pmatrix} l & i_2 & \cdots & i_p \\ k_1 & k_2 & \cdots & k_p \end{pmatrix} F$$

$$+ \sum_{\lambda=1}^{p} (-1)^{\lambda+1} \alpha_{k_\lambda} \vartheta \begin{pmatrix} i_2 & \cdots \cdots \cdots & i_p \\ k_1 & \cdots \widehat{k_\lambda} \cdots & k_p \end{pmatrix} F.$$

Proof. This is proved by the linearity of determinant and (3.84).

Similarly, by (3.83), we can prove the following lemma.

Lemma 3.10 (Operations with respect to columns).

$$\vartheta \begin{pmatrix} i_1 & i_2 & \cdots & i_p \\ k_1 & k_2 & \cdots & k_p \end{pmatrix} F = -\sum_{h \in H} \vartheta \begin{pmatrix} i_1 & i_2 & \cdots & i_p \\ h & k_2 & \cdots & k_p \end{pmatrix} F$$

$$- \sum_{\lambda=1}^{p} (-1)^{\lambda+1} \beta_{i_\lambda} \vartheta \begin{pmatrix} i_1 & \cdots & \widehat{i_\lambda} & \cdots & i_p \\ k_2 & \cdots \cdots \cdots & & k_p \end{pmatrix} F.$$

With these preparations, let us prove Proposition 3.1.

Proof (of Proposition 3.1). Let us show (1) by induction on $p = |I|$. For $p = 0$, the assertion is evident by the choice of S. Suppose that, for $|I| \leq p - 1$, we have shown $\partial_{\iota\kappa}\vartheta\binom{I}{K}F \in S$. For multi-indices I, K with $|I| = |K| = p$, to show $\partial_{\iota\kappa}\vartheta\binom{I}{K}F \in S$, we verify it by the following steps:

(i)   when $\iota \notin I$ and $\kappa \notin K$,
(ii)   when only one of $\iota \in I$, $\kappa \in K$ holds,
(iii)   when $\iota \in I$ and $\kappa \in K$ holds.

For (i), Lemma 3.8 (3) implies

$$\partial_{\iota\kappa}\vartheta\begin{pmatrix} I \\ K \end{pmatrix} F = x\begin{pmatrix} I \\ K \end{pmatrix} \partial_{\iota\kappa}\partial_{i_1 k_1} \cdots \partial_{i_p k_p} F$$

$$= \frac{x\begin{pmatrix} I \\ K \end{pmatrix}}{x\begin{pmatrix} \iota & I \\ \kappa & K \end{pmatrix}} \vartheta\begin{pmatrix} \iota & I \\ \kappa & K \end{pmatrix} F.$$

Hence, if $p \leq n - 1$, the assertion follows by the definition of S, and if $p = n$, (1) of the same lemma implies $\partial_{\iota\kappa}\vartheta\binom{I}{K}F \in S$.

For (ii), we show the case $\iota = i_1 \in I$, but the other cases can be treated similarly by Lemma 3.10. By Lemma 3.9, we have

$$\partial_{\iota_1\kappa}\vartheta\begin{pmatrix} I \\ K \end{pmatrix} F = -\sum_{l\in L} \partial_{i_1\kappa}\vartheta\begin{pmatrix} l & i_2 & \cdots & i_p \\ k_1 & k_2 & \cdots & k_p \end{pmatrix} F \tag{3.86}$$

$$+ \sum_{\lambda=1}^{p}(-1)^{\lambda+1}\alpha_{k_\lambda}\partial_{i_1\kappa}\vartheta\begin{pmatrix} i_2 & \cdots\cdots\cdots & i_p \\ k_1 & \cdots & \widehat{k_\lambda} & \cdots & k_p \end{pmatrix} F.$$

By assumption, $\kappa \notin K$, and since $L = \{0, \cdots, n\}\backslash I$, it follows from the result for (i) that the first term in the right-hand side of (3.86) is contained in S. On the other hand, the second term is also contained in S by induction hypothesis, hence, we obtain $\partial_{\iota\kappa}\vartheta\binom{I}{K}F \in S$.

For (iii), we assume $\iota = i_1$, $\kappa = k_1$ for simplicity. The other cases can be reduced to this case by the skew-symmetry of determinant. By repeated use of Lemma 3.9 and 3.10, we obtain the following formula: setting

$$I' := I\backslash\{i_1\}, \quad K' := K\backslash\{k_1\},$$

$$\vartheta\begin{pmatrix} I \\ K \end{pmatrix} F = \sum_{l \in L,\, h \in H} \vartheta\begin{pmatrix} l & I' \\ h & K' \end{pmatrix} F + \left(\sum_{k \in K} \alpha_k + \sum_{l \in L} \beta_l \right) \vartheta\begin{pmatrix} I' \\ K' \end{pmatrix} F$$

$$+ \sum_{h \in H} \sum_{\mu=2}^{p} (-1)^\mu \alpha_{k_\mu} \vartheta\begin{pmatrix} & & I' & & \\ h & k_2 & \cdots & \widehat{k_\mu} & \cdots & k_p \end{pmatrix} F \qquad (3.87)$$

$$+ \sum_{l \in L} \sum_{\lambda=2}^{p} (-1)^{\lambda+1} \beta_{i_\lambda} \vartheta\begin{pmatrix} l & i_2 & \cdots & \widehat{i_\lambda} & \cdots & i_p \\ & & & K' & & \end{pmatrix} F$$

$$+ \sum_{\lambda,\mu=2}^{p} (-1)^{\lambda+\mu} \alpha_{k_\mu} \beta_{i_\lambda} \vartheta\begin{pmatrix} i_2 & \cdots & \widehat{i_\lambda} & \cdots & i_p \\ k_2 & \cdots & \widehat{k_\mu} & \cdots & k_p \end{pmatrix} F.$$

Differentiating both sides by $\partial_{i_1 k_1}$, it follows from the result for (i) that $\partial_{i_1 k_1} \vartheta\begin{pmatrix} l & I' \\ h & K' \end{pmatrix} F \in S$. The second term to the fifth term in the right-hand side of (3.87) are all sums of $(p-1) \times (p-1)$ and $(p-2) \times (p-2)$ minors; by induction hypothesis, these terms differentiated by $\partial_{i_1 k_1}$ are contained in S. Combining these, we see that $\partial_{i_1 k_1} \vartheta\begin{pmatrix} I \\ K \end{pmatrix} F \in S$.

Next, let us show (2). Let $p \leq n$ and take a $p \times p$ minor $\vartheta\begin{pmatrix} i_1 \cdots i_p \\ k_1 \cdots k_p \end{pmatrix}$ of ϑ. If $i_1 = 0, k_1 \neq 0$, Lemma 3.9 implies

$$\vartheta\begin{pmatrix} 0 & I' \\ & K \end{pmatrix} F = -\sum_{l \in L} \vartheta\begin{pmatrix} l & I' \\ & K \end{pmatrix} F$$

$$+ \sum_{\mu=1}^{p} (-1)^{\mu+1} \alpha_{k_\mu} \vartheta\begin{pmatrix} & & I' & & \\ k_1 & \cdots & \widehat{k_\mu} & \cdots & k_p \end{pmatrix} F.$$

As I' and L does not contain 0, a $p \times p$ minor $\vartheta\begin{pmatrix} I \\ K \end{pmatrix} F$ of ϑ containing the first row can be expressed as the sum of minors of degree at most p not containing the first row. Similarly, by Lemma 3.10, a minor $\vartheta\begin{pmatrix} I \\ K \end{pmatrix} F$ of degree p containing the first column can be expressed as the sum of minors of degree at most p not containing the first column. Hence, we conclude (2).

Now, Proposition 3.1 (2) asserts that it suffices to take $\sum_{p=0}^{n} \binom{n}{p} \binom{m-n-1}{p}$ $\vartheta\begin{pmatrix} I \\ K \end{pmatrix} F$ to generate S over $\mathbb{C}(x)$. On the other hand, this sum is $\binom{m-1}{n}$ by Lemma 3.5. Hence, we have

$$\text{rank } E'(n+1, m+1; \alpha') \leq \binom{m-1}{n}.$$

This together with Theorem 3.4 implies the following theorem:

Theorem 3.8. *The hypergeometric differential equation $E'(n+1, m+1; \alpha')$, under the assumption $\alpha_j \notin \mathbb{Z}$, $0 \leq j \leq m$, admits $\binom{m-1}{n}$ linearly independent solutions. In particular, for these solutions, we can take*

$$\int_{\sigma_\nu} \prod_{j=0}^{m} (x_{0j} + x_{1j}u_1 + \cdots + x_{nj}u_n)^{\alpha_j} \, du_1 \wedge \cdots \wedge du_n, \quad 1 \le \nu \le \binom{m-1}{n}.$$

Here, σ_ν takes over a basis of $H_n(M, \mathcal{L}_\omega^\vee)$.

Remark 3.10. In [G-Z-K], the equality rank $E'(n+1, m+1; \alpha') = \binom{m-1}{n}$ was shown by computing the multiplicity of a $\mathcal{D}$-module.

3.7 Duality of $E(n+1, m+1; \alpha)$

In Theorem 3.3, we have seen that a hypergeometric series of type $(n+1, m+1)$ admits an n-fold integral representation and an $(m-n-1)$-fold integral representation. Here, we explain a relation between this fact and the duality of Grassmannians. To make the duality clear, we change notation and we consider $E(k+1, k+l+2; \alpha)$.

3.7.1 Duality of Equations

Let us rewrite what we have considered in § 3.4.3 by changing symbols.
    A matrix

$$(1_{k+1}, x') = \begin{pmatrix} 1 & & & & x_{00} & \cdots & x_{0l} \\ & 1 & & & \vdots & & \vdots \\ & & \ddots & & \vdots & & \vdots \\ & & & 1 & x_{k0} & \cdots & x_{kl} \end{pmatrix}$$

defines the integral

$$F(x'; \alpha) = \int_\sigma u_1^{\alpha_1} \cdots u_k^{\alpha_k} \prod_{j=0}^{l} (x_{0j} + x_{1j}u_1 + \cdots + x_{kj}u_k)^{\alpha_{k+1+j}} \qquad (3.88)$$

$$\cdot \, du_1 \wedge \cdots \wedge du_k,$$

which satisfies the system of differential equations

$E'(k+1, k+l+2; \alpha_0, \ldots, \alpha_{k+l+1})$

$$\begin{cases} (1) \quad \displaystyle\sum_{i=0}^{k} x_{ij}\frac{\partial F}{\partial x_{ij}} = \alpha_{j+k+1}F, \quad 0 \le j \le l, \\ (2) \quad \displaystyle\sum_{j=0}^{l} x_{ij}\frac{\partial F}{\partial x_{ij}} = -(1+\alpha_i)F, \, 0 \le i \le k, \qquad (3.89) \\ (3) \quad \dfrac{\partial^2 F}{\partial x_{ij}\partial x_{pq}} = \dfrac{\partial^2 F}{\partial x_{iq}\partial x_{pj}}, \quad 0 \le i, \, p \le k, \quad 0 \le j, \, q \le l. \end{cases}$$

Let us describe the relations between the parameters in (3.88) and in (3.89) in the matrix form:

$$\begin{pmatrix} 1 & & & x_{00} & \cdots\cdots & x_{0l} \\ & 1 & & \vdots & & \vdots \\ & & \ddots & \vdots & & \vdots \\ & & & 1 & x_{k0} & \cdots\cdots & x_{kl} \end{pmatrix}, \quad \sum_{j=0}^{k+l+1} \alpha_j + k + 1 = 0 \qquad (3.90)$$

$$\alpha_0, \alpha_1, \ldots, \alpha_k, \; \alpha_{k+1}, \ldots, \alpha_{k+l+1}$$

$$\begin{pmatrix} x_{00} & \cdots\cdots & x_{0l} \\ \vdots & & \vdots \\ x_{k0} & \cdots\cdots & x_{kl} \end{pmatrix} \begin{matrix} -(1+\alpha_0) \\ \vdots \\ -(1+\alpha_k) \end{matrix} \qquad (3.91)$$

$$\alpha_{k+1}, \; \cdots\cdots, \; \alpha_{k+l+1}$$

The differential equations (3.89) are partial differential equations with respect to the variables x_{ij}, $0 \le i \le k$, $0 \le j \le l$, we can relate these with another matrix expressed in terms of the transpose of (3.91)

$$\begin{pmatrix} 1 & & & & x_{00} & \cdots\cdots & x_{k0} \\ & 1 & & & \vdots & & \vdots \\ & & \ddots & & \vdots & & \vdots \\ & & & \ddots & \vdots & & \vdots \\ & & & & 1 & x_{0l} & \cdots\cdots & x_{kl} \end{pmatrix}, \quad \sum_{j=0}^{k+l+1}(-1-\alpha_j) + l + 1 = 0$$

$$-1 - \alpha_{k+1}, \cdots, -1 - \alpha_{k+l+1}, -1 - \alpha_0, \cdots, -1 - \alpha_k$$

$$\begin{pmatrix} x_{00} & \cdots\cdots & x_{k0} \\ \vdots & & \vdots \\ x_{0l} & \cdots\cdots & x_{kl} \end{pmatrix} \begin{matrix} \alpha_{k+1} \\ \vdots \\ \alpha_{k+l+1} \end{matrix} \qquad (3.92)$$

$$-(1+\alpha_0), \; \cdots\cdots, \; -(1+\alpha_k)$$

Here, we remark $\alpha_{k+1} = -[1 + (-1 - \alpha_{k+1})]$ etc. We have

$$E'(l+1, k+l+2; -1-\alpha_{k+1}, \ldots, -1-\alpha_{k+l+1}, -1-\alpha_0, \ldots, -1-\alpha_k)$$

$$\begin{cases} (1) & \displaystyle\sum_{j=0}^{l} x_{ij}\frac{\partial F}{\partial x_{ij}} = -(1+\alpha_i)F, \ 0 \le i \le k, \\[2mm] (2) & \displaystyle\sum_{i=0}^{k} x_{ij}\frac{\partial F}{\partial x_{ij}} = \alpha_{j+k+1}F, \quad 0 \le j \le l, \\[2mm] (3) & \dfrac{\partial^2 F}{\partial x_{ij}\partial x_{pq}} = \dfrac{\partial^2 F}{\partial x_{iq}\partial x_{pj}}, \quad 0 \le i, \ p \le k, \quad 0 \le j, \ q \le l. \end{cases} \tag{3.93}$$

Hence, by the table representing the fact that the integral (3.88) is a solution of (3.89), we see that the integral

$$F({}^t x'; -1-\alpha_{k+1}, \ldots, -1-\alpha_{k+l+1}, -1-\alpha_0, \ldots, -1-\alpha_k) \tag{3.94}$$

$$= \int_\tau u_1^{-(1+\alpha_{k+2})} \cdots u_l^{-(1+\alpha_{k+l+1})}$$

$$\cdot \prod_{i=0}^{k} (x_{i0} + x_{i1}u_1 + \cdots + x_{il}u_l)^{-(1+\alpha_i)} du_1 \wedge \cdots \wedge du_l$$

is a solution of the differential equations (3.93). Let us write the $(l+1) \times (k+l+2)$ matrix associated to the integral, following the order of the parameters, by

$$({}^t x', 1_{l+1}) = \begin{pmatrix} x_{00} & \cdots & x_{k0} & 1 & & \\ \vdots & & \vdots & & \ddots & \\ x_{0l} & \cdots & x_{kl} & & & 1 \end{pmatrix}.$$

$$-(1+\alpha_0), \ \cdots \ -(1+\alpha_k), \ -(1+\alpha_{k+1}), \ \cdots, \ -(1+\alpha_{k+l+1})$$

Summarizing, we have:

Lemma 3.11. *Two systems of hypergeometric differential equations $E'(k+1, k+l+2; \alpha_0, \cdots, \alpha_{k+l+1})$ and $E'(l+1, k+l+2; -1-\alpha_{k+1}, \cdots, -1-\alpha_{k+l+1}, -1-\alpha_0, \cdots, -1-\alpha_k)$ are the same. For each system, the integral (3.88) (resp. (3.94)) provides us the solutions. Hence, this system of differential equations admits a k-fold integral and an l-fold integral as different solutions.*

3.7.2 Duality of Grassmannians

Let $G(k+1, k+l+2)$ be the complex Grassmannian of $k+1$-dimensional subspaces of $\mathbb{C}^{k+l+2}$. Expressing an element of $\mathbb{C}^{k+l+2}$ as a row vector, $(1_{k+1}, x')$

defines a point of $G(k+1, k+l+2)$, and the set $X_{01\cdots k}$ of all points of $G(k+1, k+l+2)$ corresponding to the matrices of such form is an open subset of $G(k+1, k+l+2)$ which clearly satisfies $X_{01\cdots k} \simeq \mathbb{C}^{(k+1)(l+1)}$. For the reason we shall explain later, considering $({}^t x', -1_{l+1})$ instead of $({}^t x', 1_{l+1})$, $({}^t x', -1_{l+1})$ similarly defines a point of $G(l+1, k+l+2)$, and the set of all points $X_{k+1,\cdots,k+l+1} \simeq \mathbb{C}^{(k+1)(l+1)}$ corresponding to the matrices of such form is an open subset of $G(l+1, k+l+2)$.

Here, let us briefly explain the duality between $G(k+1, k+l+2)$ and $G(l+1, k+l+2)$. Introduce the bilinear product on $\mathbb{C}^{k+l+2}$ by

$$x \cdot y = \sum_{j=0}^{k+l+1} x_j y_j, \quad x = (x_0, \ldots, x_{k+l+1}), \quad y = (y_0, \ldots, y_{k+l+1}).$$

Let L be a $(k+1)$-dimensional subspace of $\mathbb{C}^{k+l+2}$ and set

$$L^\perp = \{y \in \mathbb{C}^{k+l+2} | (y, x) = 0 \ \forall x \in L\}.$$

Then, $L^\perp$ is an $(l+1)$-dimensional subspace of $\mathbb{C}^{k+l+2}$ and the map

$$\perp : G(k+1, k+l+2) \longrightarrow G(l+1, k+l+2)$$
$$L \longmapsto L^\perp$$

is defined. This is a biholomorphic map and this relation is called the duality of Grassmannians.

Now, considering $({}^t x', -1_{l+1})$ which is slightly different from $({}^t x', 1_{l+1})$, the corresponding integral

$$\int_\tau \prod_{j=1}^l (-u_j)^{-(1+\alpha_{k+j+1})} \cdot \prod_{i=0}^k (x_{i0} + x_{i1}u_1 + \cdots + x_{il}u_l)^{-(1+\alpha_i)} du^l,$$

$$du^l = du_1 \wedge \cdots \wedge du_l$$

only differs from (3.94) by a scalar factor, hence, it satisfies the same differential equations (3.93). By the way, as

$$(1_{k+1}, x') \begin{pmatrix} x' \\ -1_{l+1} \end{pmatrix} = 0_{k+1, l+1},$$

the image of the point of $G(k+1, k+l+2)$ corresponding to $(1_{k+1}, x')$ by the duality $\perp$ is the point of $G(l+1, k+l+2)$ corresponding to $({}^t x', -1_{l+1})$. Hence, we have shown the following lemma.

Lemma 3.12. *Under the duality map* $\perp : G(k+1, k+l+2) \xrightarrow{\sim} G(l+1, k+l+2)$, $X_{01\cdots k}$ *maps to* $X_{k+1,\cdots,k+l+1}$:

$$\perp(X_{01\cdots k}) = X_{k+1,\cdots,k+l+1}.$$

3.7.3 Duality of Hypergeometric Functions

By Lemma 3.11 and 3.12, we obtain the following theorem:

Theorem 3.9 (Duality of systems of hypergeometric differential equations). *The systems of hypergeometric differential equations $E'(k + 1, k + l + 2; \alpha_0, \cdots, \alpha_{k+l+1})$ and $E'(l + 1, k + l + 2; -1 - \alpha_{k+1}, \cdots, -1 - \alpha_{k+l+1}, -1 - \alpha_0, \cdots, -1 - \alpha_k)$, which are defined over $X_{01\cdots k} = \{[(1_{k+1}, x')]\} \subset G(k+1, k+l+2)$ and $X_{k+1,\cdots,k+l+2} = \{[({}^t x', -1_{l+1})]\} \subset G(l+1, k+l+2)$ respectively that are isomorphic to $\mathbb{C}^{(k+1)(l+1)}$ and map to each other by the duality of Grassmannians, are the same. A solution of each system has an integral representation (3.88), (3.94) associated to the matrix $(1_{k+1}, x')$, $({}^t x', -1_{l+1})$, respectively. Hence, these systems of equations admit the solutions expressed as a k-fold integral and an l-fold integral defined by the correspondence*

$$\perp : (1_{k+1}, x') \longleftrightarrow ({}^t x', -1_{l+1})$$

induced from the duality of Grassmannians.

This fact has been discovered for the first time in [Ge-Gr]. The following proof follows [Kit-Ma].

3.7.4 Duality of Integral Representations

Theorem 3.3 can be regarded as relating two integral solutions more precisely in the above consideration. The hypergeometric series

$$F(\alpha_1, \cdots, \alpha_k, \beta_1, \cdots, \beta_l, \gamma; z), \qquad z = \begin{pmatrix} z_{11} & \cdots & z_{1l} \\ \vdots & & \vdots \\ z_{k1} & \cdots & z_{kl} \end{pmatrix} \in M_{k,l}(\mathbb{C})$$

has a two integral representations, up to Γ-factors,

$$\int_{\Delta^k(\omega_1)} \prod_{i=1}^{k} u_i^{\alpha_i - 1} \cdot \left(1 - \sum_{i=1}^{k} u_i\right)^{\gamma - \sum \alpha_i - 1} \cdot \prod_{j=1}^{l} \left(1 - \sum_{i=1}^{k} z_{ij} u_i\right)^{-\beta_j} d^k u,$$

$$\tag{3.95}$$

$$\int_{\Delta^l(\omega_2)} \prod_{j=1}^{l} u_j^{\beta_j - 1} \cdot \left(1 - \sum_{j=1}^{l} u_j\right)^{\gamma - \sum \beta_j - 1} \cdot \prod_{i=1}^{k} \left(1 - \sum_{j=1}^{l} z_{ij} u_j\right)^{-\alpha_i} d^l u.$$

$$\tag{3.96}$$

Expressing the matrix corresponding to (3.95) as

$$
\begin{pmatrix}
1 & & & 1 & 1 & \cdots & 1 \\
 & -1 & & & -z_{11} & \cdots & -z_{1l} \\
 & & \ddots & & & \vdots & \\
 & 1 & -1 & & -z_{k1} & \cdots & -z_{kl}
\end{pmatrix},
$$
$$
\alpha_0 - 1, \ldots, \alpha_k - 1, \gamma - \sum \alpha_i - 1, -\beta_1, \ldots, -\beta_l
$$

Theorem 3.9 asserts that the integral

$$
\int \prod_{j=1}^{l} (-u_j)^{\beta_j - 1} \cdot \left(1 + \sum_{j=1}^{l} u_j \right)^{-\alpha_0} \cdot \prod_{i=1}^{k} \left(-1 - \sum_{j=1}^{l} z_{ij} u_j \right)^{-\alpha_i} d^l u \quad (3.97)
$$

associated to the matrix

$$
\begin{pmatrix}
1 & -1 & & -1 & -1 & & & \\
1 & -z_{11} & \cdots & -z_{k1} & & -1 & & \\
\vdots & & & & & & \ddots & \\
1 & -z_{1l} & \cdots & -z_{kl} & & & & -1
\end{pmatrix}
$$
$$
-\alpha_0, -\alpha_1, \cdots - \alpha_k, -\gamma + \sum \alpha_i, -1 + \beta_1, \cdots, -1 + \beta_l
$$

corresponds by the duality. Making the coordinate transformation

$$
u_j \longrightarrow -u_j, \qquad 1 \leq j \leq l,
$$

by the identities

$$
\sum_{i=0}^{k} (\alpha_i - 1) + \gamma - \sum_{i=1}^{k} \alpha_i - 1 + \sum_{j=1}^{l} (-\beta_j) + k + 1 = 0
$$

and $\alpha_0 = \sum_{i=0}^{l} \beta_j - \gamma + 1$, we see that (3.96) and (3.97) coincides up to a scalar factor. Hence, if we choose a special k-simplex $\Delta^k(\omega_1)$ and an l-simplex $\Delta^l(\omega_2)$ as integral domains corresponding to them respectively, the duality becomes precise and we see that the corresponding integrals coincide up to a scalar factor.

3.7.5 Example

For Gauss' hypergeometric series, we have the concrete expressions:

$$F(\alpha, \beta, \gamma; z) = \frac{\Gamma(\gamma)}{\Gamma(\gamma - \alpha)\Gamma(\alpha)} \int_0^1 u^{\alpha-1}(1-u)^{\gamma-\alpha-1}(1-zu)^{-\beta}du,$$

$$= \frac{\Gamma(\gamma)}{\Gamma(\gamma - \beta)\Gamma(\beta)} \int_0^1 u^{\beta-1}(1-u)^{\gamma-\beta-1}(1-zu)^{-\alpha}du.$$

The corresponding matrices look as follows:

$$\begin{pmatrix} 1 & 0 & 1 & 1 \\ 0 & 1 & -1 & -z \\ \beta-\gamma & \alpha-1 & \gamma-\alpha-1 & -\beta \end{pmatrix} \longleftrightarrow \begin{pmatrix} 1 & -1 & -1 & 0 \\ 1 & -z & 0 & -1 \\ \gamma-\beta-1 & -\alpha & \alpha-\gamma & \beta-1 \end{pmatrix}$$

$\updownarrow$ change of variable $u \to -u$

$$\begin{pmatrix} 1 & -1 & -1 & 0 \\ -1 & z & 0 & 1 \\ \gamma-\beta-1 & -\alpha & \alpha-\gamma & \beta-1 \end{pmatrix}.$$

Remark 3.11. The relation between $\Delta^k(\omega_1)$ and $\Delta^l(\omega_2)$ stated in § 3.7.4 can be understood as the duality of the twisted homologies corresponding to the integrals (3.95) and (3.96). For detail, see [Kit-Ma].

3.8 Logarithmic Gauss–Manin Connection Associated to an Arrangement of Hyperplanes in General Position

3.8.1 Review of Notation

Using the notation defined in § 3.4, we consider the arrangement of hyperplanes in $\mathbb{P}^n(\mathbb{C})$ defined by the $(n+1) \times (m+1)$ matrix

$$x = \begin{pmatrix} x_{01} & \cdots\cdots & x_{0m} & 1 \\ x_{11} & & x_{1m} & 0 \\ \vdots & & \vdots & \vdots \\ x_{n1} & \cdots\cdots & x_{nm} & 0 \end{pmatrix}.$$

Recall the symbols:

$u = (u_1, \ldots, u_n) \in \mathbb{C}^n, \quad$ for later use, we further set $u_0 = 1$.

$P_j(u) = x_{0j} + x_{1j}u_1 + \cdots + x_{nj}u_n, \quad 1 \le j \le m,$

$$U(u) = \prod_{j=1}^{m} P_j(u)^{\alpha_j}, \quad M = \mathbb{C}^n \setminus \bigcup_{j=1}^{m} \{P_j = 0\},$$

for $J = \{j_1, \ldots, j_p\}, \; 1 \le j_1 < \cdots < j_p \le m - 1,$

$$\varphi\langle J \rangle = \frac{dP_{j_1}}{P_{j_1}} \wedge \cdots \wedge \frac{dP_{j_p}}{P_{jp}},$$

in particular, for $|J| = n$, $\widehat{\varphi}\langle J \rangle = \displaystyle\int_{\sigma} U \cdot \varphi\langle J \rangle, \quad \sigma \in H_n(M, \mathcal{L}_\omega^\vee),$

we denote the minor obtained from extracting i_1th, $\ldots$, i_pth rows

and j_1th, $\ldots$, j_pth columns of x by $x\begin{pmatrix} i_1 & \cdots & i_p \\ j_1 & \cdots & j_p \end{pmatrix}$.

As was explained in § 3.4.1, setting

$$\overline{X} = \left\{ x = \begin{pmatrix} x_{01} & \cdots\cdots & x_{0m} & 1 \\ \vdots & & \vdots & 0 \\ \vdots & & \vdots & \vdots \\ x_{n1} & \cdots\cdots & x_{nm} & 0 \end{pmatrix}; \text{ each column vector} \ne 0 \right\},$$

$$X = \overline{X} \setminus Y,$$

where $\quad Y := \displaystyle\bigcup_{1 \le j_0 < \cdots < j_n \le m+1} \left\{ x\begin{pmatrix} 0 & \cdots\cdots & n \\ j_0 & \cdots\cdots & j_n \end{pmatrix} = 0 \right\},$

$\widehat{\varphi}\langle J \rangle$ is a multi-valued function on X. By Corollary 2.6, $\binom{m-1}{n} \widehat{\varphi}\langle J \rangle$ provides a basis of $H^n(\Omega^\bullet(*D), \nabla_\omega)$. Hence, in the exterior derivative of $\widehat{\varphi}\langle J \rangle$ with respect to x

$$d\widehat{\varphi}\langle J \rangle = \sum_{i=0}^{n} \sum_{j=1}^{m} dx_{ij} \int_{\sigma} U \left\{ \frac{1}{U} \frac{\partial (U\varphi\langle J \rangle)}{\partial x_{ij}} \right\},$$

each $\frac{1}{U} \frac{\partial (U\varphi\langle J \rangle)}{\partial x_{ij}}$ should be in principle expressed as a linear combination of $\varphi\langle K \rangle$, $K = \{k_1, \cdots, k_n\}, \; 1 \le k_1 < \cdots < k_n \le m - 1$ over $\mathbb{C}(x)$ in $H^n(\Omega^\bullet(*D), \nabla_\omega)$. Below, we compute $d\widehat{\varphi}\langle J \rangle = \sum_K \theta_{J,K}\widehat{\varphi}\langle K \rangle$ concretely and show that $\theta_{J,K}$ is a 1-form admitting logarithmic poles along Y.

3.8.2 *Variational Formula*

By simple calculation, we obtain

$$d\widehat{\varphi}\langle J \rangle = \sum_{i=0}^{n} \sum_{j_0 \notin J} dx_{ij_0} \int_{\sigma} U \frac{\alpha_{j_0} u_i}{P_{j_0}} \varphi\langle J \rangle$$

$$+ \sum_{i=0}^{n} \sum_{k=1}^{n} dx_{ij_k} \int_{\sigma} U \left\{ \frac{(\alpha_{j_k} - 1) u_i}{P_{j_k}} \varphi\langle J \rangle \right.$$

$$\left. + (-1)^{k-1} \frac{du_i}{P_{j_k}} \wedge \varphi\langle J \setminus \{j_k\}\rangle \right\}.$$

Here, $\varphi\langle J \setminus \{j_k\}\rangle = \varphi\langle j_1 \cdots j_{k-1} j_{k+1} \cdots j_n \rangle$ and $du_0 = 0$. Here and after, the symbol $\sum_{j_0 \notin J}$ signifies the sum over $1 \le j_0 \le m$ with j_0 satisfying $j_0 \notin J$. On the other hand, since we have

$$\nabla_\omega \left(\frac{(-1)^{k-1} u_i}{P_{j_k}} \varphi\langle J \setminus \{j_k\}\rangle \right)$$

$$= \frac{(-1)^{k-1}}{P_{j_k}} du_i \wedge \varphi\langle J \setminus \{j_k\}\rangle + \frac{(\alpha_{j_k} - 1) u_i}{P_{j_k}} \varphi\langle J \rangle$$

$$+ \sum_{j_0 \notin J} (-1)^{k-1} \frac{\alpha_{j_0} u_i}{P_{j_k}} \varphi\langle j_0 \cdots \widehat{j_k} \cdots j_n \rangle,$$

combining the above two formulas, we obtain

$$d\widehat{\varphi}\langle J \rangle = \int_{\sigma} U \sum_{i=0}^{n} \sum_{j_0 \notin J} dx_{ij_0} \frac{\alpha_{j_0} u_i}{P_{j_0}} \varphi\langle J \rangle$$

$$- \int_{\sigma} U \sum_{i=0}^{n} \sum_{k=1}^{n} \sum_{j_0 \notin J} (-1)^{k-1} dx_{ij_k} \frac{\alpha_{j_0} u_i}{P_{j_k}} \varphi\langle j_0 \cdots \widehat{j_k} \cdots j_n \rangle.$$

Since we obtain the first term by setting $k = 0$ in the second term of the above formula, we can further simplify it:

$$d\widehat{\varphi}\langle J \rangle = \int_{\sigma} U \sum_{\substack{i,k=0}}^{n} \sum_{\substack{j_0 \notin J \\ K := \{j_0\} \cup J}} (-1)^k dx_{ij_k} \frac{\alpha_{j_0} u_i}{P_{j_k}} \varphi\langle K \setminus \{j_k\}\rangle. \qquad (3.98)$$

Hence, the problem is reduced to compute the partial fraction expansion explained in § 2.9.1, determining all the coefficients precisely, and express the above formula with the basis $\{\varphi\langle i_1 \cdots i_n \rangle\}$.

3.8.3 Partial Fraction Expansion

Setting $K = \{j_0, \cdots, j_n\}$, $N = \{0, 1, \cdots, n\}$, we solve

$$
\begin{cases}
x_{0j_0} + x_{1j_0}u_1 + \cdots + x_{nj_0}u_n = P_{j_0}, \\
\cdots\cdots\cdots\cdots\cdots\cdots\cdots \\
x_{0j_n} + x_{1j_n}u_1 + \cdots + x_{nj_n}u_n = P_{j_n},
\end{cases}
$$

with respect to u_i by using Cramer's formula and obtain

$$
u_i = \sum_{l=0}^{n}(-1)^{i+l}\frac{P_{j_l}x\begin{pmatrix}N\setminus\{i\}\\K\setminus\{j_l\}\end{pmatrix}}{x\begin{pmatrix}N\\K\end{pmatrix}}. \tag{3.99}
$$

Here, introducing the symbols

$$
P_K := \prod_{j\in K}P_j, \quad P_{K\setminus\{j_l\}} = \prod_{j\in K\setminus\{j_l\}}P_j,
$$

by (3.99), we obtain a formula on partial fraction expansion:

$$
\frac{u_i}{P_K} = \sum_{l=0}^{n}(-1)^{i+l}\cdot\frac{x\begin{pmatrix}N\setminus\{i\}\\K\setminus\{j_l\}\end{pmatrix}}{x\begin{pmatrix}N\\K\end{pmatrix}}\cdot\frac{1}{P_{K\setminus\{j_l\}}}. \tag{3.100}
$$

3.8.4 Reformulation

Let us rewrite (3.98) to apply the formula (3.100). First, we notice that

$$
\frac{u_i}{P_{j_k}}\varphi\langle K\setminus\{j_k\}\rangle \tag{3.101}
$$

$$
= \frac{u_i}{P_K}\cdot x\begin{pmatrix}N\setminus\{0\}\\K\setminus\{j_k\}\end{pmatrix}du_1\wedge\cdots\wedge du_n
$$

$$
= \sum_{l=0}^{n}(-1)^{i+l}\cdot\frac{x\begin{pmatrix}N\setminus\{0\}\\K\setminus\{j_k\}\end{pmatrix}\cdot x\begin{pmatrix}N\setminus\{i\}\\K\setminus\{j_l\}\end{pmatrix}}{x\begin{pmatrix}N\\K\end{pmatrix}}\cdot\frac{du_1\wedge\cdots\wedge du_n}{P_{K\setminus\{j_l\}}}.
$$

Second, by using

$$\varphi\langle K \setminus \{j_l\}\rangle = \frac{1}{P_{K\setminus\{j_l\}}} dP_{j_0} \wedge \cdots \wedge \widehat{dP_{j_l}} \wedge \cdots \wedge dP_{j_n}$$

$$= \frac{x\begin{pmatrix} N \setminus \{0\} \\ K \setminus \{j_l\} \end{pmatrix}}{P_{K\setminus\{j_l\}}} \cdot du_1 \wedge \cdots \wedge du_n,$$

rewriting the right-hand side of (3.101), we finally obtain

$$\frac{u_i}{P_{j_k}}\varphi\langle K \setminus \{j_k\}\rangle$$

$$= \sum_{l=0}^{n}(-1)^{i+l} \cdot \frac{x\begin{pmatrix} N \setminus \{0\} \\ K \setminus \{j_k\} \end{pmatrix} \cdot x\begin{pmatrix} N \setminus \{i\} \\ K \setminus \{j_l\} \end{pmatrix}}{x\begin{pmatrix} N \\ K \end{pmatrix} \cdot x\begin{pmatrix} N \setminus \{0\} \\ K \setminus \{j_l\} \end{pmatrix}} \cdot \varphi\langle K \setminus \{j_l\}\rangle.$$

Rewriting the right-hand side of (3.98) with this formula, we obtain

$$d\widehat{\varphi}\langle J\rangle = \int_\sigma U \sum_{\substack{i,k,l=0}}^{n} \sum_{\substack{j_0 \notin J \\ K:=\{j_0\}\cup J}} (-1)^{i+k+l} \cdot \alpha_{j_0} dx_{ij_k} \qquad (3.102)$$

$$\cdot \frac{x\begin{pmatrix} N \setminus \{0\} \\ K \setminus \{j_k\} \end{pmatrix} \cdot x\begin{pmatrix} N \setminus \{i\} \\ K \setminus \{j_l\} \end{pmatrix}}{x\begin{pmatrix} N \\ K \end{pmatrix} \cdot x\begin{pmatrix} N \setminus \{0\} \\ K \setminus \{j_l\} \end{pmatrix}} \cdot \varphi\langle K \setminus \{j_l\}\rangle.$$

Here, we use the well-known Jacobi formula (cf. [Sata]).

Lemma 3.13 (Jacobi's formula).

$$x\begin{pmatrix} N \setminus \{0\} \\ K \setminus \{j_k\} \end{pmatrix} x\begin{pmatrix} N \setminus \{i\} \\ K \setminus \{j_l\} \end{pmatrix} - x\begin{pmatrix} N \setminus \{0\} \\ K \setminus \{j_l\} \end{pmatrix} x\begin{pmatrix} N \setminus \{i\} \\ K \setminus \{j_k\} \end{pmatrix} \qquad (3.103)$$

$$= x\begin{pmatrix} N \\ K \end{pmatrix} x\begin{pmatrix} N \setminus \{0,i\} \\ K \setminus \{j_k, j_l\} \end{pmatrix}.$$

Proof. Permuting rows and columns of (3.103), we can reduce it to the reference cited above (the formula with respect to the last two columns and rows).

We rewrite (3.102) by using this lemma:

$$d\widehat{\varphi}\langle J\rangle = \int_\sigma U \sum_{l=0}^n \sum_{\substack{j_0\notin J \\ K:=\{j_0\}\cup J}} (-1)^l$$

$$\cdot \alpha_{j_0}\left\{\frac{1}{x\binom{N}{K}}\cdot \sum_{i,k=0}^n (-1)^{i+k} x\binom{N\setminus\{i\}}{K\setminus\{j_k\}}\right) dx_{ij_k}$$

$$+ \frac{1}{x\binom{N\setminus\{0\}}{K\setminus\{j_l\}}} \sum_{i,k=0}^n (-1)^{i+k}$$

$$\cdot x\binom{N\setminus\{0,i\}}{K\setminus\{j_k,j_l\}} dx_{ij_k}\bigg\}\varphi\langle K\setminus\{j_l\}\rangle$$

$$= \int_\sigma U \sum_{l=0}^n \sum_{\substack{j_0\notin J \\ K:=\{j_0\}\cup J}} (-1)^l$$

$$\cdot \alpha_{j_0}\left\{\frac{dx\binom{N}{K}}{x\binom{N}{K}} - \frac{dx\binom{N\setminus\{0\}}{K\setminus\{j_l\}}}{x\binom{N\setminus\{0\}}{K\setminus\{j_l\}}}\right\}\cdot \varphi\langle K\setminus\{j_l\}\rangle$$

$$= \sum_{l=0}^n \sum_{\substack{j_0\notin J \\ K:=\{j_0\}\cup J}} (-1)^l\cdot \alpha_{j_0}\cdot \widehat{\varphi}\langle K-\{j_l\}\rangle d\log \frac{x\binom{N}{K}}{x\binom{N\setminus\{0\}}{K\setminus\{j_l\}}}.$$

By the above computation, we have shown the following lemma:

Lemma 3.14. $\widehat{\varphi}\langle J\rangle$, $J = \{j_1,\cdots,j_n\}$, $1\le j_1 < \cdots < j_n \le m-1$ *satisfies the relation:*

$$d\widehat{\varphi}\langle J\rangle = \sum_{\substack{j_0\notin J \\ K:=\{j_0\}\cup J}} \sum_{l=0}^n (-1)^l\cdot \alpha_{j_0}\cdot d\log \frac{x\binom{N}{K}}{x\binom{N\setminus\{0\}}{K\setminus\{j_l\}}}\cdot \widehat{\varphi}\langle K\setminus\{j_l\}\rangle.$$

Here, $N = \{0,1,\cdots,n\}$ *and* $\widehat{\varphi}\langle L\rangle$ *appearing in the right-hand side runs over* $L = \{l_1,\cdots,l_n\}$, $1\le l_1 < \cdots < l_n \le m$.

3.8.5 Example

To clarify the form of the formula obtained in Lemma 3.14, let us compute it for $n = 2$, $m = 5$.

$$x = \begin{pmatrix} x_{01} & \cdots\cdots & x_{05} & 1 \\ x_{11} & \cdots\cdots & x_{15} & 0 \\ x_{21} & \cdots\cdots & x_{25} & 0 \end{pmatrix}, \quad N = \{0, 1, 2\},$$

A basis of $H^2(\Omega^\bullet(*D), \nabla_\omega)$ is $\varphi\langle 12\rangle$, $\varphi\langle 13\rangle$, $\varphi\langle 14\rangle$, $\varphi\langle 23\rangle$, $\varphi\langle 24\rangle$, $\varphi\langle 34\rangle$.

$$d\widehat{\varphi}\langle 12\rangle = \left\{ \alpha_3 d\log \frac{x\begin{pmatrix} 0\ 1\ 2 \\ 3\ 1\ 2 \end{pmatrix}}{x\begin{pmatrix} 1\ 2 \\ 1\ 2 \end{pmatrix}} + \alpha_4 d\log \frac{x\begin{pmatrix} 0\ 1\ 2 \\ 4\ 1\ 2 \end{pmatrix}}{x\begin{pmatrix} 1\ 2 \\ 1\ 2 \end{pmatrix}} \right\} \widehat{\varphi}\langle 12\rangle$$

$$- \alpha_3 d\log \frac{x\begin{pmatrix} 0\ 1\ 2 \\ 3\ 1\ 2 \end{pmatrix}}{x\begin{pmatrix} 1\ 2 \\ 2\ 3 \end{pmatrix}} \widehat{\varphi}\langle 23\rangle + \alpha_3 d\log \frac{x\begin{pmatrix} 0\ 1\ 2 \\ 3\ 1\ 2 \end{pmatrix}}{x\begin{pmatrix} 1\ 2 \\ 1\ 3 \end{pmatrix}} \widehat{\varphi}\langle 13\rangle$$

$$- \alpha_4 d\log \frac{x\begin{pmatrix} 0\ 1\ 2 \\ 4\ 1\ 2 \end{pmatrix}}{x\begin{pmatrix} 1\ 2 \\ 2\ 4 \end{pmatrix}} \widehat{\varphi}\langle 24\rangle + \alpha_4 d\log \frac{x\begin{pmatrix} 0\ 1\ 2 \\ 4\ 1\ 2 \end{pmatrix}}{x\begin{pmatrix} 1\ 2 \\ 1\ 4 \end{pmatrix}} \widehat{\varphi}\langle 14\rangle \qquad (3.104)$$

$$+ \alpha_5 \left\{ d\log \frac{x\begin{pmatrix} 0\ 1\ 2 \\ 1\ 2\ 5 \end{pmatrix}}{x\begin{pmatrix} 1\ 2 \\ 1\ 2 \end{pmatrix}} \widehat{\varphi}\langle 12\rangle - d\log \frac{x\begin{pmatrix} 0\ 1\ 2 \\ 1\ 2\ 5 \end{pmatrix}}{x\begin{pmatrix} 1\ 2 \\ 2\ 5 \end{pmatrix}} \widehat{\varphi}\langle 25\rangle \right.$$

$$\left. + d\log \frac{x\begin{pmatrix} 0\ 1\ 2 \\ 1\ 2\ 5 \end{pmatrix}}{x\begin{pmatrix} 1\ 2 \\ 1\ 5 \end{pmatrix}} \widehat{\varphi}\langle 15\rangle \right\}.$$

Other $d\widehat{\varphi}\langle j_1 j_2\rangle$'s can be similarly computed. As can be seen from these computations, the right-hand side of Lemma 3.14 has some terms containing $\widehat{\varphi}\langle j_1 \cdots j_{n-1} m\rangle$, and this expression of differential equations is still not in the Pfaffian form. We should eliminate $\widehat{\varphi}\langle 15\rangle$, $\widehat{\varphi}\langle 25\rangle$ in the right-hand side of (3.104) by using the relations

$$\alpha_2 \widehat{\varphi}\langle 12\rangle + \alpha_3 \widehat{\varphi}\langle 13\rangle + \alpha_4 \widehat{\varphi}\langle 14\rangle + \alpha_5 \widehat{\varphi}\langle 15\rangle = 0,$$

$$\alpha_1 \widehat{\varphi}\langle 12\rangle - \alpha_3 \widehat{\varphi}\langle 23\rangle - \alpha_4 \widehat{\varphi}\langle 24\rangle - \alpha_5 \widehat{\varphi}\langle 25\rangle = 0.$$

The result is as follows:

$$
d\widehat{\varphi}\langle 12\rangle = \left\{ \alpha_1 d\log \frac{x\begin{pmatrix} 0\,1\,2 \\ 5\,1\,2 \end{pmatrix}}{x\begin{pmatrix} 1\,2 \\ 5\,2 \end{pmatrix}} + \alpha_2 d\log \frac{x\begin{pmatrix} 0\,1\,2 \\ 5\,1\,2 \end{pmatrix}}{x\begin{pmatrix} 1\,2 \\ 5\,1 \end{pmatrix}} \right.
$$

$$
\left. + \sum_{j=3}^{5} \alpha_j d\log \frac{x\begin{pmatrix} 0\,1\,2 \\ j\,1\,2 \end{pmatrix}}{x\begin{pmatrix} 1\,2 \\ 1\,2 \end{pmatrix}} \right\} \widehat{\varphi}\langle 12\rangle
$$

$$
+ \alpha_3 \left\{ d\log \frac{x\begin{pmatrix} 0\,1\,2 \\ 3\,1\,2 \end{pmatrix}}{x\begin{pmatrix} 1\,2 \\ 1\,3 \end{pmatrix}} - d\log \frac{x\begin{pmatrix} 0\,1\,2 \\ 5\,1\,2 \end{pmatrix}}{x\begin{pmatrix} 1\,2 \\ 1\,5 \end{pmatrix}} \right\} \widehat{\varphi}\langle 13\rangle
$$

$$
+ \alpha_4 \left\{ d\log \frac{x\begin{pmatrix} 0\,1\,2 \\ 4\,1\,2 \end{pmatrix}}{x\begin{pmatrix} 1\,2 \\ 1\,4 \end{pmatrix}} - d\log \frac{x\begin{pmatrix} 0\,1\,2 \\ 5\,1\,2 \end{pmatrix}}{x\begin{pmatrix} 1\,2 \\ 1\,5 \end{pmatrix}} \right\} \widehat{\varphi}\langle 14\rangle
$$

$$
+ \alpha_3 \left\{ d\log \frac{x\begin{pmatrix} 0\,1\,2 \\ 5\,1\,2 \end{pmatrix}}{x\begin{pmatrix} 1\,2 \\ 2\,5 \end{pmatrix}} - d\log \frac{x\begin{pmatrix} 0\,1\,2 \\ 3\,1\,2 \end{pmatrix}}{x\begin{pmatrix} 1\,2 \\ 2\,3 \end{pmatrix}} \right\} \widehat{\varphi}\langle 23\rangle
$$

$$
+ \alpha_4 \left\{ d\log \frac{x\begin{pmatrix} 0\,1\,2 \\ 5\,1\,2 \end{pmatrix}}{x\begin{pmatrix} 1\,2 \\ 2\,5 \end{pmatrix}} - d\log \frac{x\begin{pmatrix} 0\,1\,2 \\ 4\,1\,2 \end{pmatrix}}{x\begin{pmatrix} 1\,2 \\ 2\,4 \end{pmatrix}} \right\} \widehat{\varphi}\langle 24\rangle .
$$

3.8.6 Logarithmic Gauss–Manin Connection

Since extending the above computation to general cases is not so difficult, except that the symbols become complicated, we only state its essence. First, in the formula stated in Lemma 3.14, we distinguish the summation over j_0 into the parts with $1 \le j_0 \le m-1$ and $j_0 \notin J$ and the parts with $j_0 = m$. As the former has no problem, we consider the latter case. It takes the form:

$$\alpha_m d\log \frac{x\begin{pmatrix} N \\ \{m\}\cup J \end{pmatrix}}{x\begin{pmatrix} N\setminus\{0\} \\ J \end{pmatrix}} \widehat{\varphi}\langle J\rangle \tag{3.105}$$

$$+ \alpha_m \sum_{l=1}^{n} (-1)^l d\log \frac{x\begin{pmatrix} N \\ \{m\}\cup J \end{pmatrix}}{x\begin{pmatrix} N\setminus\{0\} \\ m, J\setminus\{j_l\} \end{pmatrix}} \cdot \widehat{\varphi}\langle m, J\setminus\{j_l\}\rangle,$$

where the multi-index $\{m, J\setminus\{j_l\}\}$ signifies $\{m, j_1, \cdots, \widehat{j_l}, \cdots, j_n\}$. By the formula

$$\nabla_\omega \varphi\langle J\setminus\{j_l\}\rangle = (-1)^{l-1}\alpha_{j_l}\varphi\langle J\rangle + \alpha_m \varphi\langle m, J\setminus\{j_l\}\rangle$$

$$+ \sum_{\substack{j_0\notin J \\ 1\le j_0 \le m-1 \\ K:=\{j_0\}\cup J}} \alpha_{j_0}\varphi\langle K\setminus\{j_l\}\rangle,$$

we obtain

$$\alpha_m \widehat{\varphi}\langle m, J\setminus\{j_l\}\rangle = (-1)^l \alpha_{j_l}\widehat{\varphi}\langle J\rangle \tag{3.106}$$

$$- \sum_{\substack{j_0\notin J \\ 1\le j_0 \le m-1 \\ K:=\{j_0\}\cup J}} \alpha_{j_0}\widehat{\varphi}\langle K\setminus\{j_l\}\rangle.$$

Rewriting the second term in (3.105) with this formula, we obtain:

$$\left(\sum_{\substack{l=0 \\ j_0=m \\ K=\{j_0\}\cup J}}^{n} \alpha_{j_l} d\log \frac{x\begin{pmatrix} N \\ K \end{pmatrix}}{x\begin{pmatrix} N\setminus\{0\} \\ K\setminus\{j_l\} \end{pmatrix}} \right) \widehat{\varphi}\langle J\rangle \tag{3.107}$$

$$+ \sum_{\substack{j_0\notin J \\ 1\le j_0 \le m-1 \\ K=\{j_0\}\cup J}} \sum_{l=1}^{n} (-1)^{l-1}\alpha_{j_0} d\log \frac{x\begin{pmatrix} N \\ \{m\}\cup J \end{pmatrix}}{x\begin{pmatrix} N\setminus\{0\} \\ m, J\setminus\{j_l\} \end{pmatrix}} \cdot \widehat{\varphi}\langle K\setminus\{j_l\}\rangle.$$

Summarizing all these computations, the Pfaffian forms satisfied by $\binom{m-1}{n}$ $\widehat{\varphi}\langle J\rangle$, $1\le j_1 < \cdots < j_n \le m-1$ can be expressed as follows: setting

$$\theta_{J,L} := \begin{cases} \displaystyle\sum_{\substack{j_0 \notin J \\ 1 \le j_0 \le m \\ K := \{j_0\} \cup J}} \alpha_{j_0} d\log \frac{x\begin{pmatrix} N \\ K \end{pmatrix}}{x\begin{pmatrix} N \setminus \{0\} \\ J \end{pmatrix}} \\[2em] \qquad + \displaystyle\sum_{l=1}^{n} \alpha_{j_l} d\log \frac{x\begin{pmatrix} N \\ \{m\} \cup J \end{pmatrix}}{x\begin{pmatrix} N \setminus \{0\} \\ m, J \setminus \{j_l\} \end{pmatrix}} \qquad \text{(for } L = J) \\[2em] (-1)^l \alpha_{j_0} \left\{ d\log \frac{x\begin{pmatrix} N \\ K \end{pmatrix}}{x\begin{pmatrix} N \setminus \{0\} \\ K \setminus \{j_l\} \end{pmatrix}} - d\log \frac{x\begin{pmatrix} N \\ \{m\} \cup J \end{pmatrix}}{x\begin{pmatrix} N \setminus \{0\} \\ m, J \setminus \{j_l\} \end{pmatrix}} \right\} \\[1em] \qquad \begin{pmatrix} 1 \le j_0 \le m - 1, \text{ setting } K := \{j_0\} \cup J, \\ L = K \setminus \{j_l\}, \quad \text{for } l \ne 0 \end{pmatrix} \\[1em] 0 \qquad \text{(otherwise)}, \end{cases}$$

arranging J, L in the lexicographic order and letting $\Theta = ((\theta_{J,L}))$ be the matrix with coefficients in this logarithmic differential forms. Let $(y_J)_J$ be the column vector arranging y_J in the lexicographic order and consider the Pfaffian form

$$(dy_J)_J = \Theta(y_L)_L. \tag{3.108}$$

As we have seen above, $(\widehat{\varphi}\langle J \rangle)_J$ becomes a solution of (3.108). Here, taking a real matrix x, we take a basis $\{\Delta_\nu(\omega)\}_{1 \le \nu \le \binom{m-1}{n}}$ of $H_n(M, \mathcal{L}_\omega^\vee)$ stated in Theorem 3.1 and set $\widehat{\varphi}_\nu\langle J \rangle := \int_{\Delta_\nu(\omega)} U \cdot \varphi\langle J \rangle$. Then, $\binom{m-1}{n}$ column vectors $(\widehat{\varphi}_\nu\langle J \rangle)_J$ become solutions of (3.108), and by Corollary 3.1, we have $\det(\widehat{\varphi}_\nu\langle J \rangle)_{\nu,J} \ne 0$. This means that the column vectors $(\widehat{\varphi}_\nu\langle J \rangle)_J$, $1 \le \nu \le \binom{m-1}{n}$ provide us with linearly independent solutions of (3.108).

Now, $x\binom{N}{K}$ etc. appearing in the expression of $\theta_{J,L}$ can be regarded as Plücker coordinates of the point of $G(n+1, m+1)$ associated to the matrix x stated in § 3.4.2. Multiplying an element g of $GL_{n+1}(\mathbb{C})$ by x from the left induces $x\binom{N}{K} \longrightarrow \det(g) \cdot x\binom{N}{K}$. In the expression of $\theta_{J,L}$, since Plücker coordinates appear in the denominator and in the enumerator of the factors in $d\log$, $\theta_{J,L}$ is invariant under the action of $GL_{n+1}(\mathbb{C})$. This means that (3.108) is a differential equation defined on the Grassmannian $G(n+1, m+1)$ omitting finite hypersurfaces. Moreover, equation (3.108) is integrable in the following sense, i.e., as a consequence of (3.108), the curvature form of Θ becomes 0:

$$d\Theta - \Theta \wedge \Theta = -\Theta \wedge \Theta = 0 \qquad (3.109)$$

since $d\Theta = 0$. In this case, we say that (3.108) defines a Gauss–Manin connection of rank $\binom{m-1}{n}$ ([De]).

Exercise 3.1 ([Ao8]). Show that the identity (3.109) can be derived from Plücker relations (cf. § 3.4.2).

Theorem 3.10. *In the open subset of $M_{n+1,m}(\mathbb{C})$*

$$\overline{X} = \left\{ x = \begin{pmatrix} x_{01} & \cdots\cdots & x_{0m} & 1 \\ x_{11} & & x_{1m} & 0 \\ \vdots & & \vdots & \vdots \\ x_{n1} & \cdots\cdots & x_{nm} & 0 \end{pmatrix}, \; each \; column \; vector \neq 0 \right\},$$

the Pfaffian form (3.108) which has logarithmic poles along

$$Y = \bigcup_{1 \leq j_0 < \cdots < j_n \leq m+1} \left\{ x \begin{pmatrix} 0 & 1 & \cdots & n \\ j_0 & & \cdots & j_n \end{pmatrix} = 0 \right\}$$

defines a Gauss–Manin connection on $X = \overline{X} \backslash Y$ and its linearly independent solutions are given by $(\widehat{\varphi}_\nu \langle J \rangle)_J$, $1 \leq \nu \leq \binom{m-1}{n}$.

Example 3.6. Let us explicitly give a Pfaffian form which provides us a Gauss–Manin connection for $n = 1$:

$$x = \begin{pmatrix} x_{01} & x_{02} & \cdots & x_{0m} & 1 \\ x_{11} & x_{12} & \cdots & x_{1m} & 0 \end{pmatrix},$$

for $1 \leq j, k \leq m-1$, $j \neq k$, we have

$$\left\{ \begin{aligned} \theta_{jj} &= \sum_{\substack{1 \leq l \leq m \\ l \neq j}} \alpha_l d\log \frac{x \begin{pmatrix} 0 & 1 \\ l & j \end{pmatrix}}{x_{1j}} + \alpha_j d\log \frac{x \begin{pmatrix} 0 & 1 \\ m & j \end{pmatrix}}{x_{1m}}, \\ \theta_{jk} &= -\alpha_k \left\{ d\log \frac{x \begin{pmatrix} 0 & 1 \\ k & j \end{pmatrix}}{x_{1k}} - d\log \frac{x \begin{pmatrix} 0 & 1 \\ m & j \end{pmatrix}}{x_{1m}} \right\}. \end{aligned} \right.$$

Hence, fixing $1 \leq j \leq m - 1$,

$$\varphi_\nu \langle j \rangle = \int_{\Delta_\nu(\omega)} U \cdot \varphi \langle j \rangle, \quad 1 \leq \nu \leq m - 1$$

gives solutions of the Pfaffian form

$$d \begin{pmatrix} y_1 \\ \vdots \\ y_{m-1} \end{pmatrix} = \begin{pmatrix} \theta_{11} & \cdots & \theta_{1, \, m-1} \\ \vdots & & \vdots \\ \theta_{m-1, \, 1} & \cdots & \theta_{m-1, \, m-1} \end{pmatrix} \begin{pmatrix} y_1 \\ \vdots \\ y_{m-1} \end{pmatrix}. \tag{3.110}$$

Remark 3.12. As a consequence of equation (3.108), we obtain the higher logarithmic expansion of hypergeometric functions as in § 1.2.5. In [Koh7], [Lin], higher logarithms are discussed from the viewpoint of K. T. Chen's iterated integral. For their applications to higher logarithmic functions, its algebraic K-theory and algebraic geometry, see, for example, [Gon] and [Ha-M]. As for an extension of the connection form (3.108) to a general arrangement of hyperplanes, there are further precise discussions on it (cf. [C-D-O], [C-O1], [C-O2], [Or-Te4]).

Remark 3.13. In this chapter, we have presented the holonomic system of differential equations $E(n + 1, m + 1; \alpha)$ on the Grassmannian $\overline{X}$ and the Gauss−Manin connection (3.108) on $\overline{X} \setminus Y$ on the other. There are several references treating the system $E(n + 1, m + 1; \alpha)$ [Har], [Iw-Kit1], [Iw-Kit2], [Kit2], [Kit3], [Kit-It], [Sas], or the geometry of arrangements of hyperplanes which reflect properties of special hypergeometric functions [Hat], [Koh5], [Or-Te1], [R-T], [Sch-Va2], [Ter1], [Ter2]. In general, a holonomic $\mathcal{D}$-module corresponds to a constructible sheaf as a solution sheaf (Riemann−Hilbert correspondence) (see [H-T-T], [Kas1], [Kas2], [Sab2], [Tan] etc.). One may say that the solution sheaf on $\overline{X} \setminus Y$ defined by (3.108), being a locally constant sheaf, gives a construcible sheaf on a one of the strata of $\overline{X}$ associated with the singularity of $E(n + 1, m + 1; \alpha)$ defined on the strata realized by arragements of hyperplanes not in general position (for example, see [C-O1], [C-O2], [Or-Te4]). It seems an interesting problem to study their interrelationship in a systematic way.

Chapter 4
Holonomic Difference Equations
and Asymptotic Expansion

As we have seen in Chapter 1, the Γ-function is a solution of a first-order difference equation which can be uniquely determined by its asymptotic behavior at infinity. This fact can be generalized to the cases of several variables that contain a finite number of unknown meromorphic functions which satisfy a holonomic system of difference equations. The hypergeometric functions discussed in Chapters 2 and 3 satisfy holonomic systems of difference equations with respect to the parameters $\alpha = (\alpha_1, \ldots, \alpha_m)$. Their asymptotic structure at infinity strongly reflects topological aspects of their twisted de Rham (co)homology.

In this chapter, we will see such an aspect of hypergeometric functions as solutions of a holonomic system of difference equations. In § 4.1, 4.2, applying classical results due to G.D. Birkhoff, we will show a fundamental result on the existence and the uniqueness of a meromorphic solution of a holonomic system of difference equations having an asymptotic expansion at a specific direction of infinity. In § 4.3, as an application of the structure of the twisted de Rham cohomology explained in Chapter 2, we will explain a way to derive a holonomic system of difference equations satisfied by hypergeometric functions and a relation between their asymptotic expansion with respect to α and the Morse theory. In § 4.4, we derive concretely and explicitly a system of equations in the case of an arrangement of hyperplanes. As § 4.1, 4.2 are not directly related to hypergeometric functions, it might be better for the reader to start reading from § 4.3 and refer to § 4.1, 4.2 from time to time.

First, we start by explaining the existence and the uniqueness theorem due to G.D. Birkhoff for the single-variable case. See [Bir2], [N], [Gelfo] as basic notion for analytic difference equations.

K. Aomoto et al., *Theory of Hypergeometric Functions*, Springer Monographs in Mathematics, DOI 10.1007/978-4-431-53938-4_4, ⓒ Springer 2011

Existence Theorem Due to G.D. Birkhoff and Infinite-Product Representation of Matrices

4.1.1 Normal Form of Matrix-Valued Function

Suppose that a square matrix $A(z) \in GL_r(\mathbb{C}(z))$ of order r with components of single-variable rational functions is given. We consider a matrix difference equation of a square matrix $\Phi(z)$

$$\Phi(z+1) = A(z)\Phi(z), \quad z \in \mathbb{C}, \tag{4.1}$$

$$\Phi(z) = \left(\left(\varphi_{ij}(z)\right)\right)_{i,j=1}^{r}.$$

Here, $\varphi_{ij}(z)$ are unknown meromorphic functions on $\mathbb{C}$. First, we look at the normal form of $A(z)$ at $z = \infty$.

Lemma 4.1. *Suppose that $A(z)$ has the Laurent expansion at $z = \infty$ of the form*

$$A(z) = z^{\mu}\left(A_0 + \frac{A_1}{z} + \frac{A_2}{z^2} + \cdots\right), \quad \mu \in \mathbb{Z} \tag{4.2}$$

where $A_0 \in GL_r(\mathbb{C})$ is semi-simple, i.e., diagonalizable and its eigenvalues are mutually different, and $A_l(l \geq 1) \in M_r(\mathbb{C})$. There exists a formal Laurent series

$$S(z) = S_0 + \frac{S_1}{z} + \frac{S_2}{z^2} + \cdots \in M_r\left(\mathbb{C}\left[\left[\frac{1}{z}\right]\right]\right) \tag{4.3}$$

with $\left(S_0 \in GL_r(\mathbb{C})\right)$ such that $A(z)$ is reduced to

$$S(z+1)^{-1}A(z)S(z) = z^{\mu}\left(\overline{A}_0 + \frac{\overline{A}_1}{z}\right) \tag{4.4}$$

($\overline{A}_0$, $\overline{A}_1$ are both diagonal matrices). Here, $\overline{A}_0$, $\overline{A}_1$ are uniquely determined up to a permutation of diagonal components. In particular, if A_0 is a diagonal matrix, then $A_0 = \overline{A}_0$ and $\overline{A}_1$ is equal to the diagonal part of A_1, and we can take $S_0 = 1$ in (4.3). In this case, $S(z)$ (4.3) satisfying (4.4) is uniquely determined.

Proof. First, since A_0 can be diagonalized by a transformation $A_0 \mapsto S_0^{-1}A_0S_0$, below, we assume that $S_0 = 1$ and A_0 is already a diagonal matrix.
Step 1. There exists

$$S'(z) = 1 + \frac{S'_1}{z} + \frac{S'_2}{z^2} + \cdots \in GL_r\left(\mathbb{C}\left[\left[\frac{1}{z}\right]\right]\right)$$

such that

$$B(z) = S'(z+1)^{-1}A(z)S'(z) \tag{4.5}$$

$$= z^\mu \left(B_0 + \frac{B_1}{z} + \frac{B_2}{z^2} + \cdots \right) \in GL_r \left(\mathbb{C} \left[\left[\frac{1}{z} \right] \right] \right)$$

becomes a diagonal matrix. Indeed, by

$$(z+1)^{-n} = z^{-n} \sum_{l=0}^{\infty} \frac{n(n+1)\cdots(n+l-1)}{l!}(-z)^{-l},$$

the relations satisfied by S'_1, S'_2, $\cdots$ are given, by comparing the coefficients of the Laurent series of the identity $S'(z+1)B(z) = A(z)S'(z)$ in z^μ, $z^{\mu-1}$,

$$A_0 = B_0, \qquad A_1 + A_0 S'_1 = S'_1 B_0 + B_1, \tag{4.6}$$

that is,

$$[A_0, S'_1] = B_1 - A_1.$$

More generally, comparing the coefficients of $z^{-n+\mu}$, we have

$$[A_0, S'_n] = -\sum_{l=1}^{n-1} A_l S'_{n-l} + B_n - A_n \tag{4.7}$$

$$+ \sum_{\substack{0 \le \sigma \le n-l \\ 1 \le l \le n-1}} S'_l B_\sigma (-1)^{n-l-\sigma} \frac{l(l+1)\cdots(n-\sigma-1)}{(n-l-\sigma)!}, \quad n \ge 1.$$

Now, for a matrix X with 0 on the diagonal, the equation $[A_0, S'_n] = X$ on S'_n always admits a solution. Choosing $B_1, B_2, \cdots$ by degrees, we can determine $S'_1, S'_2, \ldots, S'_n$ in a way that B_n becomes diagonal. As a consequence, $B_0 = A_0$, $B_1, B_2, \cdots$ can be all reduced to diagonal matrices.

Step 2.   By choosing appropriate diagonal matrices $S''_1, S''_2, \cdots$, there exists

$$S''(z) = 1 + \frac{S''_1}{z} + \frac{S''_2}{z^2} + \cdots \in GL_r \left(\mathbb{C} \left[\left[\frac{1}{z} \right] \right] \right), \tag{4.8}$$

such that $B(z)$ can be reduced to

$$S''(z+1)^{-1}B(z)S''(z) = z^\mu \left(\overline{A}_0 + \frac{\overline{A}_1}{z} \right).$$

Here, we have $\overline{A}_0 = B_0$, $\overline{A}_1 = B_1$. Indeed, when we have

$$S''(z+1)^{-1}B(z)S''(z) = \overline{A}_0 + \frac{\overline{A}_1}{z} + \frac{\overline{A}_2}{z^2} + \cdots, \tag{4.9}$$

by induction, we can choose diagonal matrices S_l'' ($l \geq 1$) in such a way that $\overline{A}_n$ ($n \geq 2$) vanish. First, we have $\overline{A}_0 = B_0$, $\overline{A}_1 = B_1$ and they are diagonal matrices. Next, by the equality

$$B_0 S_2'' + B_2 = \overline{A}_2 + (S_2'' - S_1'')B_0, \tag{4.10}$$

$$[B_0, S_2''] = 0, \tag{4.11}$$

we have $B_2 = \overline{A}_2 - S_1'' B_0$. For $\overline{A}_2$ to be 0, we set $S_1'' = -B_2 B_0^{-1}$. By induction hypothesis, assuming that we could choose $S_1'', S_2'', \ldots, S_{n-2}''$ in such a way that $\overline{A}_2 = \cdots = \overline{A}_{n-1} = 0$, the equality

$$(n-1)S_{n-1}'' B_0 = -(B_2 S_{n-2}'' + \cdots + B_{n-1} S_1'' + B_n) \tag{4.12}$$

$$+\overline{A}_n + \sum_{l=2}^{n-1}(-1)^l \frac{(n-l)\cdots(n-1)}{l!} S_{n-l}'' \overline{A}_0$$

$$+\sum_{l=1}^{n-2}(-1)^l \frac{(n-1-l)\cdots(n-2)}{l!} S_{n-l-1}'' \overline{A}_1$$

should hold, we can determine S_{n-1}'' uniquely in such a way that $\overline{A}_n = 0$.

Step 3. Setting $S(z) = S'(z)S''(z)$, we obtain a matrix $S(z)$ of the lemma.

Step 4. Assuming that A_0 is a diagonal matrix and $S_0 = 1$, we would like to show the uniqueness of $S(z)$. It is enough to show that, in (4.6) and (4.7), $A_0 = B_0 = \overline{A}_0$, $A_1 = B_1 = \overline{A}_1$, $A_l = B_l = 0$ ($l \geq 2$) implies $S_l' = 0$ ($l \geq 1$). By (4.6), $[A_0, S_1'] = 0$, i.e., S_1' is a diagonal matrix. Setting $n = 2$ in (4.7), we have

$$[A_0, S_2'] = -A_1 S_1' + S_1'(B_1 - B_0) = -S_1' B_0$$

which implies $[A_0[A_0, S_2']] = 0$. As A_0 is a diagonal matrix, this means $[A_0, S_2'] = 0$, that is, $S_1' = 0$. By induction hypothesis, assuming that $S_1' = \cdots = S_{n-2}' = 0$. By (4.7), we have $[A_0, S_{n-1}'] = 0$, i.e., S_{n-1}' is a diagonal matrix. Again by (4.7), we have

$$[A_0, S_n'] = -A_1 S_{n-1}' - (n-1)S_{n-1}' B_0 + S_{n-1}' B_1$$

$$= -(n-1)S_{n-1}' B_0$$

which implies $[A_0, [A_0, S_n']] = 0$, hence similarly as above, we obtain $[A_0, S_n'] = 0$, i.e., $S_{n-1}' = 0$. Thus, we have shown that S_l' ($l \geq 1$) are 0 and we terminate the proof of the lemma.

Remark 4.1. By a similar argument, even if the eigenvalues of A_0 are not necessarily mutually different, if A_0 is diagonalizable, one can choose $S(z)$ in such a way that $\overline{A}_0$ and $\overline{A}_1$ commute.

4.1.2 Asymptotic Form of Solutions

There exists a matrix solution of a difference equation in the normal form

$$\Phi(z+1) = z^\mu \left(\overline{A}_0 + \frac{\overline{A}_1}{z} \right) \Phi(z) \qquad (4.13)$$

having an asymptotic behavior at $\Re z \mapsto +\infty$ given as follows. It is expressed by using the asymptotic formula of the Γ-function due to Stirling:

$$\Phi(z) = \Gamma(z)^\mu (\overline{A}_0)^z \frac{\Gamma(z + (\overline{A}_0)^{-1}\overline{A}_1)}{\Gamma(z)} \qquad (4.14)$$

$$= (\overline{A}_0)^z \exp\left[(\overline{A}_0)^{-1}\overline{A}_1 \log z \right] e^{-\mu z} z^{\mu\left(z - \frac{1}{2}\right)} (2\pi)^{\frac{\mu}{2}} \left\{ 1 + O\left(\frac{1}{|z|} \right) \right\},$$

$$-\pi + \delta < \arg z < \pi - \delta \quad (\pi > \delta > 0).$$

Similarly its asymptotic behavior at $\Re z \to -\infty$ is given by

$$\Phi(z) = \left(\frac{e^{\pi\sqrt{-1}z}}{\Gamma(1-z)} \right)^\mu (\overline{A}_0)^z \frac{\Gamma(1-z)}{\Gamma(1 - z - (\overline{A}_0)^{-1}\overline{A}_1)} \qquad (4.15)$$

$$= e^{\mu(1 + (\pi\sqrt{-1}-1)z)} (-z)^{\mu\left(z - \frac{1}{2}\right)} (2\pi)^{-\frac{\mu}{2}}$$

$$\cdot (\overline{A}_0)^z \exp\left[(\overline{A}_0)^{-1}\overline{A}_1 \log(-z) \right] \left\{ 1 + O\left(\frac{1}{|z|} \right) \right\}.$$

$$\delta < \arg z < 2\pi - \delta$$

Here, $\overline{A}_0^z$, $\exp\left[(\overline{A}_0)^{-1}\overline{A}_1 \log(\pm z) \right]$, $\Gamma(z \pm (\overline{A}_0)^{-1}\overline{A}_1)$ etc. are matrix analytic functions, but since we may assume that $\overline{A}_0$, $\overline{A}_1$ are both diagonal matrices in this case, they are regarded as diagonal matrix analytic functions defined for each diagonal component.

For matrix analytic functions, see. e.g., [Gan] and [La].

Below, we set an asymptotic behavior of $\Phi(z)$ at $\Re z \mapsto \pm\infty$ by

$$\widehat{\Phi}_{0,+}(z) = (\overline{A}_0)^z \exp\left[(\overline{A}_0)^{-1}\overline{A}_1 \log z \right] e^{-\mu z} z^{\mu\left(z - \frac{1}{2}\right)} (2\pi)^{\frac{\mu}{2}}, \qquad (4.16)$$

$$-\pi + \delta < \arg z < \pi - \delta,$$

$$\widehat{\Phi}_{0,-}(z) = e^{\mu(1 + (\pi\sqrt{-1}-1)z)} (-z)^{\mu\left(z - \frac{1}{2}\right)} (2\pi)^{-\frac{\mu}{2}} \qquad (4.17)$$

$$\cdot (\overline{A}_0)^z \exp\left[(\overline{A}_0)^{-1}\overline{A}_1 \log(-z) \right], \quad \delta < \arg z < 2\pi - \delta.$$

4.1.3 Existence Theorem (i)

Next, let us state the existence theorem due to G.D. Birkhoff [Bir2]:

Theorem 4.1. *Suppose that $A_0 \in GL_r(\mathbb{C})$ has been already diagonalized and its eigenvalues $\lambda_1, \ldots, \lambda_r$ satisfy $|\lambda_1| > |\lambda_2| > \cdots > |\lambda_r|$. We denote the corresponding eigenvalues $(\overline{A}_0)^{-1}\overline{A}_1$ $(\overline{A}_0 = A_0, \overline{A}_1 = $ (the diagonal part of A_1)) by $\rho_1, \rho_2, \ldots, \rho_r$.*

(1) There uniquely exists a matrix solution $\Phi(z)$ of the difference equation (4.1) that is meromorphic on $\mathbb{C}$, holomorphic on the left domain $\Delta_R^- = \{z \in \mathbb{C}; \Re z \leq 0, |z| \geq R, \text{ or } \Re z \geq 0, |\Im z| \geq R\}$ (cf. Figure 4.1), and has an asymptotic expansion at $\Re z \mapsto -\infty$ ($|\Im z|$: bounded)

$$\Phi(z) = \widehat{\Phi}_{0,-}(z)\left(1 + O\left(\frac{1}{|z|}\right)\right), \tag{4.18}$$

(where $O\left(\frac{1}{|z|}\right)$ is the Landau symbol. cf. § 1.1.1). Here, R is a sufficiently big positive constant. Similarly,

(2) there uniquely exists a matrix solution $\Phi(z)$ of the difference equation (4.1) that is meromorphic on $\mathbb{C}$, holomorphic on the right domain $\Delta_R^+ = \{z \in \mathbb{C}; \Re z \geq 0, |z| \geq R, \text{ or } \Re z \leq 0, |\Im z| \geq R\}$, and has an asymptotic expansion at $\Re z \mapsto +\infty$ ($|\Im z|$: bounded)

$$\Phi(z) = \widehat{\Phi}_{0,+}(z)\left(1 + O\left(\frac{1}{|z|}\right)\right). \tag{4.19}$$

We denote the meromorphic solutions on $\mathbb{C}$ in (1), (2) by $\Phi_-(z)$, $\Phi_+(z)$, respectively, and when $|\Im z|$ is bounded, we call each of them the fundamental solution of (4.1) having an asymptotic expansion at $\Re z \to -\infty$ (resp. $\Re z \mapsto +\infty$) as (4.18) (resp. (4.19)).

When a column vector $(\varphi_j(z))_{j=1}^r$ satisfies the equation

$$(\varphi_j(z+1))_{j=1}^r = A(z)(\varphi_j(z))_{j=1}^r, \tag{4.20}$$

we simply call $(\varphi_j(z))_{j=1}^r$ a solution of (4.1).

The proof of Theorem 4.1 is given in § 4.1.5–4.1.13.

When $(p_j(z))_{j=1}^r$ is a periodic meromorphic function on $\mathbb{C}$, i.e.,

$$p_j(z+1) = p_j(z), \tag{4.21}$$

then, the function

$$(\varphi_j(z))_{j=1}^r = \Phi_\pm(z)(p_j(z))_{j=1}^r \tag{4.22}$$

is a solution of (4.1) and vice versa.

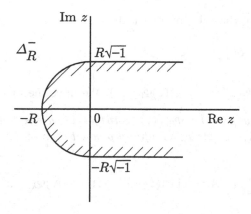

Fig. 4.1

4.1.4 Infinite-Product Representation of Matrices

For any positive integer N, by (4.1), we obtain

$$\Phi(z) = A(z-1)A(z-2)\cdots A(z-N)\Phi(z-N). \tag{4.23}$$

Furthermore, replacing $\Phi(z-N)$ with $\widehat{\Phi}_{0,-}(z-N)$, since we have

$$\lim_{N\mapsto+\infty} \widehat{\Phi}_{0,-}(z-N)^{-1}\Phi(z-N) = 1,$$

as $N \mapsto +\infty$,

$$\lim_{N\mapsto+\infty} A(z-1)A(z-2)\cdots A(z-N)\widehat{\Phi}_{0,-}(z-N), \tag{4.24}$$

may become a solution of (4.1). But in general, we do not know whether (4.24) converges or not. Instead, if we think of (4.24) as a function not with values in $GL_r(\mathbb{C})$ but with values in an appropriate quotient of $GL_r(\mathbb{C})$, then it would be convergent. We explain this below.

Let $\mathcal{B}_r$ be the maximal solvable subgroup of $GL_r(\mathbb{C})$ consisting of upper triangular matrices, i.e., a Borel subgroup, and $\mathcal{U}_r$ be the maximal unipotent subgroup of upper triangular matrices that are unipotent (all the diagonal components are 1). Then, we have $\mathcal{U}_r \subset \mathcal{B}_r$, and the quotient spaces of left classes $\mathcal{A} = GL_r(\mathbb{C})/\mathcal{U}_r$, $\mathcal{F} = GL_r(\mathbb{C})/\mathcal{B}_r$ are an $\frac{r(r+1)}{2}$-dimensional affine variety over $\mathbb{C}$ and an $\frac{r(r-1)}{2}$-dimensional projective variety over $\mathbb{C}$ (hence compact), respectively. The former is called a principal affine space, and the latter a flag manifold. Now, as we can regard $\Phi_\pm(z)$ as an element of the left classes by $\mathcal{U}_r$ and $\mathcal{B}_r$, $\Phi_\pm(z)\mathcal{U}_r$, $\Phi_\pm(z)\mathcal{B}_r$ can be regarded as an $\mathcal{A}$-valued (resp. an $\mathcal{F}$-valued) meromorphic function, more precisely a meromorphic map, on $\mathbb{C}$. Taking the kth finite part of $S(z)$

$$S^{(k)}(z) = S_0 + \frac{S_1}{z} + \cdots + \frac{S_k}{z^k}, \quad k \geq 0 \tag{4.25}$$

$(S_0 = 1)$, we set $\widehat{\Phi}_{k,\pm}(z) = S^{(k)}(z)\widehat{\Phi}_{0,\pm}(z)$. We have the following theorem.

Theorem 4.2. *Suppose that $\widehat{\Phi}_{0,\pm}(z)$ (4.16)–(4.17) are both diagonalized. For a sufficiently big k, we obtain the infinite-product representation of the matrix $\Phi_-(z)$*

$$\Phi_-(z)\mathcal{U}_r = \lim_{N \mapsto +\infty} A(z-1)A(z-2)\cdots A(z-N)\widehat{\Phi}_{k,-}(z-N)\mathcal{U}_r, \tag{4.26}$$

$$\Phi_-(z)\mathcal{B}_r = \lim_{N \mapsto +\infty} A(z-1)A(z-2)\cdots A(z-N)S^{(k)}(z-N)\mathcal{B}_r, \tag{4.27}$$

as a meromorphic function of $z(\in \mathbb{C})$. Similarly, $\Phi_+(z)$ has the infinite-product representation

$$\Phi_+(z)\mathcal{U}_r = \lim_{N \mapsto +\infty} A(z)^{-1} \cdots A(z+N-1)^{-1}\widehat{\Phi}_{k,+}(z+N)\mathcal{U}_r, \tag{4.28}$$

$$\Phi_+(z)\mathcal{B}_r = \lim_{N \mapsto +\infty} A(z)^{-1} \cdots A(z+N-1)^{-1}S^{(k)}(z+N)\mathcal{B}_r. \tag{4.29}$$

Here, the limits in the right-hand side of (4.26)–(4.29) converge uniformly on every compact subset of $\mathbb{C}$.

In particular, for $r = 2$, $GL_2(\mathbb{C})/\mathcal{B}_2$ is the projective line and (4.27), (4.29) give us countably infinite-product representations by linear fractional transformations.

4.1.5 Gauss' Decomposition

Since it suffices to prove the existence Theorem 4.1 in the case $\mu = 0$, for simplicity, we assume that $\mu = 0$.

The remaining part of § 4.1 will be devoted to the proof of the existence Theorem 4.1. Theorem 4.2 will be proved in between the process. We will construct $\Phi_\pm(z)$ via the decomposition of a matrix into the product of a lower triangular and an upper triangular matrices, that is, by the Gauss decomposition.

Step 1.   First, we show the uniqueness. Suppose that $\Phi(z)$, $\widetilde{\Phi}(z)$ are two matrix meromorphic solutions of (4.1) having the same asymptotic behavior (4.15) at $\Re z \mapsto -\infty$. The function $\Phi(z)^{-1}\widetilde{\Phi}(z)$ is periodic with respect to the translations $z \mapsto z \pm 1$. On the other hand, as we have

$\lim_{\Re z \mapsto -\infty} \Phi(z)^{-1}\widetilde{\Phi}(z) = 1$ at $\Re z \mapsto -\infty$, it follows that

$$\Phi^{-1}(z)\widetilde{\Phi}(z) = \lim_{N \mapsto +\infty} \Phi(z - N)^{-1}\widetilde{\Phi}(z - N) = 1.$$

Step 2.　Next, we show the existence. First, we construct a solution $\Phi(z)$ of (4.1) not as a $GL_r(\mathbb{C})$-valued function but as a function with its value in the quotient space $GL_r(BC)/\mathcal{U}_r$. That is, we decompose the matrix $\Phi(z)$

$$\Phi(z) = \Xi(z) \cdot H(z) \tag{4.30}$$

($\Xi(z)$ is lower triangular and $H(z)$ is a unipotent upper triangular) in the form of Gauss' decomposition, and find each of $\Xi(z)$, $H(z)$. $\Xi(z) = ((\xi_{ij}(z)))_{i,j=1}^r$ and $H(z) = ((\eta_{ij}(z)))_{i,j=1}^r$ can be uniquely expressed by the components $\varphi_{ij}(z)$ of $\Phi(z)$ as follows (cf., e.g., [Gan]).

$$\begin{cases} \xi_{ij} = 0, & i < j \\[2mm] \xi_{ij} = \dfrac{\Phi\begin{pmatrix} 1\,2\cdots j-1\,i \\ 1\,2\cdots j-1\,j \end{pmatrix}}{\Phi(1\,2\cdots j-1\,j)}, & i \geq j \end{cases} \tag{4.31}$$

$$\begin{cases} \eta_{ij} = 0 \;(i > j), \quad \eta_{ii} = 1 \\[2mm] \eta_{ij} = \dfrac{\Phi\begin{pmatrix} 1\,2\cdots i-1\,i \\ 1\,2\cdots i-1\,j \end{pmatrix}}{\Phi(1\,2\cdots i)}, & i < j. \end{cases} \tag{4.32}$$

Here, $\Phi\binom{i_1\cdots i_p}{j_1\cdots j_p}$, $1 \leq i_1 < \cdots < i_p \leq r$, $1 \leq j_1 < \cdots < j_p \leq r$ is the minor of $\Phi(z)$ obtained by extracting $i_1, \ldots, i_p$th rows and $j_1, \ldots, j_p$th columns, and $\Phi(i_1 \cdots i_p)$ signifies $\Phi\binom{i_1\cdots i_p}{i_1\cdots i_p}$. Notice that we take $\overline{A}_0$, $\overline{A}_1$ in diagonal forms. For $N \in \mathbb{Z}_{>0}$, as an approximative solution of $\Phi(z)$, we set

$$\Phi_N(z) = A(z - 1)A(z - 2) \cdots A(z - N)\widehat{\Phi}_{k,-}(z - N). \tag{4.33}$$

We rewrite it further as

$$\Phi_N(z) = \widehat{\Phi}_{k,-}(z)\Phi'_N(z), \tag{4.34}$$

$$\Phi'_N(z) = \left\{\widehat{\Phi}_{k,-}(z)^{-1}A(z - 1)\widehat{\Phi}_{k,-}(z - 1)\right\} \cdots$$

$$\cdots \left\{\widehat{\Phi}_{k,-}(z - N + 1)^{-1}A(z - N)\widehat{\Phi}_{k,-}(z - N)\right\}.$$

When we regard it as a formal Laurent series, by

$$S(z)^{-1}S^{(k)}(z) = 1 + O\left(\frac{1}{|z|^{k+1}}\right),$$

(4.4) implies

$$S^{(k)}(z+1)^{-1}A(z)S^{(k)}(z) = \overline{A}(z)\left(1 + \frac{Q_k(z)}{z^{k+1}}\right). \tag{4.35}$$

Here, $\overline{A}(z) = \overline{A}_0 + \frac{\overline{A}_1}{z}$ and $Q_k(z)$ is a meromorphic function which is holomorphic in a neighborhood of $z = \infty$. Now, we have

$$\widehat{\Phi}_{k,-}(z+1)^{-1}A(z)\widehat{\Phi}_{k,-}(z) \tag{4.36}$$

$$= \widehat{\Phi}_{0,-}(z+1)^{-1}S^{(k)}(z+1)^{-1}A(z)S^{(k)}(z)\widehat{\Phi}_{0,-}(z)$$

$$= \widehat{\Phi}_{0,-}(z+1)^{-1}\overline{A}(z)\left(1 + \frac{Q_k(z)}{z^{k+1}}\right)\widehat{\Phi}_{0,-}(z)$$

$$= \widehat{\Phi}_{0,-}(z)^{-1}\left(1 + \frac{Q_k(z)}{z^{k+1}}\right)\widehat{\Phi}_{0,-}(z).$$

Setting the (i,j)-component of $Q_k(z)$ as $q_{ij}(z)$ $(1 \leqslant i, j \leqslant r)$, as $\widehat{\Phi}_{0,-}(z)$ is a diagonal matrix (4.17), the (i,j)-component of the matrix $\widehat{\Phi}_{0,-}(z)^{-1}Q_k(z)\widehat{\Phi}_{0,-}(z)z^{-k-1}$ is expressed as

$$\delta_{ij} + \left(\frac{\lambda_j}{\lambda_i}\right)^z (-z)^{\rho_j - \rho_i}z^{-k-1}q_{ij}(z).$$

4.1.6 Regularization of the Product

Now, take $d \in \mathbb{Z}_{>0}$ in such a way that $d-1 > \max_{i \neq j}|\rho_j - \rho_i|$ and a sufficiently big $k \in \mathbb{Z}_{>0}$ such that $k - d > \max_{i \neq j}|\rho_j - \rho_i|$, and fix them. (4.36) can be expressed as

$$\widehat{\Phi}_{k,-}(z+1)^{-1}A(z)\widehat{\Phi}_{k,-}(z) = 1 + \frac{\Theta_k(z)}{z^{d+1}}, \tag{4.37}$$

$$\Theta_k(z) = z^{-k+d}\widehat{\Phi}_{0,-}(z)^{-1}Q_k(z)\widehat{\Phi}_{0,-}(z). \tag{4.38}$$

Here, the (i,j)-component $\vartheta_{ij}(z)$ of the matrix $\Theta_k(z)$ is given by $\vartheta_{ij}(z) = \left(\frac{\lambda_j}{\lambda_i}\right)^z \psi_{ij}(z)$, $\psi_{ij}(z) = (-z)^{\rho_j - \rho_i}z^{-k+d}q_{ij}(z)$. As a consequence, $\Phi'_N(z)$ can be expressed in the form of an ordered product of matrices

$$\Phi'_N(z) = \left(1 + \frac{\Theta_k(z-1)}{(z-1)^{d+1}}\right) \cdots \left(1 + \frac{\Theta_k(z-N)}{(z-N)^{d+1}}\right). \tag{4.39}$$

The $(i,1)$-component $(i \geq 1)$ of $\Phi'_N(z)$ can be expanded as

$$\delta_{i1} + \sum_{\nu=1}^{N} \sum_{1 \leq k_1 < \cdots < k_\nu} \frac{1}{(z-k_1)^{d+1} \cdots (z-k_\nu)^{d+1}} \tag{4.40}$$

$$\cdot \sum_{1 \leq l_1, \ldots, l_{\nu-1} \leq r} \left(\frac{\lambda_{l_1}}{\lambda_i}\right)^{z-k_1} \left(\frac{\lambda_{l_2}}{\lambda_{l_1}}\right)^{z-k_2} \cdots \left(\frac{\lambda_{l_{\nu-1}}}{\lambda_{l_{\nu-2}}}\right)^{z-k_{\nu-1}} \left(\frac{\lambda_1}{\lambda_{l_{\nu-1}}}\right)^{z-k_\nu}$$

$$\cdot \psi_{il_1}(z-k_1)\psi_{l_1 l_2}(z-k_2) \cdots \psi_{l_{\nu-1}1}(z-k_\nu).$$

Notice that $\psi_{ij}(z-k)$ $(k=0,1,2,\cdots)$ is bounded on any point z of Δ_R^-, i.e.,

$$|\psi_{ij}(z-k)| \leq K \quad (K \text{ is a positive constant}).$$

Hence, the estimate of the absolute value of each term in (4.40) has, by $|\lambda_1| > |\lambda_j|$, a dominant series

$$\delta_{i1} + \sum_{\nu=1}^{N} \sum_{1 \leq k_1 < \cdots < k_\nu} \sum_{1 \leq l_1, \ldots, l_{\nu-1} \leq r} \frac{K^\nu}{|z-k_1|^{d+1} \cdots |z-k_\nu|^{d+1}} \tag{4.41}$$

$$\left|\left(\frac{\lambda_1}{\lambda_i}\right)^z\right| \left|\left(\frac{\lambda_i}{\lambda_1}\right)^{k_1} \left(\frac{\lambda_{l_1}}{\lambda_1}\right)^{k_2-k_1} \cdots \left(\frac{\lambda_{l_{\nu-1}}}{\lambda_1}\right)^{k_\nu-k_{\nu-1}}\right|$$

$$\leq \delta_{i1} + \sum_{\nu=1}^{N} \sum_{1 \leq k_1 < \cdots < k_\nu} \frac{r^{\nu-1}K^\nu}{|z-k_1|^{d+1} \cdots |z-k_\nu|^{d+1}} \left|\left(\frac{\lambda_1}{\lambda_i}\right)^z\right|$$

$$\leq \delta_{i1} + \frac{1}{r} \left|\left(\frac{\lambda_1}{\lambda_i}\right)^z\right| \left\{\prod_{\nu=1}^{\infty} \left(1 + \frac{rK}{|z-\nu|^{d+1}}\right) - 1\right\}.$$

The infinite product

$$\prod_{\nu=1}^{\infty} \left(1 + \frac{rK}{|z-\nu|^{d+1}}\right), \tag{4.42}$$

for $d > 0$, converges uniformly on every compact subset of Δ_R^-. Hence, the first column of $\Phi'_N(z)$ converges uniformly on every compact subset as $N \mapsto +\infty$. Thus, so does $\lim_{N \mapsto +\infty}\{$the first column of $\Phi_N(z)\}$ on Δ_R^-.

4.1.7 Convergence of the First Column

Furthermore, since we have, for any natural number $l \leq N$,

$$\Phi_N(z) = A(z-1)A(z-2)\cdots A(z-l)\Phi_{N-l}(z-l), \qquad (4.43)$$

and for any $z \in \mathbb{C}$, $z - l \in \Delta_R^-$ for a sufficiently big l, each of

$$\xi_{i1}(z) = \lim_{N \mapsto \infty} \{\text{the } (i,1)\text{-component of } \Phi_N(z)\} \qquad (4.44)$$

$$\xi_{i1}'(z) = \lim_{N \mapsto \infty} \{\text{the } (i,1)\text{-component of } \Phi_N'(z)\} \qquad (4.45)$$

is analytically continued to a meromorphic function on the whole $\mathbb{C}$. The functions $\xi_{i1}(z)$, $\xi_{i1}'(z)$ are expanded as uniformly convergent series on every compact subset of the whole $\mathbb{C}$:

$$\xi_{i1}'(z) = \delta_{i1} + \left(\frac{\lambda_1}{\lambda_i}\right)^z \sum_{\nu=1}^{\infty} \sum_{1 \leq k_1 < \cdots < k_\nu} \sum_{1 \leq l_1,\ldots,l_{\nu-1} \leq r} \qquad (4.46)$$

$$\left(\frac{\lambda_i}{\lambda_1}\right)^{k_1} \left(\frac{\lambda_{l_1}}{\lambda_1}\right)^{k_2-k_1} \cdots \cdots \left(\frac{\lambda_{l_{\nu-1}}}{\lambda_1}\right)^{k_\nu-k_{\nu-1}}$$

$$\frac{\psi_{il_1}(z-k_1)\psi_{l_1l_2}(z-k_2)\cdots\psi_{l_{\nu-1}1}(z-k_\nu)}{(z-k_1)^{d+1}(z-k_2)^{d+1}\cdots(z-k_\nu)^{d+1}},$$

$$(\xi_{i1}(z))_{i=1}^r = \widehat{\Phi}_{k,-}(z)(\xi_{i1}'(z))_{i=1}^r. \qquad (4.47)$$

Proposition 4.1. *The function $\xi_{i1}'(z)$ has an asymptotic expansion*

$$\xi_{i1}'(z) = \begin{cases} \delta_{i1} + \left(\frac{\lambda_1}{\lambda_i}\right)^z O\left(\frac{1}{|z|^d}\right), & \Re z \leq 0 \\ \delta_{i1} + \left(\frac{\lambda_1}{\lambda_i}\right)^z O\left(\frac{1}{|\Im z|^d}\right), & \Re z \geq 0 \end{cases} \qquad (4.48)$$

at $|z| \mapsto +\infty$ on Δ_R^-.

4.1.8 Asymptotic Estimate of Infinite Product

To prove Proposition 4.1, we prepare two lemmata.

Lemma 4.2. *If $\Re z \leq 0$, we have the inequality:*

$$\sum_{k=1}^{\infty} \frac{1}{|z-k|^{d+1}} \leq \frac{1}{2}|z|^{-d} \frac{\Gamma\left(\frac{d}{2}\right)\Gamma\left(\frac{1}{2}\right)}{\Gamma\left(\frac{d+1}{2}\right)}. \qquad (4.49)$$

Proof. By $|z - k| \geq \sqrt{|z|^2 + k^2}$, we have

$$\sum_{k=1}^{\infty} \frac{1}{|z - k|^{d+1}} \leq \sum_{k=1}^{\infty} \frac{1}{(|z|^2 + k^2)^{\frac{d+1}{2}}} \tag{4.50}$$

$$\leq \int_0^{\infty} \frac{dt}{(|z|^2 + t^2)^{\frac{d+1}{2}}} = \frac{1}{2}|z|^{-d} \frac{\Gamma\left(\frac{d}{2}\right) \Gamma\left(\frac{1}{2}\right)}{\Gamma\left(\frac{d+1}{2}\right)}.$$

Lemma 4.3. *If $\Re z \geq 0$ and $|\Im z| \neq 0$, we have*

$$\sum_{k=1}^{\infty} \frac{1}{|z - k|^{d+1}} \leq \frac{2}{|\Im z|^d} \left(\frac{1}{|\Im z|} + \frac{\Gamma\left(\frac{d}{2}\right)\Gamma\left(\frac{1}{2}\right)}{2\Gamma\left(\frac{d+1}{2}\right)} \right). \tag{4.51}$$

Proof. Decomposing the sum in (4.51) into two parts $|z - k| < 1$ and $|z - k| \geq 1$, the result immediately follows.

By Lemma 4.2 and 4.3, there exists a positive constant C such that

$$\sum_{k=1}^{\infty} \frac{1}{|z - k|^{d+1}} \leq \begin{cases} C \frac{1}{|z|^d}, & \Re z \leq 0 \\ C \frac{1}{|\Im z|^d}, & \Re z \geq 0 \end{cases} \tag{4.52}$$

for $z \in \Delta_R^-$. Hence, we obtain the estimate

$$\left(1 + \frac{rK}{|z - 1|^{d+1}} \right) \left(1 + \frac{rK}{|z - 2|^{d+1}} \right) \cdots$$

$$\leq \begin{cases} \exp\left(\frac{rKC}{|z|^d} \right), & \Re z \leq 0 \\ \exp\left(\frac{rKC}{|\Im z|^d} \right), & \Re z \geq 0. \end{cases}$$

Thus, Proposition 4.1 follows from (4.41).

On the sectorial domain $\delta < \arg z < 2\pi - \delta$ (δ a small positive number), since we have $|\Im z| \geq \frac{|z|}{\sin \delta}$ when $\Re z \geq 0$, Proposition 4.1 implies:

Corollary 4.1. *On the sectorial domain $\delta < \arg z < 2\pi - \delta$, an asymptotic behavior of $\xi'_{i1}(z)$ at $|z| \mapsto +\infty$ is given by*

$$\xi'_{i1}(z) = \delta_{i1} + \left(\frac{\lambda_1}{\lambda_i} \right)^z O\left(\frac{1}{|z|^d} \right) \tag{4.53}$$

for $1 \leq i \leq r$.

By the transformation

$$\Phi(z) = S^{(k)}(z) \Phi_k(z), \tag{4.54}$$

the difference equation (4.1) is rewritten in the form

$$\Phi_k(z+1) = S^{(k)}(z+1)^{-1}A(z)S^{(k)}(z)\Phi_k(z). \tag{4.55}$$

Equation (4.55) is the same type of equation as (4.1), and if Theorem 4.1 is shown for (4.55), via the transformation (4.54), it is also shown for (4.1). Hence below, we assume $k = 0$ and show Theorem 4.1. Now, by Proposition 4.1 and (4.47), we have:

Corollary 4.2. *For $k = 0$, we have an asymptotic expansion at $|z| \mapsto +\infty$ ($|\Im z|$: bounded)*

$$\xi_{i1}(z) = \begin{cases} \lambda_1^z(-z)^{\rho_1}\left(\delta_{i1} + O\left(\frac{1}{|z|}\right)\right), & \Re z \leqslant 0, \\ \lambda_1^z(-z)^{\rho_i}\left(\delta_{i1} + O\left(\frac{1}{|\Im z|^d}\right)\right), & \Re z \geq 0. \end{cases} \tag{4.56}$$

Proof. The equality

$$(\xi_{i1}(z))_{i=1}^r = \widehat{\Phi}_{0,-}(z)(\xi_{i1}'(z))_{i=1}^r \tag{4.57}$$

holds, and by the choice of d in § 4.1.6, $(-z)^{\rho_i - \rho_1}z^{-d+1}$ is bounded at $|z| \to +\infty$ ($|\Im z|$: bounded), hence, we obtain (4.56).

4.1.9 Convergence of Lower Triangular Matrices

In place of $A(z)$ and $\Phi(z)$, we consider the exterior product of $A(z)$ and $\Phi(z)$ of degree s ($1 \leq s \leq r$) and the difference equations of $\binom{r}{s} \times \binom{r}{s}$ matrices:

$$\underbrace{\Phi \wedge \cdots \wedge \Phi}_{s\text{th exterior product}}(z) = \Phi^{(s)}(z), \tag{4.58}$$

$$\underbrace{A \wedge \cdots \wedge A}_{s\text{th exterior product}}(z) = A^{(s)}(z), \tag{4.59}$$

$$\Phi^{(s)}(z+1) = A^{(s)}(z)\Phi^{(s)}(z). \tag{4.60}$$

Each component of the first column of the matrix $\Phi^{(s)}$ is, for an s-tuple of integers $i_1, i_2, \ldots, i_s$ ($1 \leq i_1 < \cdots < i_s \leq r$), a difference equation on the minor $\Phi\binom{i_1 \cdots i_s}{1 \cdots s}$ of degree s, and a component $\Phi_N\binom{i_1 \cdots i_s}{1 \cdots s}$ of the first column of $\Phi_N^{(s)} = \underbrace{\Phi_N \wedge \cdots \wedge \Phi_N}_{s\text{th}}(z)$, similarly to (4.2), has the Laurent expansion

$$A^{(s)}(z) = z^{\mu s}\left(A_0^{(s)} + \frac{A_1^{(s)}}{z} + \cdots\right), \tag{4.61}$$

where we set $(A_0^{(s)} = \underbrace{A_0 \wedge \cdots \wedge A_0}_{s\text{th}})$. The eigenvalues of $A_0^{(s)}$ are the s dif-ferent products $\lambda_{i_1}, \ldots, \lambda_{i_s}$, $1 \le i_1 < \cdots < i_s \le r$ of the eigenvalues λ_j of A_0. Since, under the assumption of Theorem 4.1, $A^{(s)}(z)$ again satisfies the assumption of Theorem 4.1 or the assumption in Remark 4.1, by an argument similar to § 4.1.8,

$$\Xi \begin{pmatrix} i_1 \cdots i_s \\ 1 \cdots s \end{pmatrix} = \lim_{N \mapsto +\infty} \Phi_N \begin{pmatrix} i_1 \cdots i_s \\ 1 \cdots s \end{pmatrix} \tag{4.62}$$

converges uniformly on every compact subset of Δ_R^-, hence, on the whole $\mathbb{C}$.

4.1.10 Asymptotic Estimate of Lower Triangular Matrices

Now, an element $\Phi_N(z)\mathcal{U}_r$ of the left coset by the maximal unipotent sub-group $\mathcal{U}_r$ is expressed as $\Xi_N(z)\mathcal{U}_r$ by a lower triangular matrix

$$\Xi_N(z) = \begin{pmatrix} \xi_{11}^{(N)} & & & \\ \xi_{21}^{(N)} & \xi_{22}^{(N)} & & \\ \vdots & & \ddots & \ddots \\ \xi_{r1}^{(N)} & \cdots & & \xi_{rr}^{(N)} \end{pmatrix}, \tag{4.63}$$

and this expression is unique. Here, $\xi_{ij}^{(N)}$ is given, from the formula of Gauss's decomposition (4.31), by

$$\xi_{ij}^{(N)}(z) = \frac{\Phi^{(N)} \begin{pmatrix} 1\ 2 \cdots j-1\ i \\ 1\ 2 \cdots j-1\ j \end{pmatrix}}{\Phi^{(N)}(1\ 2 \cdots j-1)}, \quad i \ge j \tag{4.64}$$

hence on Δ_R^-, the limit

$$\xi_{ij}(z) = \lim_{N \mapsto +\infty} \xi_{ij}^{(N)}(z) \tag{4.65}$$

exists. In this sense, $\lim_{N \mapsto +\infty} \Phi_N(z)\mathcal{U}_r$ converges to a meromorphic function (in fact, a map) $\Xi(z)\mathcal{U}_r$, $\Xi(z) = ((\xi_{ij}(z)))_{i,j=1}^r$ ($\xi_{ij}(z) = 0$, $i < j$) on Λ_R^- with values in the principal affine space $GL_r(\mathbb{C})/\mathcal{U}_r$. Moreover, since the equality

$$\Xi(z+1)\mathcal{U}_r = A(z)\Xi(z)\mathcal{U}_r \tag{4.66}$$

is satisfied, $\Xi(z)\mathcal{U}_r$ extends analytically to a meromorphic function on $\mathbb{C}$ with values in $GL_r(\mathbb{C})/\mathcal{U}_r$.

Proposition 4.2. *On Δ_R^-, $\xi_{ij}(z)$ $(i \geq j)$ has an asymptotic expansion at $|z| \mapsto +\infty$ $(|\Im z|$: bounded)*

$$\xi_{ij}(z) = \begin{cases} \lambda_j^z(-z)^{\rho_j} \left(\delta_{ij} + O\left(\frac{1}{|z|}\right)\right), & \Re z \leq 0, \\ \lambda_j^z(-z)^{\rho_i} \left(\delta_{ij} + O\left(\frac{1}{|z|}\right) + O\left(\frac{1}{|\Im z|^d}\right)\right), & \Re z \geq 0. \end{cases} \tag{4.67}$$

Here, $d \in \mathbb{Z}_{>0}$ satisfies $d - 1 > r \max_{i \neq j} |\rho_i - \rho_j|$.

Proof. For $j = 1$, the proof has already been given. The sth exterior product of $\Phi_N'(z)$ is expressed as

$$\Phi_N'^{(s)}(z) = \left(1 + \frac{\Theta^{(s)}(z-1)}{(z-1)^{d+1}}\right) \cdots \left(1 + \frac{\Theta^{(s)}(z-N)}{(z-N)^{d+1}}\right), \tag{4.68}$$

where each component of $\Theta^{(s)}(z)$ is given by

$$\theta_{I,J}^{(s)}(z) = \left(\frac{\lambda_{j_1} \cdots \lambda_{j_s}}{\lambda_{i_1} \cdots \lambda_{i_s}}\right)^z \psi_{I,J}(z), \tag{4.69}$$

$I = \{i_1 < \cdots < i_s\}$, $J = \{j_1 < \cdots < j_s\}$ and $\psi_{I,J}(z)$ is bounded on Δ_R^-. By an argument similar to the case of $s = 1$, the asymptotic expansion of

$$\lim_{N \mapsto +\infty} \Phi_N \begin{pmatrix} i_1 \cdots i_s \\ 1 \cdots s \end{pmatrix} = \Xi \begin{pmatrix} i_1 \cdots i_s \\ 1 \cdots s \end{pmatrix} \tag{4.70}$$

at $|z| \to +\infty$ $(|\Im z|$: bounded) is given by

$$\Xi \begin{pmatrix} i_1 \cdots i_s \\ 1 \cdots s \end{pmatrix} \tag{4.71}$$

$$= \begin{cases} (\lambda_1 \cdots \lambda_s)^z (-z)^{\rho_1 + \cdots + \rho_s} \left(\delta_{i_1 1} \cdots \delta_{i_s s} + O\left(\frac{1}{|z|}\right)\right), & \Re z \leq 0 \\ (\lambda_1 \cdots \lambda_s)^z (-z)^{\rho_{i_1} + \cdots + \rho_{i_s}} \left(\delta_{i_1 1} \cdots \delta_{i_s 1} + O\left(\frac{1}{|\Im z|^d}\right)\right), & \Re z \geq 0. \end{cases}$$

Hence, at $|z| \mapsto +\infty$ $(|\Im z|$: bounded), we have

$$\xi_{ii}(z) = \begin{cases} \lambda_i^z(-z)^{\rho_i} \left(1 + O\left(\frac{1}{|z|}\right)\right), & \Re z \leq 0, \\ \lambda_i^z(-z)^{\rho_i} \left(1 + O\left(\frac{1}{|\Im z|^d}\right)\right), & \Re z \geq 0. \end{cases} \tag{4.72}$$

Similarly, rewriting

$$\xi_{ij}(z) = \frac{\Xi \begin{pmatrix} 1 \cdots j-1 \ i \\ 1 \cdots j-1 \ j \end{pmatrix}}{\Xi(1 \cdots j-1)} \tag{4.73}$$

with (4.71), we obtain (4.67).

4.1.11 Difference Equation Satisfied by Upper Triangular Matrices

Since we have constructed the lower triangular part $\Xi(z)$ of the Gauss decomposition (4.30) of a solution $\Phi(z)$ in the previous subsection, in this subsection, we construct the upper triangular part $H(z)$. In (4.30), we may assume that $\Xi(z)$ already satisfies (4.66). By the uniqueness of the expression in (4.30), $H(z)$ satisfies

$$\begin{cases} H(z+1) = B(z)H(z). \\ B(z) = \Xi(z+1)^{-1}A(z)\Xi(z). \end{cases} \tag{4.74}$$

Here, $B(z) \in \mathcal{U}_r$, that is, $B(z)$ is a unipotent upper triangular matrix.

Now, we should solve (4.74) with respect to the unipotent upper triangular matrix $H(z)$. Writing (4.74) for each component, by the fact that $B(z)$ is a unipotent upper triangular matrix, we have

$$\eta_{ij}(z+1) - \sum_{\nu=i+1}^{j-1} b_{i\nu}(z)\eta_{\nu j}(z) - b_{ij}(z) = \eta_{ij}(z), \quad i < j. \tag{4.75}$$

First, for $i = j - 1$, we find a solution $\eta_{j-1,j}(z)$ of the difference equation

$$\eta_{j-1,j}(z+1) = b_{j-1,j}(z) + \eta_{j-1,j}(z). \tag{4.76}$$

Fixing j, we solve $\eta_{j-1,j}(z)$, $\eta_{j-2,j}(z), \ldots, \eta_{ij}(z)$ $(i < j)$ step by step. By (4.74), since $b_{ij}(z)$ $(i < j)$ is expressed as

$$b_{ij}(z) = \sum_{\substack{1 \le \nu \le i \\ j \le \sigma \le r}} \widetilde{\xi}_{i\nu}(z+1)a_{\nu\sigma}(z)\xi_{\sigma j}(z), \tag{4.77}$$

(where $\widetilde{\Xi}(z) = ((\widetilde{\xi}_{ij}(z)))_{i,\ j=1}^{r}$ is the inverse matrix of $\Xi(z)$), by Lemma 4.1 and Proposition 4.2, we obtain the following estimate.

Lemma 4.4. On $z \in \Delta_R^-$, at $|z| \mapsto +\infty$ $(|\Im z|$: bounded), for $i < j$, we have

$$b_{ij}(z) = \begin{cases} \left(\frac{\lambda_j}{\lambda_i}\right)^z (-z)^{\rho_j - \rho_i} z^{-1} \left(1 + O\left(\frac{1}{|z|}\right)\right), & \Re z \le 0 \\ \left(\frac{\lambda_j}{\lambda_i}\right)^z O(|z|^{\rho_* - 1}) \left(1 + O\left(\frac{1}{|\Im z|^d}\right)\right), & \Re z \ge 0, \end{cases} \tag{4.78}$$

where we set $\rho_* = \max_{i \ne j} |\rho_j - \rho_i|$.

Proof. In fact, since we have

$$\widetilde{\xi}_{i\nu}(z) = \begin{cases} \lambda_i^{-z}(-z)^{-\rho_i}\left(\delta_{i\nu} + O\left(\frac{1}{|z|}\right)\right), & \Re z \leqslant 0 \\ \lambda_i^{-z}(-z)^{-\rho_\nu}\left(\delta_{i\nu} + O\left(\frac{1}{|\Im z|^d}\right)\right), & \Re z \geq 0 \end{cases} \tag{4.79}$$

as we see from (4.67), and $a_{\nu\sigma}(z) = O(|z|^{-1})$ $(\nu < \sigma)$ (see (4.2)), this together with (4.67) and (4.77) imply (4.78).

4.1.12 Resolution of Difference Equations

To solve (4.75), (4.76), we prepare the next lemma.

Lemma 4.5. *Let $\beta, \gamma \in \mathbb{C}$ be such that $|\beta| < 1$. Suppose that $g(z)$ is meromorphic on $\mathbb{C}$ and holomorphic on Δ_R^- which has an asymptotic behavior $\Re z \mapsto -\infty$ ($|\Im z|$: bounded) given by*

$$g(z) = \begin{cases} \beta^z(-z)^\gamma O(1), & \Re z \leq 0, \\ \beta^z O(1), & \Re z \geq 0. \end{cases} \tag{4.80}$$

Then, there exists a meromorphic function $f(z)$ on $\mathbb{C}$ which is holomorphic on Δ_R^- and has an asymptotic at $\Re z \mapsto -\infty$ ($|\Im z|$ bounded)

$$f(z) = \begin{cases} \beta^z(-z)^\gamma O(1), & \Re z \leqslant 0, \\ O(1), & \Re z \geq 0, \end{cases} \tag{4.81}$$

and satisfies the difference equation

$$f(z+1) = g(z) + f(z). \tag{4.82}$$

Proof. First, we decompose $\mathbb{C}$ into three strips and define the functions $f_+(z)$, $f_0(z)$, $f_-(z)$:

$$f_+(z) = -\sum_{n=0}^{\infty} g(z+n), z \in \Delta_R^- \cap \{z \in \mathbb{C}; \Im z \geq R\}, \tag{4.83}$$

$$f_0(z) = -\sum_{\substack{n=0 \\ \Re z + n \leq -R}}^{\infty} g(z+n), z \in \Delta_R^- \cap \{z \in \mathbb{C}; |\Im z| \leq 2R\}, \tag{4.84}$$

$$f_-(z) = -\sum_{n=0}^{\infty} g(z+n), z \in \Delta_R^- \cap \{z \in \mathbb{C}; \Im z \leq -R\}. \tag{4.85}$$

As can be seen from (4.80), $f_+(z)$, $f_0(z)$, $f_-(z)$ are well-defined and each of them satisfies (4.81). Indeed, for $R \leq \Im z$, if $\Re \gamma \geq 0$, we have

$$|f_+(z)| \le C_1 \sum_{n=0}^{\infty} |\beta^{z+n}|\, |z+n|^{\Re\gamma}$$

$$\stackrel{\le}{{}} C_1 |\beta^z| \sum_{n=0}^{\infty} n^{\Re\gamma}(|z|+1)^{\Re\gamma}|\beta|^n \quad \text{(because } n(1+|z|) \ge n + |z|)$$

$$\le C_2 |\beta^z|\, (1+|z|)^{\Re\gamma},$$

and if $\Re\gamma < 0$, we have

$$|f_+(z)| \le C_1 |\beta^z| \sum_{n=0}^{\infty} |z+n|^{\Re\gamma}|\beta|^n$$

$$= C_1 |\beta^z| \left\{ \sum_{n < \frac{1}{2}|z|} + \sum_{n \ge \frac{1}{2}|z|} \right\} \left(|z+n|^{\Re\gamma}|\beta^n| \right)$$

$$= C_1 |\beta^z| \left\{ C_3 |z|^{\Re\gamma} + \frac{|\beta|^{\frac{1}{2}|z|}}{1-|\beta|} \right\}$$

$$\le C_4 |\beta^z|\, |z|^{\Re\gamma}.$$

Here, C_1, C_2, C_3, C_4 are appropriate positive constants. Similar estimates hold for $f_0(z)$, $f_-(z)$ and they satisfy (4.81). Furthermore, $f_+(z)$, $f_0(z)$, $f_-(z)$ satisfy (4.82). Hence, on the intersection $\{z \in \mathbb{C}; R \le \Im z \le 2R\}$, the difference $f_+(z) - f_0(z)$ is a periodic holomorphic function, that is,

$$f_+(z+1) - f_0(z+1) = f_+(z) - f_0(z). \tag{4.86}$$

Similarly, on $\{z \in \mathbb{C}; -2R \le \Im z \le -R\}$, $f_0(z) - f_-(z)$ is a periodic holomorphic function, that is,

$$f_0(z+1) - f_-(z+1) = f_0(z) - f_-(z). \tag{4.87}$$

Hence, for each $f_+(z) - f_0(z)$, $f_0(z) - f_-(z)$, we may set

$$f_+(z) - f_0(z) = f_{+,0}(e^{2\pi\sqrt{-1}z}), \quad R \le \Im z \le 2R$$

$$f_0(z) - f_-(z) = f_{0,-}(e^{2\pi\sqrt{-1}z}), \quad -2R \le \Im z \le -R.$$

Here, $f_{+,0}(w)$, $f_{0,-}(w)$ are single-valued holomorphic functions on $e^{-2\pi R} < |w| < e^{-\pi R}$, $e^{\pi R} < |w| < e^{2\pi R}$, respectively. Consider the Cousin integrals (cf. [Ka])

$$\frac{1}{2\pi\sqrt{-1}} \int_{|w|=e^{-\frac{3}{2}\pi R}} \frac{f_{+,0}(w)dw}{w - e^{2\pi\sqrt{-1}z}} = \begin{cases} \psi_+(z), & \Im z > \frac{3}{2}R \\ \psi_0(z), & \Im z < \frac{3}{2}R \end{cases} \tag{4.88}$$

$$\frac{1}{2\pi\sqrt{-1}} \int_{|w|=e^{\frac{3}{2}\pi R}} \frac{f_{0,-}(w)dw}{w - e^{2\pi\sqrt{-1}z}} = \begin{cases} \psi_0'(z), & \Im z > -\frac{3}{2}R \\ \psi_-(z), & \Im z < -\frac{3}{2}R. \end{cases} \tag{4.89}$$

Then, each function extends analytically to a periodic function: on $\Im z \geq R$ for $\psi_+(z)$, on $\Im z \leq 2R$ for $\psi_0(z)$, on $\Im z \geq -2R$ for $\psi_0'(z)$, and on $\Im z \leq -R$ for $\psi_-(z)$, respectively. Hence, by setting,

$$f(z) = \begin{cases} f_+(z) - \psi_+(z) - \psi_0'(z), & \Im z \geq R, \\ f_0(z) - \psi_0(z) - \psi_0'(z), & 2R \geq \Im z \geq -2R, \\ f_-(z) - \psi_0(z) - \psi_-(z), & -R \geq \Im z, \end{cases} \tag{4.90}$$

we see that $f(z)$ is defined consistently and is meromorphic on $\mathbb{C}$. As $\psi_+(z)$, $\psi_0(z)$, $\psi_0'(z)$, $\psi_-(z)$ are bounded, $f(z)$ satisfies (4.81) and (4.82).

4.1.13 Completion of the Proof

Since we have $|\lambda_i| > |\lambda_j|$ for $i < j$, by applying Lemma 4.5 to (4.78), we see that there exists a solution $\eta_{j-1,j}(z)$ of (4.76) which is meromorphic on $\mathbb{C}$ and is holomorphic on Δ_R^- and which has an asymptotic expansion at $\Re z \mapsto -\infty$ ($|\Im z|$: bounded)

$$\eta_{j-1,j}(z) = \begin{cases} \left(\frac{\lambda_j}{\lambda_{j-1}}\right)^z (-z)^{\rho_j - \rho_{j-1}} z^{-1} \left(1 + O\left(\frac{1}{|z|}\right)\right), & \Re z \leqslant 0, \\ O(1), & \Re z \geq 0. \end{cases} \tag{4.91}$$

By applying Lemma 4.5 successively to the difference equation (4.75), we obtain the following lemma.

Lemma 4.6. *There exists a meromorphic function $\eta_{ij}(z)$ $(i < j)$ which is meromorphic on $\mathbb{C}$ and holomorphic on Δ_R^- that has an asymptotic expansion at $\Re z \mapsto -\infty$ ($|\Im z|$: bounded)*

$$\eta_{i,j}(z) = \begin{cases} \left(\frac{\lambda_j}{\lambda_i}\right)^z (-z)^{\rho_j - \rho_i} z^{-1} \left(1 + O\left(\frac{1}{|z|}\right)\right), & \Re z \leqslant 0 \\ O(1), & \Re z \geq 0 \end{cases} \tag{4.92}$$

and satisfies the difference equation (4.75).

Proof. Fixing j, we show the lemma by an descending induction on i $(< j)$. For $i = j - 1$, it is proved as we stated above. Next, assume that $\eta_{\nu j}(z)$ $(i < \nu < j)$ satisfies (4.92). Since we have

$$b_{i\nu}(z)\eta_{\nu j}(z) = \begin{cases} \left(\frac{\lambda_j}{\lambda_i}\right)^z (-z)^{\rho_j - \rho_i} z^{-2} \left(1 + O\left(\frac{1}{|z|}\right)\right), & \Re z \leq 0 \\ \left(\frac{\lambda_\nu}{\lambda_i}\right)^z O(1), & \Re z \geq 0 \end{cases}$$

for $|z| \mapsto +\infty$ ($|\Im z|$: bounded) by (4.78), setting $\alpha = \frac{\lambda_j}{\lambda_i}, \beta = \frac{\lambda_{i+1}}{\lambda_i}, \gamma = \rho_j - \rho_i$ and applying Lemma 4.5, we see that there exists a solution of (4.92) which is meromorphic on $\mathbb{C}$ and homolorphic on Δ_R^-.

Set

$$\eta_{ii}(z) = 1, \quad \eta_{ij}(z) = 0, \quad i > j. \tag{4.93}$$

Let $H(z)$ be the unipotent upper triangular matrix whose (i, j)-component is $\eta_{ij}(z)$. Then, $H(z)$ is meromorphic on $\mathbb{C}$ and is a solution of the difference equation (4.74). Defining $\Phi(z)$ as the product (4.30), for the matrix $\Phi(z) = ((\varphi_{ij}(z)))_{i, j=1}^r$, we have

$$\varphi_{ij}(z) = \sum_{1 \leq h \leq \min(i, j)} \xi_{ih}(z)\eta_{hj}(z), \tag{4.94}$$

and by the asymptotic estimates (4.67) and (4.92), we obtain an asymptotic estimate of $\varphi_{ij}(z)$ at $\Re z \mapsto -\infty$ ($|\Im z|$: bounded)

$$\varphi_{ij}(z) = \lambda_j^z (-z)^{\rho_j} \left(\delta_{ij} + O\left(\frac{1}{|z|}\right)\right). \tag{4.95}$$

Thus, Theorem 4.1 (1) is proved. The proof of Theorem 4.1 (2) can be managed completely in the same way as for (1).

Remark 4.2. The difference equation (4.1) is called of regular singular type when $\mu = 0$, and of irregular singular type when $\mu \neq 0$.

Remark 4.3. An expression of the form (4.30) is sometimes called the Birkhoff decomposition.

4.2 Holonomic Difference Equations in Several Variables and Asymptotic Expansion

4.2.1 Holonomic Difference Equations of First Order

Let $\{e_1, \ldots, e_m\}$ be a basis of a rank m lattice L in $\mathbb{C}^m$ and consider a difference equation of a meromorphic square matrix $\Phi(z)$ of order r on $\mathbb{C}^m$

$$\Phi(z + e_j) = A_j(z)\Phi(z), \quad z \in \mathbb{C}^m. \tag{4.96}$$

Here, $A_j(z)$ is a $GL_r(\mathbb{C}(z))$-valued rational function. Since we have

$$\begin{aligned}\Phi(z + e_i + e_j) &= A_i(z + e_j)A_j(z)\Phi(z) \\ &= A_j(z + e_i)A_i(z)\Phi(z)\end{aligned}$$

for any i, j as the compatibility condition on (4.96), the following equality should be satisfied:

$$A_i(z + e_j)A_j(z) = A_j(z + e_i)A_i(z). \tag{4.97}$$

A system of difference equations (4.96) satisfying (4.97) is called a holonomic[1] system of difference equations, to be more precise, a rational holonomic system of difference equations of first order. Below, with the basis $\{e_j\}_{1 \leq j \leq m}$, we identify L with $\mathbb{Z}^m$. The following argument mainly follows [Ao3]. See also [Pr].

When $\eta \in \mathbb{Z}^m \setminus \{0\}$ satisfies the following two conditions, we say that η is a "regular direction" with respect to the system (4.96). For $N \in \mathbb{Z}_{>0}$, each $A_j(z)$ has a Laurent expansion of the form

$$A_j(z + N\eta) = A_{j,\eta}^{(0)}(z)N^{\mu_j} + A_{j,\eta}^{(1)}(z)N^{\mu_j - 1} + \cdots \tag{4.98}$$

$(\mu_j \in \mathbb{Z})$, the conditions are:

(i) $A_{j,\eta}^{(0)}(z)$ does not depend on z. Below, we simply denote it as $A_{j,\eta}^{(0)}$.
(ii) $\det A_{j,\eta}^{(0)} \neq 0$.

Except for very special $\{A_j(z)\}_{1 \leq j \leq m}$, it can be easily checked that such a direction η exists. In fact, each $A_j(z)$ can be expressed as

$$A_j(z) = \frac{\mathcal{A}_j(z)}{f_j(z)}$$

$(f_j(z) \in \mathbb{C}[z], \mathcal{A}_j(z) \in GL_r(\mathbb{C}[z]))$. Moreover, we assume that the highest homogeneous component $\mathcal{A}_{j,0}(z)$ (set $h_j = \deg \mathcal{A}_j(z) = \deg \mathcal{A}_{j,0}(z)$) of $\mathcal{A}_j(z)$

---

[1] This naming is credited to Mikio Sato.

with respect to z satisfies $\det \mathcal{A}_{j,0}(z) \neq 0$. We have $\mu_j = h_j - \deg f_j$, and if $\mathcal{A}_{j,0}(z + N\eta), f_j(z + N\eta)$ have degree h_j, $\deg f_j$ as polynomials in N respectively, then (i), (ii) are satisfied.

If $\eta, \eta' \in \mathbb{Z}^m \setminus \{0\}$ are mutually proportional, i.e., $\eta' = \sigma\eta$ ($\sigma \in \mathbb{Q}_{>0}$), then η and η' gives the same direction.

4.2.2 Formal Asymptotic Expansion

Now, by choosing appropriate coordinates $(z_1, \ldots, z_m)$, we assume that the direction of the coordinate z_1, namely, e_1 is a regular direction. Then, we have a Laurent expansion with respect to z_1:

$$A_j(z) = A_j^{(0)}(z')z_1^{\mu_j} + A_j^{(1)}(z')z_1^{\mu_j - 1} + \cdots, \tag{4.99}$$

$$z' = (z_2, \ldots, z_m), \quad 1 \leqslant j \leqslant m,$$

but since z_1 is a regular direction, $A_j^{(0)}(z')$ is independent of z', so we set $A_j^{(0)}(z') = A_j^{(0)}$. We have $\det A_j^{(0)} \neq 0$, and furthermore, we impose the following assumption:

(iii)   The eigenvalues $\lambda_1, \ldots, \lambda_r$ (we assume $|\lambda_1| \geq |\lambda_2| \geq \cdots \geq |\lambda_r|$) of $A_1^{(0)}$ are different to each other.

$A_j^{(l)}(z')$ is a polynomial of degree at most l. We have $A_j^{(l)}(z') \in \mathbb{C}[z']$, $\deg A_j^{(l)}(z') \leq l$, and below, we suppose that $A_1^{(0)}$ is already diagonalized.

Then, we have the following theorem as a version for several variables of Theorem 4.1.

Theorem 4.3 ([Ao3]).

(1) *There exists a formal Laurent series* $S(z) \in GL_r\left(\mathbb{C}[z'] \otimes \mathbb{C}\left(\left(\frac{1}{z_1}\right)\right)\right)$ *in* z_1

$$S(z) = 1 + \frac{S^{(1)}(z')}{z_1} + \frac{S^{(2)}(z')}{z_1^2} + \cdots, \tag{4.100}$$

$$S^{(\nu)}(z') \in M_r(\mathbb{C}[z']), \deg S^{(\nu)}(z') \leqslant \nu,$$

such that a formal Laurent series

$$S(z + e_j)^{-1}A_j(z)S(z) = B_j(z), \quad 1 \leqslant j \leqslant m \tag{4.101}$$

can be expressed in the form:

$$B_1(z) = z_1^{\mu_1}\left(A_1^{(0)} + \frac{B_1^{(1)}(z')}{z_1}\right), \tag{4.102}$$

$$B_j(z) = z_1^{\mu_j}\left(A_j^{(0)} + \frac{B_j^{(1)}(z')}{z_1} + \frac{B_j^{(2)}(z')}{z_1^2} + \cdots\right), \quad 2 \leqslant j \leqslant m. \tag{4.103}$$

(2) $A_j^{(0)}(1 \leqslant j \leqslant m)$ are constant and $B_1^{(1)}(z')$ can be expressed in the form:

$$B_1^{(1)}(z') = A_1^{(1)} + \left(\sum_{j=2}^{m}\mu_j z_j\right)A_1^{(0)}, \tag{4.104}$$

where $B_j^{(\nu)}(z') \in M_r(\mathbb{C}[z'])$ and $A_1^{(1)}$ is a constant matrix.

(3) The matrices $A_j^{(0)}$, $B_j^{(\nu)}(z')(1 \leqslant j \leqslant m)$ are all diagonal.

(4) The matrices $B_j^\nu(z')$ $(2 \leqslant j \leqslant m, 1 \leqslant \nu)$ satisfying (1), (2), (3) can be uniquely expressed in terms of $A_1^{(0)}$, $A_1^{(1)}$, $A_j^{(0)}$.

(5) In particular, for $\mu_1 = \mu_2 = \cdots = \mu_m = 0$, i.e., if (4.96) is of regular singular type, then each of (4.102), (4.103) can be reduced to the form:

$$B_1(z) = A_1^{(0)} + \frac{A_1^{(1)}}{z_1}, \tag{4.105}$$

$$B_j(z) = A_j^{(0)}, \quad 2 \leqslant j \leqslant m. \tag{4.106}$$

Proof. By (4.97), in particular, we have

$$A_j(z + e_1)A_1(z) = A_1(z + e_j)A_j(z), \quad j \geq 2. \tag{4.107}$$

From this and (4.99), we have $A_j^{(0)}A_1^{(0)} = A_1^{(0)}A_j^{(0)}$. We apply Lemma 4.1 to $A_1(z)$ as a function of z_1. There exists a formal Laurent series (4.100) such that $B_j(z)$ defined by (4.101) have an expression of the form (4.102), (4.103). For $B_j(z)$, a formula similar to (4.107) holds:

$$B_j(z + e_1)B_1(z) = B_1(z + e_j)B_j(z), \quad j \geq 2. \tag{4.108}$$

By (4.108), $A_j^{(0)}$ and $B_1^{(1)}(z')$ commute. As we have

$$[A_1^{(0)}, B_j^{(1)}(z')] = A_j^{(0)}\{-B_1^{(1)}(z' + e_j) + B_1^{(1)}(z') + \mu_j A_1^{(0)}\} \tag{4.109}$$

by comparing the coefficients of $z_1^{\mu_1 + \mu_j - 1}$ in (4.108), it follows that

$$\left[A_1^{(0)}, [A_1^{(0)}, B_j^{(1)}(z')]\right] = 0. \tag{4.110}$$

Since $A_1^{(0)}$ is a diagonal matrix, this implies

$$[A_1^{(0)}, B_j^{(1)}(z')] = 0. \tag{4.111}$$

By (4.109), we have

$$B_1^{(1)}(z' + e_j) - B_1^{(1)}(z') = \mu_j A_1^{(0)}, \quad j \geq 2 \tag{4.112}$$

and since $B_1^{(1)}(z')$ is a rational function, it has the form of (4.104). We have already proved (3), so we prove (4). By Lemma 4.1, $B_1^{(1)}(z')$ is uniquely determined and $S(z)$ satisfying

$$S(z + e_1)^{-1} A_1(z) S(z) = B_1(z), \quad (S_0 = 1)$$

is also uniquely determined. As $B_j(z)$ is given by (4.101), $B_j^{(\nu)}(z')$ is also uniquely determined. When $B_1^{(1)}(z')$ of (4.102) is expressed in the form of (4.104), by the compatibility condition (4.108) for $B_j(z)$ ($1 \leq j \leq m$), we see that $B_j^{(\nu)}(z')$ of (4.103) is uniquely expressed in terms of $A_1^{(0)}, A_1^{(1)}, A_j^{(0)}$ ($2 \leq j \leq m$). Moreover, if $\mu_1 = \mu_2 = \cdots = \mu_m = 0$, (4.105) clearly holds. $B_j(z)$ satisfies the identity (4.108), and since we also have $B_1(z+e_j) = B_1(z)$ and the commutativity $B_j(z + e_1)B_1(z) = B_1(z)B_j(z + e_1)$, it follows that $B_j(z+e_1) = B_j(z)$. Since $B_j(z)$ is rational, it follows that $B_j(z)$ is a constant. Hence, (4.106) is proved.

Corollary 4.3. *The difference equation (4.96) admits a formal solution in the form of a matrix Laurent series in z_1*

$$\widehat{\Phi}(z) = \frac{e^{\pi\sqrt{-1}\mu_1 z_1}}{\Gamma(1 - z_1)^{\mu_1}} S(z) \frac{\Gamma(1 - z_1)}{\Gamma(1 - z_1 - (A_1^{(0)})^{-1}A_1^{(1)} - \sum_{j=2}^m \mu_j z_j)} \tag{4.113}$$

$$\cdot \exp\left(\sum_{j=1}^m z_j \log A_j^{(0)}\right).$$

Indeed, it satisfies $S(z+e_1)^{-1}\widehat{\Phi}(z+e_1) = B_1(z)S(z)^{-1}\widehat{\Phi}(z)$, $S(z+e_j)^{-1}\widehat{\Phi}(z+e_j) = B_j(z)S(z)^{-1}\widehat{\Phi}(z)$ ($2 \leq j \leq m$).

4.2.3 Normal Form of Asymptotic Expansion

When $\{A_j(z)\}_{1 \leq j \leq m}$ satisfy (4.97), one can uniquely define $\{A_\nu(z)\}_{\nu \in \mathbb{Z}^m}$ in such a way that, for any $\nu, \nu' \in \mathbb{Z}^m$, they statisfy

$$A_{\nu+\nu'}(z) = A_\nu(z + \nu')A_{\nu'}(z), \tag{4.114}$$

$$A_{e_j}(z) = A_j(z), \quad A_{-e_j}(z) = A_j(z - e_j)^{-1}. \tag{4.115}$$

Below, we also denote $A_{-e_j}(z)$ by $A_{-j}(z)$. The identity

$$\Phi(z + \nu) = A_\nu(z)\Phi(z) \tag{4.116}$$

is satisfied. $\{A_\nu(z)\}_{\nu \in \mathbb{Z}^m}$ is a 1-cocycle on $\mathbb{Z}^m$ with values in $GL_r(\mathbb{C}(z))$, that is, it defines an element of 1-cohomology $H^1(\mathbb{Z}^m, GL_r(\mathbb{C}(z)))$.

Corollary 4.4. $A_\nu(z)$ has the Laurent expansion at $z_1 = \infty$ written in the form

$$S(z + \nu)^{-1}A_\nu(z)S(z) = z_1^{\mu_\nu}\left(A_\nu^{(0)} + \frac{B_\nu^{(1)}(z')}{z_1} + \cdots\right) \tag{4.117}$$

such that, for $\nu = \sum_{i=1}^m \nu_i e_i$, we have

$$\mu_\nu = \sum_{i=1}^m \nu_i \mu_i, \tag{4.118}$$

$$A_\nu^{(0)} = (A_1^{(0)})^{\nu_1} \cdots (A_m^{(0)})^{\nu_m} \tag{4.119}$$

and that $B_\nu^{(l)}(z') \in M_r(\mathbb{C}[z'])$ and $A_1^{(0)}$ commute.

Now, setting

$$\Phi(z) = \Gamma\left(\sum_{j=1}^m \mu_j z_j\right)\Phi_0(z) \tag{4.120}$$

and considering the difference equation on $\Phi_0(z)$

$$(4.96)' \qquad \Phi_0(z + e_j) = \left(\sum_{j=1}^n \mu_j z_j; \mu_j\right)^{-1} A_j(z)\Phi_0(z)^1, \quad 1 \leqslant j \leqslant n$$

equivalent to (4.96), the index μ_j in the expression (4.99) is reduced to 0. Hence, below, in (4.96), we assume that

$$\mu_1 = \mu_2 = \cdots = \mu_r = 0, \tag{4.121}$$

i.e., (4.96) is of regular singular type. Then, (4.113) and (4.117) are reduced to the following forms:

$$\widehat{\Phi}(z) = S(z)\frac{\Gamma(1 - z_1)}{\Gamma(1 - z_1 - (A_1^{(0)})^{-1}A_1^{(1)})} \cdot \exp\left(\sum_{j=1}^m z_j \log A_j^{(0)}\right), \tag{4.122}$$

---

[1] For the symbol $(\alpha; k)$, see the footnote in § 1.1, Chapter 1.

$$S(z+\nu)^{-1}A_\nu(z)S(z)=\begin{cases} A_\nu^{(0)}\displaystyle\prod_{\sigma=0}^{\nu_1-1}\left(1+\frac{(A_1^{(0)})^{-1}A_1^{(1)}}{z_1+\sigma}\right), & \nu_1\geqq 0,\\[4mm] A_\nu^{(0)}\displaystyle\prod_{\sigma=\nu_1}^{-1}\left(1+\frac{(A_1^{(0)})^{-1}A_1^{(1)}}{z_1+\sigma}\right)^{-1}, & \nu_1<0. \end{cases} \tag{4.123}$$

4.2.4 Existence Theorem (ii)

By applying the existence Theorem 4.1 due to G.D. Birkhoff, we prove the existence theorem for the case of several variables.

Theorem 4.4 ([Ao3]). *Under the assumptions (i), (ii), (iii) stated in §4.2.1–4.2.2, there uniquely exists a solution $\Phi(z)$ of (4.96) with values in $GL_r(\mathbb{C})$ that is meromorphic on $\mathbb{C}^m$ and that has an asymptotic expansion at $\Re z_1 \mapsto -\infty$ ($|\Im z_1|$: bounded) given by (4.122).*

Proof. By Theorem 4.1, there is a meromorphic function $\Phi(z)$ of z which satisfies

$$\Phi(z+e_1) = A_1(z)\Phi(z) \tag{4.124}$$

and which has an asymptotic expansion given by (4.122). Here, $A_j(z)^{-1}\Phi(z+e_j)$ also has the same property. Indeed, by the compatibility condition (4.97), we have

$$A_j(z+e_1)^{-1}\Phi(z+e_1+e_j) \tag{4.125}$$

$$= A_1(z)A_j(z)^{-1}A_1(z+e_j)^{-1}\Phi(z+e_1+e_j)$$

$$= A_1(z)A_j(z)^{-1}\Phi(z+e_j).$$

Moreover, by the fact that $S_0, A_j^{(0)}, A_1^{(1)}$ are all diagonal matrices, we see that $A_j(z)^{-1}\Phi(z+e_j)$ has an asymptotic expansion at $\Re z_1 \mapsto -\infty$ ($|\Im z_1|$: bounded)

$$A_j(z)^{-1}\Phi(z+e_j) \sim S_0 \exp\{(A_1^{(0)})^{-1}A_1^{(1)}\log(-z_1)\} \tag{4.126}$$

$$\cdot \exp\left(\sum_{j=1}^m z_j \log A_j^{(0)}\right),$$

and has the same principal part as $\Phi(z)$. Hence, by the uniqueness of the solution of (4.1), we should have

$$A_j(z)^{-1}\Phi(z+e_j) = \Phi(z), \tag{4.127}$$

namely,

$$\Phi(z + e_j) = A_j(z)\Phi(z), \quad 2 \leqslant j \leqslant m. \tag{4.128}$$

This $\Phi(z)$ is the solution of (4.96) we are looking for.

4.2.5 Connection Problem

From the above theorem, if $\eta \in \mathbb{Z}^m \setminus \{0\}$ is a regular direction, there uniquely exists a solution $\Phi(z) = \Phi_\eta(z)$ of the difference equation (4.96) which has an asymptotic expansion in the direction of η. If $\eta' \in \mathbb{Z}^m \setminus \{0\}$ and η is in the same direction, the solution $\Phi_{\eta'}(z)$ of (4.96) having an asymptotic expansion in the direction of η' coincides with $\Phi_\eta(z)$. Hence, we may restrict ourselves to the case when $\eta \in \mathbb{Z}^m \setminus \{0\}$ is a primitive element, i.e., to such an element that if $\eta = k\eta' (k \in \mathbb{Z})$, then $\eta' = \pm\eta$.

If η is primitive, then there exists a unimodular transformation $(w_{ij} \in \mathbb{Z})$

$$z_i = \sum_{j=1}^m w_{ij}\zeta_j \tag{4.129}$$

such that $w_{i1} = \eta_i$. Below, we denote this transformation as $z = W \cdot \zeta$. Setting $\Psi(\zeta) = \Phi(W \cdot \zeta)$, (4.96) can be rewritten, by using the shift operators with respect to $\zeta_1, \zeta_2, \ldots, \zeta_m$

$$T_j^{*\pm1}\psi(\zeta_1, \ldots, \zeta_j, \ldots, \zeta_m) = \psi(\zeta_1, \ldots, \zeta_j \pm 1, \ldots, \zeta_m)$$

as a system of difference equations

$$T_j^{*\pm1}\Psi(\zeta) = A_{\pm j}^*(\zeta)\Psi(\zeta). \tag{4.130}$$

If the direction of η is regular with respect to (4.96), then the direction of ζ_1 is regular with respect to (4.130). By using a formal Laurent series

$$S^*(\zeta) = 1 + \frac{S^{*(1)}(\zeta')}{\zeta_1} + \frac{S^{*(2)}(\zeta')}{\zeta_1^2} + \cdots \tag{4.131}$$

$(\zeta' = (\zeta_2, \ldots, \zeta_m))$, we can normalize as

$$A_1^*(\zeta) = S^*(\zeta + e_1^*)^{-1}\left(A_1^{*(0)} + \frac{A_1^{*(1)}}{\zeta_1}\right)S^*(\zeta), \tag{4.132}$$

where we set $(e_1^* = (1, 0, \ldots, 0))$. Moreover, when each of $A_1^{*(0)}, A_1^{*(1)}$ is expressed as diagonal matrices

$$A_1^{*(0)} = \mathrm{Diag}[\lambda_1^*, \ldots, \lambda_r^*], \tag{4.133}$$

$$A_1^{*(1)} = \text{Diag}[\rho_1^*, \dots, \rho_r^*], \tag{4.134}$$

we obtain $2r$-tuple of numbers $\{\lambda_1^*, \dots, \lambda_r^*, \rho_1^*, \dots, \rho_r^*\}$. $\{\lambda_1^*, \dots, \lambda_r^*\}$ is called the first characteristic index and $\{\rho_1^*, \dots, \rho_r^*\}$ is called the second characteristic index. In (4.98), $A_1^{*(0)}$ can be expressed as a product of the principal part $A_{j,\eta}^{(0)}$ ($\mu_j = 0$) of the Laurent expansion of $A_j(z)$ at the direction of η as

$$A_1^{*(0)} = (A_{1,\eta}^{(0)})^{\eta_1} \cdots (A_{m,\eta}^{(0)})^{\eta_m}. \tag{4.135}$$

In fact, $A_1^*(\zeta)$ can be expressed as an ordered product of $\{A_j(z)\}_{j=1}^m$ as follows. When $\eta_1 \geq 0, \dots, \eta_m \geq 0$, it is of the form

$$A_1^*(\zeta) = G_m \cdot G_{m-1} \cdots G_1, \tag{4.136}$$

$$G_j = A_j\left(z + (\eta_j - 1)e_j + \sum_{k=1}^{j-1} \eta_k e_k\right) \cdots A_j\left(z + \sum_{k=1}^{j-1} \eta_k e_k\right).$$

By Theorem 4.3, $A_{j,\eta}^{(0)}$ ($1 \leq j \leq m$) are mutually commutative, and the principal part of (4.136) at $N \mapsto +\infty$ is reduced to (4.135). When $\eta_1 \geq 0, \dots, \eta_m \geq 0$ are not satisfied and $\eta_j < 0$, by considering $A_j(z - e_j)^{-1}$ in place of $A_j(z)$, we again have (4.135).

4.2.6 Example

As a simple example, we consider the Euler beta function. Euler's beta function $\Phi(z)$ is defined by the integral:

$$\Phi(z_1, z_2) = \int_0^1 u^{z_1 - 1}(1 - u)^{z_2 - 1} du \tag{4.137}$$

$$= \frac{\Gamma(z_1)\Gamma(z_2)}{\Gamma(z_1 + z_2)}, \quad \Re z_1 > 0, \ \Re z_2 > 0.$$

$\Phi(z)$ satisfies the system of difference equations of rank 1

$$\begin{cases} \Phi(z + e_1) = \frac{z_1}{z_1 + z_2}\Phi(z), \\ \Phi(z + e_2) = \frac{z_2}{z_1 + z_2}\Phi(z). \end{cases} \tag{4.138}$$

e_1, e_2 are not regular directions. Now, setting $\eta = \eta_1 e_1 + \eta_2 e_2$, we have

$$\begin{cases} A_1(z + N\eta) \sim \frac{\eta_1}{\eta_1 + \eta_2}\left(1 + O\left(\frac{1}{N}\right)\right), \\ A_2(z + N\eta) \sim \frac{\eta_2}{\eta_1 + \eta_2}\left(1 + O\left(\frac{1}{N}\right)\right), \end{cases} \tag{4.139}$$

and if it satisfies $(\eta_1 + \eta_2)\eta_1\eta_2 \neq 0$, then η gives a regular direction. In particular, setting $\eta_1 = \eta_2 = 1$, i.e., $\eta = e_1 + e_2$, we have

$$\Phi(z + \eta) = \frac{z_1 z_2}{(z_1 + z_2)(z_1 + z_2 + 1)}\Phi(z), \qquad (4.140)$$

or in other words,

$$\Phi(z) = \frac{(z_1 - 1)(z_2 - 1)}{(z_1 + z_2 - 1)(z_1 + z_2 - 2)}\Phi(z - \eta). \qquad (4.140)'$$

The infinite product

$$\Phi(z) = \lim_{N \mapsto +\infty}\left\{\prod_{j=1}^{N}\frac{(z_1 - j)(z_2 - j)}{(z_1 + z_2 - 2j + 1)(z_1 + z_2 - 2j)}\right\}\frac{2^{2N - z_1 - z_2}}{\sqrt{\pi N}} \qquad (4.141)$$

$$= \frac{\Gamma(1 - z_1 - z_2)}{\Gamma(1 - z_1)\Gamma(1 - z_2)}$$

is a unique solution of (4.138) having an asymptotic expansion in the direction of $-\eta$ (Stirling's formula in § 1.1)

$$\Phi(z - N\eta) \sim \frac{2^{2N}}{\sqrt{N\pi}}2^{-z_1 - z_2}. \qquad (4.142)$$

For more general η, when $\eta_1 > 0, \eta_2 > 0$, the solution $\Phi_{-\eta}(z)$ having an asymptotic expansion in the direction of $-\eta$ does not depend on η and is equal to (4.141). In the η-plane $\mathbb{R}^2$, for each η belonging to one of the six domains, separated by the three lines $\eta_1 = 0, \eta_2 = 0, \eta_1 + \eta_2 = 0, \Delta_1 : \eta_1 > 0, \eta_2 > 0; \Delta_2 : \eta_2 > 0, \eta_1 + \eta_2 > 0; \Delta_3 : \eta_2 > 0, \eta_1 + \eta_2 < 0; \Delta_4 : \eta_1 < 0, \eta_2 < 0; \Delta_5 : \eta_1 > 0, \eta_1 + \eta_2 < 0; \Delta_6 : \eta_1 + \eta_2 > 0, \eta_2 < 0$, the solution of (4.138) having an asymptotic expansion in the direction of η exists:

For $\Delta_1 \ni \eta$, $\quad \Phi_\eta(z) = \dfrac{\Gamma(z_1)\Gamma(z_2)}{\Gamma(z_1 + z_2)}$,

$\Delta_2 \ni \eta$, $\quad \Phi_\eta(z) = \dfrac{e^{\pi\sqrt{-1}z_1}\Gamma(z_2)}{\Gamma(1 - z_1)\Gamma(z_1 + z_2)}$,

$\Delta_3 \ni \eta$, $\quad \Phi_\eta(z) = e^{\pi\sqrt{-1}z_2}\dfrac{\Gamma(z_2)\Gamma(1 - z_1 - z_2)}{\Gamma(1 - z_1)}$,

$\Delta_4 \ni \eta$, $\quad \Phi_\eta(z) = \dfrac{\Gamma(1 - z_1 - z_2)}{\Gamma(1 - z_1)\Gamma(1 - z_2)}$,

$\Delta_5 \ni \eta$, $\quad \Phi_\eta(z) = e^{\pi\sqrt{-1}z_1}\dfrac{\Gamma(z_1)\Gamma(1 - z_1 - z_2)}{\Gamma(1 - z_2)}$,

$\Delta_6 \ni \eta$, $\quad \Phi_\eta(z) = e^{\pi\sqrt{-1}z_2}\dfrac{\Gamma(z_1)}{\Gamma(1 - z_2)\Gamma(z_1 + z_2)}$.

4.2.7 Remark on 1-Cocyles

For $r = 1$, a set of rational functions $\{A_i(z)\}_{i=1}^m$ in (4.96), or in other owrds, $\{A_\nu(z)\}_{\nu \in \mathbb{Z}^m}$ defines a $\mathbb{C}^*(z) = \mathbb{C}(z) \setminus \{0\}$-valued 1-cocycle on $\mathbb{Z}^m$. In this case, the structure of $\{A_\nu(z)\}_{\nu \in \mathbb{Z}^m}$ is completely understood and is called the Bernstein–Sato b-function, or simply a b-function (cf. Appendix A). And the solution of (4.96) is expressed by the Γ-functions of several variables (see Appendix of [S-S-M]). On the contrary, for $r \geq 2$, its structure is not well studied. The result of C. Sabbah [Sab1] imposing a condition on the poles of $A_\nu(z)$ is fairly interesting, but its proof requires a complicated technique such as Quillen's theorem on projective modules, so a simpler proof is expected. In § 4.4, we discuss systems of differential equations for hypergeometric functions of type $(n + 1, m + 1; \alpha)$ in detail.

4.2.8 Gauss' Contiguous Relations

A typical example of a difference equation for $r = 2$ is the Gauss difference equation stated in § 1.4.1.

Another solution of (1.41) (1.42) is given by an integral representation

$$\widetilde{F}(\alpha, \beta, \gamma; x) = \frac{\Gamma(\gamma)}{\Gamma(\alpha)\Gamma(\gamma - \alpha)} \int_1^{\frac{1}{x}} u^{\alpha-1}(1 - u)^{\gamma-\alpha-1}(1 - ux)^{-\beta}du \quad (4.143)$$

for $\Re(\gamma - \alpha) > 0, \Re\alpha > 0$ ([AAR], [Ol], [W-W]). Namely, the vector

$$\begin{pmatrix} \widetilde{F}(\alpha, \beta+1, \gamma+1; x) \\ \widetilde{F}(\alpha, \beta, \gamma; x) \end{pmatrix}$$

satisfies the difference equations (1.41)–(1.42). Hence, the square matrix of order 2

$$\Phi(\alpha, \beta, \gamma) = \begin{pmatrix} F(\alpha, \beta+1, \gamma+1; x) & \widetilde{F}(\alpha, \beta+1, \gamma+1; x) \\ F(\alpha, \beta, \gamma; x) & \widetilde{F}(\alpha, \beta, \gamma; x) \end{pmatrix} \qquad (4.144)$$

satisfies the difference equation (a contiguous relation) with respect to the shift in the direction $\eta = (1, 1, 2)$ given by

$$\Phi(\alpha, \beta, \gamma) = A(\alpha, \beta, \gamma)\Phi(\alpha+1, \beta+1, \gamma+2), \qquad (4.145)$$

$$A(\alpha, \beta, \gamma) = \begin{pmatrix} 0 & 1 \\ -\frac{\alpha(\gamma-\beta)}{\gamma(\gamma+1)}x & 1 \end{pmatrix} \begin{pmatrix} 0 & 1 \\ -\frac{(\beta+1)(\gamma+1-\alpha)}{(\gamma+1)(\gamma+2)} & 1 \end{pmatrix} \qquad (4.146)$$

$$= \begin{pmatrix} -\frac{(\beta+1)(\gamma+1-\alpha)}{(\gamma+1)(\gamma+2)}x & 1 \\ -\frac{(\beta+1)(\gamma+1-\alpha)}{(\gamma+1)(\gamma+2)}x & 1 - \frac{\alpha(\gamma-\beta)}{\gamma(\gamma+1)}x \end{pmatrix}.$$

More generally, for $t \in \mathbb{C}$, $\Phi(\alpha+t, \beta+t, \gamma+2t)$ (below, we denote it by $\Phi_\eta(t)$) satisfies the difference equation with respect to the shift in the direction η:

$$\Phi_\eta(t) = A_\eta(t)\Phi_\eta(t+1), \qquad (4.147)$$

$$A_\eta(t) = \begin{pmatrix} -\frac{(\beta+1+t)(\gamma+1-\alpha+t)}{(1+\gamma+2t)(2+\gamma+2t)}x & 1 \\ -\frac{(\beta+1+t)(\gamma+1-\alpha+t)}{(\gamma+1+2t)(\gamma+2+2t)}x & 1 - \frac{(\alpha+t)(\gamma-\beta+t)}{(\gamma+2t)(\gamma+1+2t)}x \end{pmatrix}. \qquad (4.148)$$

By an asymptotic expansion of $A_\eta(t)$ at $t = \infty$

$$A_\eta(t) = \begin{pmatrix} -\frac{x}{4} & 1 \\ -\frac{x}{4} & 1-\frac{x}{4} \end{pmatrix} + \frac{x}{4t}\begin{pmatrix} \alpha-\beta-\frac{1}{2} & 0 \\ \alpha-\beta-\frac{1}{2} & -\alpha+\beta+\frac{1}{2} \end{pmatrix} + \cdots \qquad (4.149)$$

$$= C\left\{ \begin{pmatrix} \lambda_1 & \\ & \lambda_2 \end{pmatrix} + \frac{1}{4t}\frac{\alpha-\beta-\frac{1}{2}}{\lambda_2-\lambda_1}\begin{pmatrix} 0 & \frac{x}{2}+2\lambda_2-1 \\ -\frac{x}{2}-2\lambda_1+1 & 0 \end{pmatrix} + \cdots \right\}C^{-1},$$

$$C = \begin{pmatrix} 1 & 1 \\ \frac{1+\sqrt{1-x}}{2} & \frac{1-\sqrt{1-x}}{2} \end{pmatrix}, \qquad (4.150)$$

$$\lambda_1 = \frac{1}{4}(1+\sqrt{1-x})^2, \quad \lambda_2 = \frac{1}{4}(1-\sqrt{1-x})^2,$$

equation (4.147) admits a unique meromorphic matrix solution $\Phi^*(t)$ on $\mathbb{C}$ (Theorem 4.3) that has an asymptotic expansion at $\Re t \mapsto +\infty$ given by

$$\Phi^*(t) \sim C \cdot S(t) \cdot \widehat{\Phi}_0(t), \tag{4.151}$$

$$\widehat{\Phi}_0(t) = \begin{pmatrix} \lambda_1^{-t} & \\ & \lambda_2^{-t} \end{pmatrix},$$

$$S(t) = 1 + \frac{S_1}{t} + \frac{S_2}{t^2} + \cdots . \tag{4.152}$$

4.2.9 Convergence

Setting

$$\widehat{\Phi}_0(t)\Phi^*(t)^{-1} = S(t)^{-1}C^{-1} = \begin{pmatrix} \psi_{11}(t) & \psi_{12}(t) \\ \psi_{21}(t) & \psi_{22}(t) \end{pmatrix}, \tag{4.153}$$

by $\lim_{N \mapsto +\infty} \widehat{\Phi}_0(t+N)\Phi^*(t+N)^{-1} = C^{-1}$, taking the left coset of $\mathcal{B}_2$, we have

$$\lim_{N \mapsto +\infty} \widehat{\Phi}_0(t+N)\Phi^*(t+N)^{-1}\mathcal{B}_2 = \lim_{N \mapsto +\infty} \begin{pmatrix} 1 & 0 \\ \frac{\psi_{21}(t+N)}{\psi_{11}(t+N)} & 1 \end{pmatrix} \mathcal{B}_2 \tag{4.154}$$

$$= \begin{pmatrix} 1 & 0 \\ \frac{\widetilde{c_{21}}}{\widetilde{c_{11}}} & 1 \end{pmatrix} \mathcal{B}_2,$$

where we set $C^{-1} = \begin{pmatrix} \widetilde{c_{11}} & \widetilde{c_{12}} \\ \widetilde{c_{21}} & \widetilde{c_{22}} \end{pmatrix}$.

Now, when $x \notin [1, \infty)$, it follows from $|\lambda_1| > |\lambda_2|$ that $\lim_{N \mapsto +\infty} \left(\frac{\lambda_2}{\lambda_1}\right)^N = 0$. Hence, we obtain

$$\lim_{N \mapsto +\infty} \widehat{\Phi}_0(t+N)^{-1}\widehat{\Phi}_0(t+N)\Phi^*(t+N)^{-1}\mathcal{B}_2 \tag{4.155}$$

$$= \lim_{N \mapsto +\infty} \begin{pmatrix} 1 & 0 \\ \lambda_2^{t+N}\frac{\psi_{21}(t+N)}{\lambda_1^{t+N}\psi_{11}(t+N)} & 1 \end{pmatrix} \mathcal{B}_2 = \mathcal{B}_2,$$

namely,

$$\lim_{N \mapsto +\infty} A(t)A(t+1)\cdots A(t+N-1)\mathcal{B}_2 = \lim_{N \mapsto +\infty} \Phi^*(t)\Phi^*(t+N)^{-1}\mathcal{B}_2$$

$$= \lim_{N \mapsto +\infty} \Phi^*(t)\widehat{\Phi}_0(t+N)^{-1}\widehat{\Phi}_0(t+N)\Phi^*(t+N)^{-1}\mathcal{B}_2$$

$$= \Phi^*(t)\mathcal{B}_2 = \begin{pmatrix} 1 & 0 \\ \frac{\varphi_{21}^*(t)}{\varphi_{11}^*(t)} & 1 \end{pmatrix}\mathcal{B}_2, \tag{4.156}$$

where we set $\Phi^*(t) = \begin{pmatrix} \varphi_{11}^*(t) & \varphi_{12}^*(t) \\ \varphi_{21}^*(t) & \varphi_{22}^*(t) \end{pmatrix}$.

4.2.10 Continued Fraction Expansion

Now, when we let a matrix $\begin{pmatrix} a & b \\ c & d \end{pmatrix} \in GL_2(\mathbb{C})$ act on the projective line $GL_2/\mathcal{B}_2 \cong \mathbb{P}^1(\mathbb{C})$, it induces a linear fractional transformation

$$\begin{pmatrix} 1 \\ \zeta' & 1 \end{pmatrix}\mathcal{B}_2 = \begin{pmatrix} a & b \\ c & d \end{pmatrix}\begin{pmatrix} 1 \\ \zeta & 1 \end{pmatrix}\mathcal{B}_2, \tag{4.157}$$

$$\zeta' = \frac{c+d\zeta}{a+b\zeta}, \qquad (\zeta, \zeta' \in \mathbb{C}).$$

Then, the transformation (1.46) is expressed as

$$\Phi(\alpha,\beta,\gamma)\mathcal{B}_2$$
$$= A_\eta(0)A_\eta(1)\cdots A_\eta(\nu-1)\Phi(\alpha+\nu,\beta+\nu,\gamma+2\nu)\mathcal{B}_2. \tag{4.158}$$

Moreover, as the right-hand side of (1.49) can be written as

$$\lim_{N \mapsto +\infty} A_\eta(0)A_\eta(1)\cdots A_\eta(N-1)\mathcal{B}_2, \tag{4.159}$$

(4.156) signifies that $\frac{\varphi_{21}^*(0)}{\varphi_{11}^*(0)}$ has the continued fraction expansion (1.49).

4.2.11 Saddle Point Method and Asymptotic Expansion

Let us see what kind of asymptotic behavior the function $\Phi(\alpha,\beta,\gamma)$ (4.144) has in the direction of $\eta = (1,1,2)$ at infinity, by analyzing an integral representation. By (1.40), we have

$$F(\alpha + N, \beta + N, \gamma + 2N; x) \tag{4.160}$$

$$= \frac{\Gamma(\gamma + 2N)}{\Gamma(\alpha + N)\Gamma(\gamma - \alpha + N)} \int_0^1 u^{\alpha - 1}(1 - u)^{\gamma - \alpha - 1}(1 - xu)^{-\beta}$$

$$\cdot \left\{ \frac{u(u - 1)}{1 - ux} \right\}^N du$$

$$= \frac{\Gamma(\gamma + 2N)}{\Gamma(\alpha + N)\Gamma(\gamma - \alpha + N)} \int_0^1 u^{\alpha - 1}(1 - u)^{\gamma - \alpha - 1}(1 - xu)^{-\beta}$$

$$\cdot \exp[NF(u)]du,$$

$$F(u) = \log u + \log(1 - u) - \log(1 - ux). \tag{4.161}$$

Let us compute an asymptotic behavior by the steepest descent method (saddle point method). This method for the higher-dimensional case will be explained in § 4.3 in detail. Now, we are going to find the zeros of the real gradient vector field

$$\mathbf{V} = \mathrm{grad}(\Re F) = \left(\frac{\partial \Re F}{\partial \Re x}, \frac{\partial \Re F}{\partial \Im x} \right) \tag{4.162}$$

on $\mathbb{C} \setminus \{0, 1, \frac{1}{x}\}$, that is, the critical points of $\Re F$, or in other words, the saddle points of $\Re F$. Here, we have

$$\mathbf{V} = 0 \quad \Leftrightarrow \quad \frac{1}{u} + \frac{1}{u - 1} - \frac{x}{ux - 1} = 0, \tag{4.163}$$

and the two solutions ξ_1, ξ_2 of (4.163) with respect to u are given by

$$\begin{cases} \xi_1 = \frac{1 - \sqrt{1 - x}}{x} = \frac{1}{1 + \sqrt{1 - x}}, \\ \xi_2 = \frac{1 + \sqrt{1 - x}}{x} = \frac{1}{1 - \sqrt{1 - x}}. \end{cases} \tag{4.164}$$

For simplicity, we suppose that $0 < x < 1$. Then, we have $0 < \xi_1 < 1 < \frac{1}{x} < \xi_2$. The orbits of $\mathbf{V}$ passing through ξ_1, ξ_2, in other words, the contracting cycles $\sigma_{\xi_1}, \sigma_{\xi_2}$ of $\mathbf{V}$ passing through ξ_1, ξ_2, respectively, are given as in Figure 4.2:

$$\sigma_{\xi_1} : \ 0 \leq x \leq 1, \tag{4.165}$$

$$\sigma_{\xi_2} : \ \{\xi_1 \leq x \leq 1\} \cup \{u \in \mathbb{C} | 0 = \arg(u) + \arg(1 - u) - \arg(1 - ux)\}.$$

$\Re F$ is maximal at ξ_1 on σ_{ξ_1} and is maximal at ξ_2 on σ_{ξ_2}. At ξ_1, we have

The phase diagram of **V**

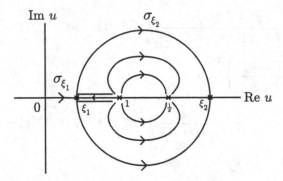

Fig. 4.2

$$\frac{d^2F}{du^2} = \frac{-2x^2}{(1-\sqrt{1-x})^2\sqrt{1-x}}, \qquad u \in \mathbb{R},$$

and at ξ_2, we have

$$\frac{d^2F}{du^2} = \frac{2x^2}{(1+\sqrt{1-x})^2\sqrt{1-x}}, \qquad u \in \mathbb{R}.$$

We take a branch of the function $U = u^{\alpha-1}(1-u)^{\gamma-\alpha-1}(1-xu)^{-\beta}$ in such a way that it becomes positive at ξ_1, ξ_2. On σ_{ξ_1}, σ_{ξ_2}, $\Im F$ is constant. So, we have

$$\int_{\sigma_{\xi_1}} u^{\alpha-1}(1-u)^{\gamma-\alpha-1}(1-xu)^{-\beta}du \qquad (4.166)$$

$$= \int_0^1 u^{\alpha-1}(1-u)^{\gamma-\alpha-1}(1-xu)^{-\beta}du,$$

$$\int_{\sigma_{\xi_2}} u^{\alpha-1}(1-u)^{\gamma-\alpha-1}(1-ux)^{-\beta}du \qquad (4.167)$$

$$= 2\sqrt{-1}\sin\beta \int_1^{\frac{1}{x}} u^{\alpha-1}(1-u)^{\gamma-\alpha-1}(1-ux)^{-\beta}du.$$

At $u = \xi_1$, ξ_2, $u^{\alpha+N-1}(1-u)^{\gamma-\alpha-1+N}(1-xu)^{-\beta-1-N}$ becomes $U(\xi_1)\xi_1^{2N}$, $U(\xi_2)\xi_2^{2N}$, respectively. On the other hand, by Stirling formula (1.5), we remark that

$$\frac{\Gamma(\gamma+2N)}{\Gamma(\alpha+N)\Gamma(\gamma-\alpha+N)} = \frac{2^{\gamma-1}}{\sqrt{\pi}}e^{2N\log 2}\left(1+O\left(\frac{1}{N}\right)\right),$$

$$\frac{\Gamma(\gamma+1+2N)}{\Gamma(\alpha+N)\Gamma(\gamma-\alpha+N)} = \frac{2^{\gamma}}{\sqrt{\pi}}e^{2N\log 2}\left(1+O\left(\frac{1}{N}\right)\right).$$

Since we have, by the saddle point method,

$$\int_{\sigma_{\xi_1}} u^{\alpha+N-1}(1-u)^{\gamma-\alpha+N-1}(1-xu)^{-\beta-N-1}du \tag{4.168}$$

$$= \xi_1^{2N} N^{-\frac{1}{2}}\left\{\frac{(1-\sqrt{1-x})(1-x)^{\frac{1}{4}}}{x}\right\}\xi_1^{\alpha-1}(1-\xi_1)^{\gamma-\alpha-1}$$

$$\cdot (1-\xi_1 x)^{-\beta-1}\sqrt{\pi}\left\{1+O\left(\frac{1}{N}\right)\right\},$$

$$\int_{\sigma_{\xi_2}} u^{\alpha+N-1}(1-u)^{\gamma-\alpha+N-1}(1-xu)^{-\beta+N-1}du \tag{4.169}$$

$$= \xi_2^{2N} N^{-\frac{1}{2}}\left\{\frac{(1+\sqrt{1-x})(1-x)^{\frac{1}{4}}}{x}\right\}\xi_2^{\alpha-1}(1-\xi_2)^{\gamma-\alpha-1}$$

$$\cdot (1-\xi_2 x)^{-\beta-1}\sqrt{\pi}\left\{1+O\left(\frac{1}{N}\right)\right\},$$

we obtain

$$F(\alpha+N,\beta+N,\gamma+2N;x) \tag{4.170}$$

$$= 4^N\left(\frac{1-\sqrt{1-x}}{x}\right)^{2N}\frac{(1-\sqrt{1-x})(1-x)^{\frac{1}{4}}}{x}\cdot 2^{\gamma-1}$$

$$\cdot \xi_1^{\alpha-1}(1-\xi_1)^{\gamma-\alpha-1}(1-\xi_1 x)^{-\beta-1}\left\{1+O\left(\frac{1}{N}\right)\right\},$$

$$F(\alpha+N,\beta+1+N,\gamma+1+2N;x) \tag{4.171}$$

$$= 4^N\left(\frac{1-\sqrt{1-x}}{x}\right)^{2N}\frac{(1-\sqrt{1-x})(1-x)^{\frac{1}{4}}}{x}\cdot 2^{\gamma}$$

$$\cdot \xi_1^{\alpha}(1-\xi_1)^{\gamma-\alpha-1}(1-\xi_1 x)^{-\beta-1}\left\{1+O\left(\frac{1}{N}\right)\right\},$$

$$\widetilde{F}(\alpha + N, \beta + N, \gamma + 2N; x) \tag{4.172}$$

$$= 4^N \left(\frac{1 + \sqrt{1-x}}{x}\right)^{2N} \frac{(1 + \sqrt{1-x})(1-x)^{\frac{1}{4}}}{x} 2^{\gamma-1}$$

$$\cdot \xi_2^{\alpha-1}(1 - \xi_2)^{\gamma-\alpha-1}(1 - \xi_2 x)^{-\beta-1} \left\{1 + O\left(\frac{1}{N}\right)\right\},$$

$$\widetilde{F}(\alpha + N, \beta + N + 1, \gamma + 1 + 2N; x) \tag{4.173}$$

$$= 4^N \left(\frac{1 + \sqrt{1-x}}{x}\right)^{2N} \frac{(1 + \sqrt{1-x})(1-x)^{\frac{1}{4}}}{x} 2^{\gamma-1}$$

$$\cdot \xi_2^{\alpha}(1 - \xi_2)^{\gamma-\alpha-1}(1 - \xi_2 x)^{-\beta-1} \left\{1 + O\left(\frac{1}{N}\right)\right\}.$$

Hence, at $N \mapsto +\infty$, we obtain

$$\Phi(\alpha + N, \beta + N, \gamma + 2N) \tag{4.174}$$

$$\sim 2^{\gamma} \frac{(1-x)^{\frac{1}{4}}}{x} \begin{pmatrix} \frac{1}{\frac{1+\sqrt{1-x}}{2}} & \frac{1}{\frac{1-\sqrt{1-x}}{2}} \end{pmatrix} \begin{pmatrix} \lambda_1^{-N} & \\ & \lambda_2^{-N} \end{pmatrix}$$

$$\begin{pmatrix} U(\xi_1)\xi_1(1 - \sqrt{1-x}) & \\ & U(\xi_2)\xi_2(1 + \sqrt{1-x}) \end{pmatrix}.$$

On the other hand, since we have

$$\Phi^*(\alpha + N, \beta + N, \gamma + 2N) \tag{4.175}$$

$$\sim \begin{pmatrix} \frac{1}{\frac{1+\sqrt{1-x}}{2}} & \frac{1}{\frac{1-\sqrt{1-x}}{2}} \end{pmatrix} \begin{pmatrix} \lambda_1^{-N} & \\ & \lambda_2^{-N} \end{pmatrix}$$

by (4.151), we obtain the relations

$$\Phi(\alpha, \beta, \gamma) = \Phi^*(\alpha, \beta, \gamma)G(\alpha, \beta, \gamma), \tag{4.176}$$

$$G(\alpha, \beta, \gamma) = 2^{\gamma} \frac{(1-x)^{\frac{1}{4}}}{x} \begin{pmatrix} U(\xi_1)\frac{(1-\sqrt{1-x})^2}{x} & \\ & U(\xi_2)\frac{(1+\sqrt{1-x})^2}{x} \end{pmatrix}. \tag{4.177}$$

Thus, by (4.156), we have

$$\Phi(\alpha, \beta, \gamma)\mathcal{B}_2 = \Phi^*(\alpha, \beta, \gamma)\mathcal{B}_2 \tag{4.178}$$

$$= \lim_{N \mapsto +\infty} A_{\eta}(0)A_{\eta}(1) \cdots A_{\eta}(N-1)\mathcal{B}_2,$$

namely,

$$\frac{F(\alpha, \beta, \gamma; x)}{F(\alpha, \beta + 1, \gamma + 1; x)} = \frac{\varphi_{21}(0)}{\varphi_{11}(0)} = \frac{\varphi_{21}^*(0)}{\varphi_{11}^*(0)} \tag{4.179}$$

has the continued fraction expansion of the right-hand side of (1.49). In this way, Gauss' formula is again obtained by a general method for difference equations.

4.3 Contracting (Expanding) Twisted Cycles and Asymptotic Expansion

Let us show how the example treated in § 4.2.11 is generalized to the case of several variables.

4.3.1 Twisted Cohomology

In Chapter 2, we have discussed the finite dimensionality of twisted cohomologies. As a consequence, we can obtain a difference equation with respect to shifts of the exponents $\alpha = (\alpha_1, \cdots, \alpha_m)$. In § 2.2, we have considered the integral of a multiplicative function $U(u) = P_1^{\alpha_1} \cdots P_m^{\alpha_m}$ on $\mathbb{C}^n$ over an n-dimensional twisted cycle σ

$$\widehat{du_1 \wedge \cdots \wedge du_n} = \int_\sigma U(u) du_1 \wedge \cdots \wedge du_n. \tag{4.180}$$

As we have seen in § 2.8, the twisted de Rham cohomology $H^*(\Omega^\bullet(*D), \nabla_\omega)$ of an affine variety $M = \mathbb{C}^n \setminus D$ is of finite dimension, and under some assumptions, we have

$$H^p(\Omega^\bullet(*D), \nabla_\omega) \cong 0, \quad p \neq n, \tag{4.181}$$

$$\dim H^n(\Omega^\bullet(*D), \nabla_\omega) = (-1)^n \chi(M), \tag{4.182}$$

where $\chi(M)$ is the Euler characterisitc of M. Now, for almost all α, there exists a basis $[\varphi_1], \cdots, [\varphi_r](\varphi_1, \varphi_2, \quad , \varphi_r \subset \Omega^n(*D))$ (here we set $r = \dim H^n(\Omega^\bullet(*D), \nabla_\omega)$) of $H^n(\Omega^\bullet(*D), \nabla_\omega)$ which does not depend on α. The integral

$$\widehat{\varphi_k(\alpha)} = \int_\sigma U\varphi_k \tag{4.183}$$

satisfies, for the shifting operator $T_j^{\pm 1}$: $(\alpha_1, \ldots, \alpha_m) \mapsto (\alpha_1, \ldots, \alpha_j \pm 1, \ldots, \alpha_m)$ with respect to each component of $\alpha = (\alpha_1, \ldots, \alpha_m)$,

$$T_j \widehat{\varphi_k(\alpha)} = \widehat{\varphi}(\alpha_1, \ldots, \alpha_j + 1, \ldots, \alpha_m) \tag{4.184}$$

$$= \int_\sigma U P_j \varphi_k,$$

$$T_j^{-1} \widehat{\varphi_k(\alpha)} = \int_\sigma U P_j^{-1} \varphi_k, \tag{4.185}$$

(remark that, here, instead of the variables z_j used in Chapter 2, we use α_j). Since $[\varphi_1], \ldots, [\varphi_r]$ form a basis of $H^n(\Omega^\bullet(*D)\nabla_\omega)$, as elements of $H^n(\Omega^\bullet(*D), \nabla_\omega)$, we have the expressions:

$$[P_j \varphi_k] = \sum_{l=1}^r a_{kl}^{(j)}(\alpha)[\varphi_l], \tag{4.186}$$

$$[P_j^{-1} \varphi_k] = \sum_{l=1}^r \widetilde{a}_{kl}^{(j)}(\alpha)[\varphi_l]. \tag{4.187}$$

Here, $a_{kl}^{(j)}$ and $\widetilde{a}_{kl}^{(j)}$ are rational functions of $\alpha_1, \ldots, \alpha_m$ and the coefficients of P_ν.

This is because, as we stated in Remark 2.5, there exist rational $(n-1)$-forms $\psi_\pm \in \Omega^{n-1}(*D)$ on the rational function field of $\alpha_1, \alpha_2, \ldots, \alpha_m$ such that

$$P_j \varphi_k - \sum_{l=1}^r a_{kl}^{(j)}(\alpha)\varphi_l = \nabla_\omega \psi_+,$$

$$P_j^{-1} \varphi_k - \sum_{l=1}^r \widetilde{a}_{kl}^{(j)}(\alpha)\varphi_l = \nabla_\omega \psi_-.$$

Passing from the cohomology classes to the integrals, we obtain the difference relations:

$$T_j \widehat{\varphi_k}(\alpha) = \sum_{l=1}^r a_{kl}^{(j)}(\alpha)\widehat{\varphi_l}(\alpha), \tag{4.188}$$

$$T_j^{-1} \widehat{\varphi_k}(\alpha) = \sum_{l=1}^r \widetilde{a}_{kl}^{(j)}(\alpha)\widehat{\varphi_l}(\alpha), \tag{4.189}$$

or in the vector representation, we have

$$T_j^{\pm}(\widehat{\varphi_k}(\alpha))_k = A_{\pm j}(\alpha)(\widehat{\varphi_k}(\alpha))_k, \tag{4.190}$$

where $A_j = ((a_{kl}^{(j)}))_{k,l=1}^r$, $A_{-j} = ((\widetilde{a}_{kl}^{(j)}))_{k,l=1}^r$ and $(\widehat{\varphi_k}(\alpha))_k$ is the column vector with components $\widehat{\varphi_1}(\alpha), \ldots, \widehat{\varphi_r}(\alpha)$. For a more general shifting operator $T^\nu = T_1^{\nu_1} \cdots T_m^{\nu_m}$, $\nu = (\nu_1, \ldots, \nu_m) \in \mathbf{Z}^m$, it transforms the relations among cohomology classes in the integrals without any change, and the relation

$$T^\nu(\widehat{\varphi_k}(\alpha))_k = A_\nu(\alpha)(\widehat{\varphi_k}(\alpha))_k \tag{4.191}$$

holds. Here, the set of $r \times r$ matrices $\{A_\nu(\alpha)\}_{\nu \in \mathbf{Z}^m}$ defines a 1-cocycle on $\mathbf{Z}^n$ with values in $GL_r(\mathbb{C}(\alpha))$. Namely, the cocycle conditions

$$A_{\nu+\nu'}(\alpha) = A_\nu(\alpha + \nu')A_{\nu'}(\alpha), \tag{4.192}$$

$(\nu, \nu' \in \mathbf{Z}^m)$,

$$A_0(\alpha) = 1, \tag{4.193}$$

hold. In particular, for the standard basis $e_j = (0, \ldots, \overset{j}{1}, \ldots, 0)$ of $\mathbf{Z}^m$, we have $A_{\pm e_j}(\alpha) = A_{\pm j}(\alpha)$ and $\{A_j(\alpha)\}_{j=1}^m$ satisfy the compatibility conditions (4.97). Conversely, when $\{A_j(\alpha)\}_{j=1}^m$ satisfying (4.97) are given, they uniquely extend to a 1-cocycle $\{A_\nu(\alpha)\}_{\nu \in \mathbf{Z}^m}$ and equation (4.191) defines a holonomic system of rational difference equations (4.96) of first order and $(\widehat{\varphi_j}(\alpha))_j$ are solutions of it.

By (4.192), (4.193), we clearly have

$$A_j(\alpha - e_j)A_{-j}(\alpha) = A_{-j}(\alpha + e_j)A_j(\alpha) = 1. \tag{4.194}$$

In particular, for a regular direction η, we obtain

$$A_{j,\eta}^{(0)} A_{-j,\eta}^{(0)} = 1 \tag{4.195}$$

in its Laurent expansion (4.98).

4.3.2 Saddle Point Method for Multi-Dimensional Case

Now, we fix $\eta \in \mathbf{Z}^m \setminus \{0\}$. For the shift operator T in the direction of η, we have

$$T^{N\eta}\widehat{\varphi}(\alpha) = \widehat{\varphi}(\alpha + N\eta) \tag{4.196}$$

$$= \int_\sigma U(P_1^{\eta_1} \cdots P_m^{\eta_m})^N du_1 \wedge \cdots \wedge du_n$$

$$= \int_\sigma U \exp(N \sum_{j=1}^m \eta_j \log P_j) du_1 \wedge \cdots \wedge du_n$$

$$= \int_\sigma U \exp(NF) du_1 \wedge \cdots \wedge du_n.$$

Here, $F(u)$ represents the function:

$$F(u) = \sum_{j=1}^m \eta_j \log P_j. \tag{4.197}$$

To compute an asymptotic expansion of $\widehat{\varphi}(\alpha + N\eta)$ at $N \mapsto +\infty$, we use the saddle point method, which is a long-established general method ([Bru],[Ar-GZ-V]). This is also called the steepest descent method or the Laplace method. On the other hand, the construction of a twisted cycle has a strong relation to the Morse theory in geometry. We explain this below.

4.3.3 Complete Kähler Metric

We define the gradient vector field $\mathbf{V} = \mathrm{grad}\Re F$ on M with level function $\Re F$. Since $\mathbf{V}$ is defined by an appropriate Kähler metric on M, we start by introducing a complete Kähler metric on M[1]. We owe the following lemma to T. Ohsawa.

Lemma 4.7. *Let M be any non-compact connected complex manifold. Assume that $\psi_\nu(u)(1 \leqslant \nu \leqslant N)$ are smooth real-valued functions defined on M such that $\max_{1 \leqslant \nu \leqslant N} |\psi_\nu(u)|$ increases to $+\infty$ at infinity. Moreover, we assume that each hermitian matrix*

$$\left(\left(\psi_\nu(\zeta) \frac{\partial^2 \psi_\nu(\zeta)}{\partial \zeta^j \partial \overline{\zeta}^k} \right) \right)_{j,k=1}^n \qquad (\zeta^1, \ldots, \zeta^n \text{ are local charts}) \tag{4.198}$$

is everywhere positive semi-definite. Now, let ds_0^2 be any, not necessarily complete, Kähler metric on M. Then, the sum of ds_0^2 and the pseudo-Kähler metric (which is positive semi-definite, but not necessarily definite)

---

[1] The ordinary gradient vector field $\left(\frac{\partial \mathrm{Re} F}{\partial u_1}, \ldots, \frac{\partial \mathrm{Re} F}{\partial u_n} \right), (u_1, \ldots, u_n) \in \mathbb{R}^n$ on $\mathbb{R}^n$ is the restriction of the above gradient vector field to its real subspace defined by the flat Kähler metric $\sum_{j=1}^n |du_j|^2$ on the complexification $\mathbb{C}^n$ of $\mathbb{R}^n$.

$$ds_1^2 = \sum_{\nu=1}^{N} \sum_{j,k=1}^{n} \frac{\partial^2 \psi_\nu^2}{\partial \zeta^j \partial \overline{\zeta}^k} d\zeta^j d\overline{\zeta}^k \qquad (4.199)$$

with potential given by a plurisubharmonic function $\sum_{\nu=1}^{N} \psi_\nu^2$, $ds^2 = ds_0^2 + ds_1^2$, defines a complete Kähler metric on M.

Proof. Since we have an expression

$$\frac{\partial^2 (\psi_\nu^2)}{\partial \zeta^j \partial \overline{\zeta}^k} = 2 \frac{\partial \psi_\nu}{\partial \zeta^j} \frac{\partial \psi_\nu}{\partial \overline{\zeta}^k} + \psi_\nu \frac{\partial^2 \psi_\nu}{\partial \zeta^j \partial \overline{\zeta}^k}$$

in local coordinates $(\zeta^1, \ldots, \zeta^n)$, it follows that

$$\sum_{j,\,k=1}^{n} \frac{\partial^2 (\psi_\nu^2)}{\partial \zeta^j \partial \overline{\zeta}^k} d\zeta^j d\overline{\zeta}^k = 2|\partial \psi_\nu|^2 + \sum_{j,k=1}^{n} \psi_\nu \frac{\partial^2 \psi_\nu}{\partial \zeta^j \partial \overline{\zeta}^k} d\zeta^j d\overline{\zeta}^k \qquad (4.200)$$

$$\geq 2|\partial \psi_\nu|^2.$$

Here, we set $\partial \psi_\nu = \sum_{j=1}^{n} \frac{\partial \psi_\nu}{\partial \zeta^j} d\zeta^j$. Now, for the total derivative

$$d\psi_\nu = \sum_{j=1}^{n} \left(\frac{\partial \psi_\nu}{\partial \zeta^j} d\zeta^j + \frac{\partial \psi_\nu}{\partial \overline{\zeta}^j} d\overline{\zeta}^j \right)$$

of ψ_ν, by

$$|d\psi_\nu|^2 = \left\{ \sum_{j=1}^{n} \left(\frac{\partial \psi_\nu}{\partial \zeta^j} d\zeta^j + \frac{\partial \psi_\nu}{\partial \overline{\zeta}^j} d\overline{\zeta}^j \right) \right\}^2 \qquad (4.201)$$

$$= 4 \left\{ \sum_{j=1}^{n} \mathrm{Re} \left(\frac{\partial \psi_\nu}{\partial \zeta^j} d\zeta^j \right) \right\}^2$$

$$\leq 4 \left| \sum_{j=1}^{n} \frac{\partial \psi_\nu}{\partial \zeta^j} d\zeta^j \right|^2 = 4|\partial \psi_\nu|^2,$$

it follows from the Cauchy–Schwarz inequality that

$$ds^2 \geq 2 \sum_{\nu=1}^{N} |\partial \psi_\nu|^2 \geq \frac{1}{2} \sum_{\nu=1}^{N} |d\psi_\nu|^2 \geq \frac{1}{2N} \left(\sum_{\nu=1}^{N} |d\psi_\nu| \right)^2. \qquad (4.202)$$

For any two points p, q on M, we consider the length $s(p, q)$ of any piecewise smooth curve connecting p and q. Integrating the above inequality along γ, it satisfies

$$s(p,q) \geq \frac{1}{\sqrt{2N}} \sum_{\nu=1}^{N} |\psi_\nu(p) - \psi_\nu(q)|. \tag{4.203}$$

Fixing p and letting q be at infinity, we must have $s(p,q) \mapsto +\infty$. In this way, ds^2 gives a complete Kähler metric on M.

Returning to the previous situation, we let M be $\mathbb{C}^n \setminus D$. Setting $\psi_\nu = \log|P_\nu|$ $(1 \leqslant \nu \leqslant m)$, it satisfies the properties of the lemma. Since we have

$$\frac{\partial^2 \psi_\nu^2}{\partial u_j \partial \overline{u}_k} = \frac{1}{2} \frac{1}{|P_\nu|^2} \frac{\partial P_\nu}{\partial u_j} \frac{\partial \overline{P}_\nu}{\partial \overline{u}_k}, \tag{4.204}$$

the metric

$$ds^2 = |du_1|^2 + \cdots + |du_n|^2 + \frac{1}{2} \sum_{\nu=1}^{m} |d\log P_\nu|^2 \tag{4.205}$$

defines a complete Kähler metric on M.

4.3.4 Gradient Vector Field

Now, $M = \mathbb{C}^n \setminus D$ becomes a complete Kähler manifold. For local coordinates $(\zeta^1, \ldots, \zeta^n)$ of a neighborhood of each point of M, the metric can be written as follows:

$$ds^2 = \sum_{i,j=1}^{n} g_{i\overline{j}} d\zeta^i d\overline{\zeta}^j, \tag{4.206}$$

where $(g_{i\overline{j}})_{i,j=1}^{n}$ is a positive definite hermitian matrix. Since the gradient vector field $\mathbf{V} = \mathrm{grad} \Re F$ on M is expressed as

$$\mathbf{V} = \left(\sum_{j=1}^{n} g^{1\overline{j}} \frac{\partial \Re F}{\partial \overline{\zeta}^j}, \ldots, \sum_{j=1}^{n} g^{n\overline{j}} \frac{\partial \Re F}{\partial \overline{\zeta}^j} \right)$$

with respect to local coordiantes $(\zeta^1, \ldots, \zeta^n)$, the differential equation that defines the trajectory of $\mathbf{V}$ is given by

$$\frac{d\zeta^i}{dt} = \sum_{j=1}^{n} \frac{\partial \Re F}{\partial \overline{\zeta}^j} g^{i\overline{j}}, \tag{4.207}$$

where $(g^{i\bar{j}})^n_{i,j=1}$ is the inverse matrix of $(g_{i,\bar{j}})^n_{i,j=1}$. Since the vector field $\mathbf{V}$ does not depend on the choice of local coordinates, it is a global vector field defined on M. By the Cauchy–Riemann equation,

$$\frac{\partial \Re F}{\partial \zeta^i} = \sqrt{-1}\frac{\partial \Im F}{\partial \zeta^i}, \tag{4.208}$$

$$\frac{\partial \Re F}{\partial \overline{\zeta}^i} = -\sqrt{-1}\frac{\partial \Im F}{\partial \overline{\zeta}^i},$$

equation (4.207) can be rewritten as

$$\frac{d\zeta^i}{dt} = -\sqrt{-1}\sum_{j=1}^{n}\frac{\partial \Im F}{\partial \overline{\zeta}^j}g^{i\bar{j}}. \tag{4.209}$$

Namely, $\mathbf{V}$ is the Hamiltonian flow with the Hamiltonian function $\Im F$ (see [Ar]). In particular, we have

$$\frac{d\Im F}{dt} = \sum_{j=1}^{n}\left(\frac{\partial \Im F}{\partial \zeta^j}\frac{d\zeta^j}{dt} + \frac{\partial \Im F}{\partial \overline{\zeta}^j}\frac{d\overline{\zeta}^j}{dt}\right) \tag{4.210}$$

$$= -2\Re\left\{\sqrt{-1}\sum_{i,j=1}^{n}\frac{\partial \Im F}{\partial \zeta^i}\frac{\partial \Im F}{\partial \overline{\zeta}^j}g^{i\bar{j}}\right\}$$

$$= 0,$$

and $\Im F$ is constant along the flow.

Now, along each trajectory of $\mathbf{V}$, the length $s(u, u')$ of the section with the boundary $u, u' \in M$ in a trajectory satisfies the equality

$$s(u, u') = \frac{1}{2}|\Re F(u) - \Re F(u')|. \tag{4.211}$$

In fact, we let u move along a trajectory of (4.207) with time t, and we have

$$\left(\frac{ds(u, u')}{dt}\right)^2 = \sum_{i,j=1}^{n}g_{i\bar{j}}\frac{d\zeta^i}{dt}\frac{d\overline{\zeta}^j}{dt} \tag{4.212}$$

$$-\sum_{i,j=1}^{n}\frac{\partial \Re F}{\partial \zeta^i}\frac{\partial \Re F}{\partial \overline{\zeta}^j}g^{i\bar{j}} = \frac{1}{2}\frac{d\Re F}{dt} \geq 0.$$

Here, by specializing the parameter t as $t = \frac{1}{2}\Re F$, we have

$$\left(2\frac{ds(u,u')}{d\Re F(u)}\right)^2 = 1, \tag{4.213}$$

namely, $2\frac{ds(u,u')}{d\Re F(u)} = \pm 1$ and we obtain (4.211). From this, fixing u', having $s(u,u') = +\infty$ on the same trajectory and having $|\Re F(u)| = +\infty$ are equivalent. In other words, a trajectory of the vector field $\mathbf{V}$ and u rest in a compact subset of M if $\Re F(u)$ is finite. Thus, the following proposition is proved.

Proposition 4.3. *The gradient vector field $\mathbf{V}$ defined by (4.207) is complete on M.*

4.3.5 Critical Points

On M, a point where $\mathbf{V}$ vanishes, that is, a singularity of $\mathbf{V}$, is a point where all of the first derivatives of $\Re F$ vanish:

$$\frac{\partial \Re F}{\partial u_j} = \frac{\partial \Re F}{\partial \overline{u}_j} = 0, \tag{4.214}$$

in other words, it is a critical point of the level function $\Re F$. Hence, we also have

$$\frac{\partial F}{\partial u_l} = \sum_{j=1}^m \frac{\eta_j}{P_j}\frac{\partial P_j}{\partial u_l} = 0, \quad 1 \leqslant l \leqslant n. \tag{4.215}$$

This is equivalent to saying that the holomorphic 1-form

$$\omega_\eta = \sum_{j=1}^m \eta_j d \log P_j \tag{4.216}$$

on M vanishes. Thus, to find a singularity of $\mathbf{V}$, it suffices to find the zeros of ω_η.

4.3.6 Vanishing Theorem (iii)

The zeros of ω_η are determined by the zeros of a system of n algebraic equations (4.215). Here, we impose the following assumption.

Assumption 3. *The system of equations (4.215) admits at most a finite number of zeros $c^{(1)}, c^{(2)}, \ldots, c^{(\kappa)}$ in M, and at each point $c^{(j)}$, the symmetric matrix $\left(\left(\frac{\partial^2 F}{\partial u_j \partial u_k}\right)\right)_{j,k=1}^n$ (this matrix is called the Hessian matrix of F) of order n is non-degenerate.*

Assumption 3 is a restriction not only on $P_l(1 \leqslant l \leqslant m)$ but also in the direction η; nevertheless, this holds for almost all η. In this case, $\Re F$ is said to be a Morse function as a level function on M ([Mi]). As can be seen from (4.212), we always have $\frac{d\Re F}{dt} \geq 0$ and $\frac{d\Re F}{dt} = 0$ can happen only if $\frac{\partial \Re F}{\partial \zeta^i} = 0$ $(1 \leqslant i \leqslant n)$.

We set the value of $\Re F$ at each critical point $c^{(j)}$ by $\Re F(c^{(j)}) = \beta_j \in \mathbf{R}$, and we assume that they are arranged as follows:

Assumption 4. $\beta_1 > \beta_2 > \cdots > \beta_\kappa$.

Now, we denote the hyperplane at infinity in $\mathbb{P}^n(\mathbb{C})$ by H_∞. $T(D^\cup H_\infty)$ signifies an appropriate tubular neighborhood of $D^\cup H_\infty$ in $\mathbb{P}^n(\mathbb{C})$. The pull back $\iota^* \mathcal{L}_\omega$ of the local system $\mathcal{L}_\omega$ on M via the inclusion $\iota : T(D^\cup H_\infty) \setminus D^\cup H_\infty \subset M$ defines a local system on $T(D^\cup H_\infty) \setminus D^\cup H_\infty$. Here, we further assume:

Assumption 5.

$$H^p(T(D^\cup H_\infty) \setminus D^\cup H_\infty, \iota^* \mathcal{L}_\omega) \cong 0, \ p \geq 0. \tag{4.217}$$

Then, we have the following vanishing theorem.

Theorem 4.5 ([Ao2]).

$$H^p(M, \mathcal{L}_\omega) \cong 0, \ p \neq n. \tag{4.218}$$

Next, we provide a more comprehensible sufficient condition to satisfy Assumption 5.

By the Hironaka theorem of resolution of singularities, there exists a finite number of repetitions of appropriate blow-ups $(T(\widetilde{D^\cup H_\infty}), \widetilde{D^\cup H_\infty})$ of $(T(D^\cup H_\infty), D^\cup H_\infty)$ such that a proper morphism (regular map)

$$f : (T(\widetilde{D^\cup H_\infty}), (\widetilde{D^\cup H_\infty})) \to (T(D^\cup H_\infty), D^\cup H_\infty), \tag{4.219}$$

$$f : T(\widetilde{D^\cup H_\infty}) \setminus \widetilde{D^\cup H_\infty} \cong T(D^\cup H_\infty) \setminus D^\cup H_\infty \ \text{(isomorphic)} \tag{4.220}$$

is defined and $\widetilde{D^\cup H_\infty}$ is a desingularization of $D^\cup H_\infty$, that is, $\widetilde{D^\cup H_\infty}$ is normal crossing in $T(\widetilde{D^\cup H_\infty})$. At any point Q of $\widetilde{D^\cup H_\infty}$, we assume that the defining equation of $\widetilde{D^\cup H_\infty}$ is expressed as $(\zeta^1 = 0)^\cup \cdots ^\cup (\zeta^l = 0)$ in terms of local coordinates $(\zeta^1, \ldots, \zeta^n)$. The lifting of the logarithmic 1-form $\omega = \sum_{j=1}^m \alpha_j d \log P_j$, defined in § 2.5, to $T(\widetilde{D^\cup H_\infty}) \setminus \widetilde{D^\cup H_\infty}$ is expressed in the form of

$$\widetilde{\omega} = \sum_{j=1}^l \lambda_j \frac{d\zeta^j}{\zeta^j} + \text{(holomorphic)}, \quad \lambda_j \in \mathbb{C} \tag{4.221}$$

at Q. Here, we assume that:

Assumption 6. *At any point Q,*

$$\lambda_1 + \cdots + \lambda_l \notin \mathbb{Z}. \tag{4.222}$$

By the morphism (4.219), each P_h can be expressed as

$$P_h(f(\zeta)) = \zeta_1^{r_1} \zeta_2^{r_2} \cdots \zeta_l^{r_l} \cdot G(\zeta) \tag{4.223}$$

$(r_1, r_2, \ldots, r_l \in \mathbb{Z}, G(\zeta)$ is a holomorphic function satisfying $G(0) \neq 0)$ on a neighborhood of $\zeta = 0$, $\lambda_j \in \sum_{k=1}^{m} \mathbb{Q}\alpha_k$. Hence, we have $\lambda_1 + \cdots + \lambda_l \in \sum_{k=1}^{m} \mathbb{Q}\alpha_k$. In (4.223), since we have $r_1 + \cdots + r_l > 0$ at least for one h, Assumption 6 is satisfied by the following assumption:

Assumption 7. *If $\sum_{j=1}^{m} |s_j| > 0$, then the linear combination $\sum_{j=1}^{m} s_j \alpha_j$ of $\alpha_1, \alpha_2, \ldots, \alpha_m$ does not belong to $\mathbb{Q}$.*

It can be easily shown by the Mayer–Vietoris sequence that Assumption 6 follows from Assumption 5. In fact, there exists a finite family $\{W_s\}_{s \in S}$ of sub-manifolds of $\widetilde{D^{\cup} H_\infty}$ satisfying

$$\widetilde{D^{\cup} H_\infty} = \bigcup_{s \in S} W_s, \tag{4.224}$$

$$T(\widetilde{D^{\cup} H_\infty}) \setminus \widetilde{D^{\cup} H_\infty} = \bigcup_{s \in S} (T(W_s) \setminus W_s) \tag{4.225}$$

and each tubular neighborhood $T(W_s) \setminus W_s$ is topologically homeomorphic to the trivial bundle $W_s \times \Delta_s^*$, $\Delta_s^* \cong (\mathbb{C}^*)^l$ $(\dim W_s = n - l)$, i.e., there is the local trivialization:

$$\rho_s : W_s \times \Delta_s^* \xrightarrow{\sim} T(W_s) \setminus W_s.$$

Similarly, for any $s_1, s_2, \ldots, s_k \in S$ $(1 \leqslant k \leqslant n)$, the k intersections

$$(T(W_{s_1}) \setminus W_{s_1}) \cap \cdots \cap (T(W_{s_k}) \setminus W_{s_k})$$

are either empty or there exists a sub-manifold $W_{s_1,\ldots,s_k}$ of $D^{\cup} H_\infty$ that allows us to have the topologically local trivialization:

$$\rho_{s_1,\ldots,s_k} : W_{s_1,\ldots,s_k} \times \Delta_{s_1,\ldots,s_k}^* \xrightarrow{\sim} \bigcap_{j=1}^{k} (T(W_{s_j}) \setminus W_{s_j}),$$

$$\Delta_{s_1,\ldots,s_k}^* = (\mathbb{C}^*)^l, \quad \dim W_{s_1,\ldots,s_k} = n - l.$$

If we further pull back the pull-back $\iota_{s_1,\ldots,s_k}^* \mathcal{L}_\omega$ of $\mathcal{L}_\omega$ via the inclusion

$$\iota_{s_1,\dots,s_k} : \bigcap_{j=1}^{k} (T(W_{s_j}) \setminus W_{s_j}) \hookrightarrow M$$

by $\rho_{s_1,\dots,s_k}$, it defines a local system on $W_{s_1,\dots,s_k} \times \Delta^*_{s_1,\dots,s_k}$ obtained by the fundamental group (a free abelian group) of $\Delta^*_{s_1,\dots,s_k}$. By (4.222), we have

$$H^\bullet(\Delta^*_{s_1,\dots,s_k}, \rho^*_{s_1,\dots,s_k}\iota^*_{s_1,\dots,s_k}\mathcal{L}_\omega) \cong 0.$$

Hence, by the Künneth formula, we obtain

$$H^\bullet(W_{s_1,\dots,s_k} \times \Delta^*_{s_1,\dots,s_k}, \rho^*_{s_1,\dots,s_k}\iota^*_{s_1,\dots,s_k}\mathcal{L}_\omega) \cong 0,$$

namely,

$$H^\bullet\left(\bigcap_{j=1}^{r} (T(W_{s_j}) \setminus W_{s_j}), \iota^*_{s_1,\dots,s_k}\mathcal{L}_w\right) \cong 0. \tag{4.226}$$

From (4.225) and (4.226), by the Mayer–Vietoris exact sequence of cohomology ([Br]), we obtain (4.217).

4.3.7 Application of the Morse Theory

Theorem 4.5 corresponds to the vanishing theorem stated in § 2.8. In Chapter 2, we derived it algebraically, and here, we derive this theorem topologically by using the Morse theory (cf. [Mi]).

Lemma 4.8. *Let $\mathcal{Z}$ be a p-dimensional ($p < n$) twisted cycle. Then, there exists a proper twisted cycle $\mathcal{Z}^*$ in $T(D^\cup H_\infty) \setminus (D^\cup H_\infty)$ of the same dimension that is homologous to $\mathcal{Z}$.*

Proof. Set $M^h = \{u \in M ; \Re F \le h\}$ and we may assume that $\mathcal{Z}$ is contained in $M^h (h > \beta_1)$. As F is non-degenerate at each critical point $c^{(1)}, c^{(2)}, \dots, c^{(\kappa)}$ and F is regular, it follows that the Morse index of $\Re F$ at $c^{(j)}$ is just n, that is, the signature of the Hessian of $\Re F$, the real symmetric matrix of order $2n$

$$\left(\left(\begin{array}{cc} \frac{\partial^2 \Re F}{\partial \Re \zeta^k \partial \Re \zeta^l}, & \frac{\partial^2 \Re F}{\partial \Re \zeta^k \partial \Im \zeta^l} \\ \frac{\partial^2 \Re F}{\partial \Im \zeta^k \partial \Re \zeta^l}, & \frac{\partial^2 \Re F}{\partial \Im \zeta^k \partial \Im \zeta^l} \end{array}\right)\right)^n_{k,l-1},$$

is (n, n). Now, we show, by induction on i, that $\mathcal{Z}$ is homologous to a twisted cycle $\mathcal{Z}_i$ contained in $M^{\beta_i - \varepsilon}$ (ε is a small positive number). Indeed, this is true for $i = 0$, so by induction hypothesis, we may assume that $\mathcal{Z}$ is already homologous to a twisted cycle $\mathcal{Z}_{i-1}(\subset M^{\beta_{i-1}-\varepsilon})$. Since $\mathcal{Z}_{i-1}$ has a compact support, by the retraction defined by the one-parameter subgroup generated by the vector field $-\mathbf{V}$, it is homologous to a twisted cycle $\widetilde{\mathcal{Z}}_{i-1}$ in $M^{\beta_i+\varepsilon}$. On the other hand, as the Morse theory shows, when $\dim \widetilde{\mathcal{Z}}_{i-1} < n$, we can let $\widetilde{\mathcal{Z}}_{i-1}$ move a little bit homotopically and let it flow along a trajectory of $-\mathbf{V}$ on avoiding the critical point β_i of $\mathbf{V}$ (cf. Figure 4.3). In this way, there exists a twisted cycle $\mathcal{Z}_i$ near $\widetilde{\mathcal{Z}}_{i-1}$ which lies in $M^{\beta_i-\varepsilon}$ and which is homologous to $\widetilde{\mathcal{Z}}_{i-1}$. Namely, $\mathcal{Z}$ and $\mathcal{Z}_i$ are homologous. For a sufficiently big positive number L, M^{-L} belongs to $T(D^\cup H_\infty)$. This is because, we have $\Re F \le -L$ at each point of M^{-L}, i.e., the twisted cycle $\mathcal{Z}_\kappa$ is retracted to a cycle contained in $T(D^\cup H_\infty) \setminus D^\cup H_\infty$. Thus, we have proved the lemma.

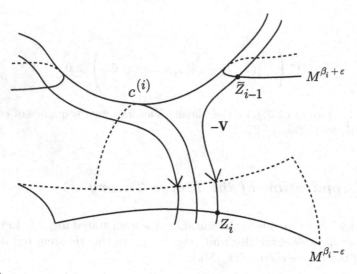

Fig. 4.3

Proof (of Theorem 4.5). Let $\mathcal{Z}$ be a p-dimensional $(p < n)$ twisted cycle in M. By the previous lemma, there exists a twisted cycle $\mathcal{Z}^*$ which is homologous to $\mathcal{Z}$ and which belongs to $T(D^\cup H_\infty) \setminus D^\cup H_\infty$. Since we have $H_p(T(D^\cup H_\infty) \setminus D^\cup H_\infty, (\iota^* \mathcal{L}_\omega)^\vee) \cong 0$ by Assumption 5, there exists a twisted chain τ which belongs to $T(D^\cup H_\infty) \setminus D^\cup H_\infty$ satisfying $\mathcal{Z}^* = \partial_\omega \tau$. This means that $\mathcal{Z}$ is homologous to 0, which implies Theorem 4.5.

4.3.8 n-Dimensional Lagrangian Cycles

Next, to construct n-dimensional twisted cycles, we explain an important local fact about trajectories of the gradient vector field $\mathbf{V}$. To avoid the complexity of the indices of coordinates, in this subsection, we take local coordinates $(z_1, \ldots, z_n)$ instead of $(\zeta^1, \ldots, \zeta^n)$, and suppose that the Kähler metric is given in the form $ds^2 = \sum_{i,j=1}^n g_{i\bar{j}}(z) dz_i d\bar{z}_j$.

Theorem 4.6 ([Ao2]). *Let $f(z)$ be a holomorphic function defined on a neighborhood of the origin of $(z_1, \ldots, z_n)$. We assume that the origin is a non-degenerate critical point of $\Re f$. Consider the equation of the trajectories (4.207)*

$$\frac{dz_i}{dt} = \sum_{j=1}^n \frac{\partial \Re f}{\partial \bar{z}_j} g^{i\bar{j}}(z) \tag{4.227}$$

of the gradient vector field $\mathbf{V}$. The set of all trajectories that have the origin as the limit point at $t = \pm\infty$ form a real n-dimensional manifold $\mathcal{M}^\pm$, respectively. $\mathcal{M}^+$ and $\mathcal{M}^-$ intersect only at the origin, and they are mutually transversal. Moreover, $\mathcal{M}^\pm$ satisfy $\Omega|_{\mathcal{M}^\pm} = 0$ with respect to the symplectic form

$$\Omega = \sqrt{-1} \sum_{i,j=1}^n g_{i\bar{j}}(z) dz_i \wedge d\bar{z}_j. \tag{4.228}$$

Such a sub-manifold $\mathcal{M}^\pm$ is called a Lagrangian submanifold (cf. [Ar], [Gui-St]).

Definition 4.1. $\mathcal{M}^+$ is called a contracting manifold with respect to $\mathbf{V}$, and $\mathcal{M}^-$ is called an expanding manifold with respect to $\mathbf{V}$. Hence, $\mathcal{M}^-, \mathcal{M}^+$ is also a contracting manifold (resp. an expanding manifold) with respect to $-\mathbf{V}$.[1]

Proof. The first half of the theorem is a special case of a well-konwn fact (cf. p. 113 in [Sma]). It also follows automatically from the proof of the latter half given below. To prove the latter half of the theorem, let us prepare two lemmata.

Lemma 4.9. *At the origin, there exists real analytic coordinates $(\xi_1, \ldots, \xi_n, \eta_1, \ldots, \eta_n)$ that admit expressions of the form*

$$\Omega = \sum_{i=1}^n d\xi_i \wedge d\eta_i, \tag{4.229}$$

$$\Im f = \sum_{j=1}^n \lambda_j \xi_j \eta_j + (higher\text{-}degree\ terms), \tag{4.230}$$

with $(\lambda_j > 0, \ 1 \leqslant j \leqslant n)$.

---

[1] Originally, S. Smale called them a stable manifold and an unstable manifold, respectively, but here we followed V.I. Arnold ([Ar], p.287). See also [Ar-Av].

Proof. Since we can transform as $g_{i\bar{j}}(0) = \delta_{ij}$ by an appropriate complex coordinate transformation

$$z_i = \sum_{j=1}^{n} a_{ij} w_j \quad (a_{i,j} \in \mathbb{C}), \tag{4.231}$$

we may assume that $g_{i\bar{j}}(0) = \delta_{ij}$ from the very beginning. Since the square matrix of order $2n$

$$S = \left(\begin{pmatrix} \frac{\partial^2 \Im f(0)}{\partial x_i \partial x_j}, & \frac{\partial^2 \Im f(0)}{\partial x_i \partial y_j} \\ \frac{\partial^2 \Im f(0)}{\partial y_i \partial x_j}, & \frac{\partial^2 \Im f(0)}{\partial y_i \partial y_j} \end{pmatrix} \right)_{i,j=1}^{n}, \quad z_i = x_i + \sqrt{-1} y_i$$

is real symmetric and infinitesimally symplectic, that is, it satisfies

$$S = {}^t S, \quad S \begin{pmatrix} 0 & 1_n \\ -1_n & 0 \end{pmatrix} + \begin{pmatrix} 0 & 1_n \\ -1_n & 0 \end{pmatrix} {}^t S = 0,$$

via the transformation (4.231) by a unitary matrix $((a_{ij}))_{i,j=1}^{n}$, S can be transformed to the form

$$S \mapsto \begin{pmatrix} 0 & \vdots & \text{diagonal matrix} \\ \cdots\cdots\cdots\cdots\cdots & \vdots & \cdots\cdots\cdots\cdots\cdots \\ \text{diagonal matrix} & \vdots & 0 \end{pmatrix}$$

such that all of the diagonal components are positive. Namely, by changing coordinates appropriately, with respect to new coordinates (x, y),

$$\Omega = \sum_{j=1}^{n} dx_j \wedge dy_j + (\text{higher-degree terms}), \tag{4.232}$$

$$\Im f = \sum_{j=1}^{n} \lambda_j x_j y_j + (\text{higher-degree terms}) \tag{4.233}$$

are expanded in convergent series of the above forms. Moreover, by an appropriate real analytic coordinate transformation $(\xi_1, \ldots, \xi_n, \eta_1, \ldots, \eta_n) \to z$

$$\begin{cases} x_j = \xi_j + (\text{higher-degree terms}) \\ y_j = \eta_j + (\text{higher-degree terms}) \end{cases}, \quad 1 \leqslant j \leqslant n,$$

Ω is reduced to the form (4.229) (Theorem 6.1 in Chap. 3 of [Ster]), and we obtain (4.229), (4.230).

Lemma 4.10. *Assume that the Hamiltonian function $H(\xi, \eta)$ is expanded to a convergent series in the form of the right-hand side of (4.230) with respect to the above local coordinates $(\xi, \eta) = (\xi_1, \ldots, \xi_n, \eta_1, \ldots, \eta_n)$ at the origin. Then, there exists a unique solution $\psi = \psi(\xi)$ of the Hamilton–Jacobi equation ([Ar])*

$$H\left(\xi, \frac{\partial \psi}{\partial \xi}\right) = 0, \tag{4.234}$$

$\frac{\partial \psi}{\partial \xi} = \left(\frac{\partial \psi}{\partial \xi_1}, \ldots, \frac{\partial \psi}{\partial \xi_n}\right)$ *being real analytic at the origin.*

Proof. We use the method of majorant series. In general, when a formal series in p variables $(t_1, \ldots, t_p)$

$$A(t) = \sum_{s_1, \ldots, s_p \geq 0} a_{s_1, \ldots, s_p} t_1^{s_1} \cdots t_p^{s_p}$$

and a series of positive terms

$$B = \sum_{s_1, \ldots, s_p \geq 0} b_{s_1, \ldots, s_p}$$

satisfy

$$|a_{s_1, \ldots, s_p} t_1^{s_1} \cdots t_p^{s_p}| \leq b_{s_1, \ldots, s_p}$$

for any $s_1, \ldots, s_p$, we say that B dominates $A(t)$, or B is a majorant series of $A(t)$. In such a case, we denote it as $A(t) \prec\prec B$. For $t \in \Delta^p(0; \delta)$ (δ a positive number), if we have $A(t) \prec\prec B$ and B converges, then so does $A(t)$ and it is real analytic. Now, assume that $H(\xi, \eta)$ has a power series expansion on $|\xi_j| \leq \rho$, $|\eta_j| \leq \rho$, and denote it by

$$H(\xi, \eta) = \sum_{j=1}^{n} \lambda_j \xi_j \eta_j + \sum_{|\mu|+|\sigma| \geq 3} h_{\mu, \sigma} \xi^\mu \eta^\sigma, \tag{4.235}$$

$$\mu = (\mu_1, \ldots, \mu_n), \quad \sigma = (\sigma_1, \ldots, \sigma_n), \quad |\mu| = \sum_{j=1}^{n} \mu_j,$$

$$|\sigma| = \sum_{j=1}^{n} \sigma_j, \quad \xi^\mu = \xi_1^{\mu_1} \cdots \xi_n^{\mu_n}, \quad \eta^\sigma = \eta_1^{\sigma_1} \cdots \eta_n^{\sigma_n}.$$

If a formal series in ξ

$$\psi(\xi) = \sum_{|\mu| \geq 3} \beta_\mu \xi^\mu \tag{4.236}$$

formally satisfies (4.234), then β_μ is uniquely determined and is given by the recursive formula:

$$\beta_\mu = -\frac{\text{the coefficient of } \xi^\mu \text{ in } \left[\sum_{|\mu|+|\sigma|\geq 3} h_{\mu,\sigma}\xi^\mu \left(\frac{\partial\psi}{\partial\xi}\right)^\sigma\right]}{\sum_{j=1}^n \lambda_j \mu_j}. \quad (4.237)$$

Here, $\frac{\partial\psi}{\partial\xi} = \left(\frac{\partial\psi}{\partial\xi_1}, \ldots, \frac{\partial\psi}{\partial\xi_n}\right)$ signifies the formal series

$$\frac{\partial\psi}{\partial\xi_j} = \sum_{|\mu|\geq 3} \mu_j \beta_\mu \xi^{\mu-e_j}, \quad e_j = (0,\ldots,\overset{j}{1},\ldots,0).$$

The right-hand side of (4.237) depends at most on $\beta_k(|k| \leq |\mu| - 1)$.
Since there exists a positive number K satisfying

$$|h_{\mu,\sigma}| \leq K\rho^{-(|\mu|+|\sigma|)},$$

we obtain an estimate as a formal power series in (ξ, η)

$$\sum_{|\mu|+|\sigma|\geq 3} h_{\mu,\sigma}\xi^\mu \eta^\sigma \prec\prec K\frac{\rho^{-3}(\sum_{j=1}^n |\xi_j| + |\eta_j|)^3}{1 - \rho^{-1}\sum_{j=1}^n(|\xi_j| + |\eta_j|)}.$$

We take $0 < \rho_1(\leq \rho)$ and consider $\xi \in \Delta^n(0;\rho_1)$. For a formal power series in ρ_1

$$\widetilde{\psi}(\rho_1) = \sum_{|\mu|\geq 3} |\mu|\,|\beta_\mu|\rho_1^{|\mu|-1}, \quad (4.238)$$

as we have

$$|\mu| \leq \sum_{j=1}^n \lambda_j\mu_j/(\min_{1\leq j\leq n} \lambda_j), \qquad \frac{\partial\psi}{\partial\xi_j} \prec\prec \widetilde{\psi}(\rho_1),$$

(4.237) implies

$$\widetilde{\psi}(\rho_1) \prec\prec \frac{K\rho_1^{-1}\rho^{-3}(n(\rho_1 + \widetilde{\psi}(\rho_1)))^3}{(\min_{1\leq j\leq n} \lambda_j)\left[1 - \rho^{-1}n(\rho_1 + \widetilde{\psi}(\rho_1))\right]} \quad (4.239)$$

as a formal power series in ρ_1. Now, setting $\psi^*(\rho_1) = \rho_1^2\omega(\rho_1)$, we can uniquely determine a convergent series of positive terms $\omega(\rho_1)$ $(0 \leq \rho_1 < \delta)$ satisfying the algebraic equation of degree 3:

$$\psi^*(\rho_1) = \frac{K\rho_1^{-1}\rho^{-3}}{(\min_{1\leq j\leq n} \lambda_j)} \frac{(n(\rho_1 + \psi^*(\rho_1)))^3}{[1 - n\rho^{-1}(\rho_1 + \psi^*(\rho_1))]},$$

in other words,

$$\omega(\rho_1) = \frac{K\rho^{-3}n^3}{\min_{1\leq j\leq n}\lambda_j}\frac{(1+\rho_1\omega(\rho_1))^3}{[1-n\rho_1\rho^{-1}(1+\rho_1\omega(\rho_1))]} \tag{4.240}$$

and $\omega(0) = \frac{K\rho^{-3}n^3}{\min_{1\leq j\leq n}\lambda_j}$. Since each coefficient of $\omega(\rho_1)$ in (4.240) can be uniquely determined, by expanding the right-hand side, from the coefficients of lower-degree terms, by induction on ρ_1^l ($l \geq 3$), we obtain

$$\widetilde{\psi}(\rho_1) \prec\prec \psi^*(\rho_1). \tag{4.241}$$

As $\psi^*(\rho_1)$ converges, so does $\widetilde{\psi}(\rho_1)$. In this way, we see that $\frac{\partial\psi}{\partial\xi_j}$ converges on $\xi \in \Delta^n(0;\rho_1)$. By

$$\psi(\xi) = \int_0^\xi \sum_{j=1}^n \frac{\partial\psi(\xi)}{\partial\xi_j}d\xi_j,$$

$\psi(\xi)$ also converges for $\xi \in \Delta^n(0;\rho_1)$. It is clear from (4.237) that $\psi(\xi)$ satisfies (4.234), hence, the proof of Lemma 4.10 is completed.

Now, let us terminate the proof of Theorem 4.6. Suppose that Ω is in the form (4.229), using the function $\psi(x)$ obtained by Lemma 4.10, we apply the canonical transformation (see, e.g., [Ar], [Ster]):

$$\begin{cases} \widetilde{\eta}_i = -\frac{\partial\psi}{\partial\xi_i} + \eta_i \\ \widetilde{\xi}_i = \xi_i \end{cases}, \quad 1 \leqslant i \leqslant n. \tag{4.242}$$

We have $\Omega = \sum_{i=1}^n d\widetilde{\xi}_i \wedge d\widetilde{\eta}_i$ and the expression:

$$H(\xi,\eta) = \sum_{j=1}^n \lambda_j\widetilde{\xi}_j\widetilde{\eta}_j + H'(\widetilde{\xi},\widetilde{\eta}). \tag{4.243}$$

Here, $H'(\widetilde{\xi},\widetilde{\eta})$ is the sum of higher-degree terms and satisfies $H'(\widetilde{\xi},0) = 0$. Hence, for any positive ε, when $(\widetilde{\xi},\widetilde{\eta}) \in \Delta^n(0;\delta)$ ($\delta > 0$ sufficiently small), we have

$$\left|\sum_{j=1}^n \widetilde{\eta}_j\frac{\partial H'(\widetilde{\xi},\widetilde{\eta})}{\partial\widetilde{\xi}_j}\right| < \varepsilon\sum_{j=1}^n \widetilde{\eta}_j^2.$$

In particular, if we take as $2\varepsilon < \min_{1\leq j\leq n}\lambda_j$, along a trajectory $(\widetilde{\xi}(t),\widetilde{\eta}(t))$ ($t \in \mathbf{R}$) of $\mathbf{V}$, we have

$$\frac{d}{dt}\left(\sum_{j=1}^{n} \widetilde{\eta}_j{}^2\right) = 2\sum_{j=1}^{n} \widetilde{\eta}_j \frac{d\widetilde{\eta}_j}{dt} \tag{4.244}$$

$$= -2\left\{\sum_{j=1}^{n} \lambda_j \widetilde{\eta}_j{}^2 + \sum_{j=1}^{n} \widetilde{\eta}_j \frac{\partial H'}{\partial \widetilde{\xi}_j}\right\} < -2\varepsilon \sum_{j=1}^{n} \widetilde{\eta}_j{}^2,$$

which implies

$$\sum_{j=1}^{n} \widetilde{\eta}_j{}^2 \leq \operatorname{Const} e^{-2\varepsilon t}.$$

Now, suppose that one of the trajectories $\Gamma : (\widetilde{\xi}(t), \widetilde{\eta}(t))$ of $\mathbf{V}$ has the origin as the limit point at $t = -\infty$, and $\widetilde{\eta}(t_0) \neq 0$ at $t = t_0$. Then, there exists $t_1(< t_0)$ such that

$$\sum_{j=1}^{n} \widetilde{\eta}_j{}^2(t_1) < \sum_{j=1}^{n} \widetilde{\eta}_j{}^2(t_0).$$

Applying the mean value theorem to the interval $[t_1, t_0]$, there exists $t_2(t_1 < t_2 < t_0)$ such that $\frac{d}{dt}\sum_{j=1}^{n} \widetilde{\eta}_j{}^2(t_2) > 0$. But this contradicts (4.244). Hence, any trajectory of $\mathbf{V}$ which converges to the origin at $t = -\infty$ satisfies $\widetilde{\eta}_1(t) = \cdots = \widetilde{\eta}_n(t) = 0$, i.e.,

$$\mathcal{M}^- \subset \{\widetilde{\eta}_1 = \cdots = \widetilde{\eta}_n = 0\}.$$

Conversely, if $\widetilde{\eta} = 0$, we have

$$\frac{d\widetilde{\xi}_j}{dt} = \lambda_j \widetilde{\xi}_j, \quad 1 \leqslant j \leqslant n,$$

and any trajectory of $\mathbf{V}$ converges to the origin at $t = -\infty$. Namely, $\mathcal{M}^-$ coincides with the n-dimensional manifold $\{\widetilde{\eta}_1 = \cdots = \widetilde{\eta}_n = 0\}$. It can be similarly proved for $\mathcal{M}^+$. The tangent spaces $\mathbb{T}(\mathcal{M}^\pm)$ of $\mathcal{M}^\pm$ are defined by

$$\mathbb{T}\mathcal{M}^+ : \ d\xi_1 = \cdots = d\xi_n = 0,$$

$$\mathbb{T}\mathcal{M}^- : \ d\eta_1 = \cdots = d\eta_n = 0,$$

hence, $\mathcal{M}^+$ and $\mathcal{M}^-$ intersect at the origin mutually transversally. Thus, Theorem 4.6 has been completely proved.

Remark 4.4. In [Bir1], it was shown that $H(\xi, \eta)$ can be reduced to a function depending only on $\xi_1\eta_1, \ldots, \xi_n\eta_n$ as a formal series, but it is not clear whether it converges or not. If the Hamiltonian function $H(\xi, \eta)$ defines a completely integrable Hamiltonian flow in a neighborhood of the origin, this fact is valid (see [It]). In such a case, Lemma 4.10 follows from this fact. In [Mo], a similar theorem was proved by using Newton's method.

4.3.9 n-Dimensional Twisted Cycles

For simplicity, we further impose the following assumption.

Assumption 8. $\Im F(c^{(1)}), \ldots, \Im F(c^{(\kappa)})$ *are all mutually different.*

Under this assumption, by (4.210), the trajectory of $\mathbf{V}$ which has $c^{(j)}$ as a limit point never passes through other critical points $c^{(k)}(k \neq j)$. In a neighborhood of $c^{(j)}$, for the intersection $\mathcal{Z}_j$ of the Lagrangian sub-manifold of all trajectories of $-\mathbf{V}$ having $c^{(j)}$ as the limit point at $t = -\infty$ with $M \setminus T(D^{\cup} H_{\infty})$, the support of $\partial_\omega \mathcal{Z}_j$ is contained in $\partial T(D^{\cup} H_{\infty})$, that is, $\partial_\omega \mathcal{Z}_j = \mathcal{Z}_j \cap \partial T(D^{\cup} H_{\infty})$. By Assumption 5, there exists a twisted chain $\tau_j \in T(D^{\cup} H_{\infty}) \setminus D^{\cup} H_{\infty}$ such that $\partial_\omega \tau_j = \partial_\omega \mathcal{Z}_j$. Hence, if we set $\sigma_j = \mathcal{Z}_j - \tau_j$, we have $\partial_\omega \sigma_j = 0$, and the intersection of the support of σ_j and $M \setminus T(D^{\cup} H_{\infty})$ is a Lagrangian sub-manifold contained in the real hypersurface: $\Im F(u) = \Im F(c^{(j)})$. These κ twisted cycles $\sigma_1, \sigma_2, \ldots, \sigma_\kappa$ belongs to $H_n(M, \mathcal{L}_\omega^\vee)$. According to the Morse theory ([Mi]), we have

$$(-1)^n \kappa = \chi(M). \tag{4.245}$$

Moreover, since the homotopy type of $M^{\beta_i + \varepsilon}$ coincides with $M^{\beta_i - \varepsilon}$ attached by the n-dimensional cycle σ_i, any n-dimensional twisted cycle of M is a linear combination of these κ twisted cycles. Combining this fact with Theorem 4.5, by Lemma 2.6, passing to the cohomology, we obtain the equality

$$\kappa = \dim H^n(M, \mathcal{L}_\omega). \tag{4.246}$$

Thus, we have the following theorem.

Theorem 4.7. *Under Assumptions 3–5 and 8, (4.245) and (4.246) hold.*

As was discussed in Chapter 2, combining

$$H^\bullet(\Omega^\bullet(*D)), \nabla_\omega) \cong H^\bullet(M, \mathcal{L}_\omega) \tag{4.247}$$

with Theorem 4.7, we see that

$$r = \dim H^n(\Omega^\bullet(*D), \nabla_\omega) = \kappa. \tag{4.248}$$

Although Theorem 2.8 and Theorem 4.5 are compatible, as the conditions of these theorems are different, they are not completely equivalent.

Remark 4.5. As in Theorem 4.5, the vanishing theorem of cohomology with coefficients in a local system was first established by Y. Matsushima and

S. Murakami ([M-M] I, II). Nowadays, it is called a theorem of Matsushima–Murakami type and there are several trials to generalize it. For example, see [Ao1] and [Koh1]. For recent developments in the case of arrangements of hyperplanes, see [Or-Te3].

4.3.10 Geometric Meaning of Asymptotic Expansion

Next, we explain a relation between n-dimensional contracting twisted cycle and an asymptotic expansion of the integral (4.180) (for detail, see [Ar-GZ-V]). We assume that the function $F(u)$ defined in (4.197) satisfies Assumptions 3 and 4. Concerning the vector field $\mathbf{V}$, the integral over n-dimensional contracting twisted cycles $\sigma_1, \ldots, \sigma_r$ that pass through the critical points $c^{(1)}, \ldots, c^{(r)}$, respectively, decomposes as follows:

$$\int_{\sigma_j} U(u; \alpha + N\eta) du_1 \wedge \cdots \wedge du_n \tag{4.249}$$

$$= \int_{\sigma_j \cap (M \setminus T(D^{\cup} H_\infty))} U(u; \alpha + N\eta) du_1 \wedge \cdots \wedge du_n$$

$$+ \int_{\sigma_j \cap T(D^{\cup} H_\infty)} U(u; \alpha + N\eta) du_1 \wedge \cdots \wedge du_n.$$

As we may regard $\mathrm{Sup}_{u \in \sigma_j \cap T(D^{\cup} H_\infty)} \Re F(u) \ll \Re F(c^{(j)})$, at $N \mapsto +\infty$, the increasing order of the second term in the right-hand side of (4.249) is negligible in comparison with that of the first term. On the other hand, we have

$$\int_{\sigma_j \cap (M \setminus T(D^{\cup} H_\infty))} U(u; \alpha + N\eta) du_1 \wedge \cdots \wedge du_n \tag{4.250}$$

$$= \exp(N \sqrt{-1} \Im F(c^{(j)}))$$

$$\cdot \int_{\sigma_j \cap (M \setminus T(D^{\cup} H_\infty))} U(u; \alpha) \exp(N \Re F(u)) du_1 \wedge \cdots \wedge du_n,$$

and on $\sigma_j \cap (M \setminus T(D^{\cup} H_\infty))$, $\Re F(u)$ takes its maximal value $\Re F(c^{(j)})$ at $u = c^{(j)}$. In addition, we have $\Re F(c^{(j)}) > \Re F(u)$ $(u \in \sigma_j \cap (M \setminus T(D^{\cup} H_\infty)))$. We can choose local coordinates $(\zeta^1, \ldots, \zeta^n)$ in such a way that $c^{(j)}$ is the origin and, on σ_j, all of $\zeta^1, \ldots, \zeta^n$ become real. Then, in a neighborhood of $\zeta = 0$, $F(u)$ has a Taylor expansion of the form

$$F(u(\zeta)) = F(c^{(j)}) - \sum_{k,l=1}^{n} v_{kl}\zeta^k\zeta^l + \cdots, \tag{4.251}$$

$$-v_{kl} = \frac{1}{2}\frac{\partial^2 F(c^{(j)})}{\partial\zeta^k\partial\zeta^l},$$

the matrix $V_j = ((v_{kl}))_{k,l=1}^{n}$ is real symmetric and positive definite. By the saddle point method ([Bru], [Ar-GZ-V]), we have:

Lemma 4.11. *At $N \mapsto +\infty$, we have an asymptotic formula*

$$\widehat{du_1 \wedge \cdots \wedge du_n} \tag{4.252}$$

$$= U(c^{(j)}; \alpha + N\eta)\frac{N^{-\frac{n}{2}}\pi^{\frac{n}{2}}}{(\det V_j)^{\frac{1}{2}}}\left(1 + O\left(\frac{1}{N}\right)\right).$$

Proof. As can be seen from (4.249), (4.250), to obtain an estimate (4.252), it is sufficient to provide an estimate of the integral (4.250) on σ_j in a neighborhood of $\zeta = 0$. Now, let $\psi(\zeta)$ be any smooth function defined on $|\zeta^1| \leq 1$, $\ldots, |\zeta^n| \leq 1$, $(\zeta^1, \ldots, \zeta^n) \in \mathbb{R}^n$. Since we have an expression

$$\psi(\zeta) = \psi(0) + \sum_{k=1}^{n}\zeta^k\frac{\partial\psi(0)}{\partial\zeta^k} + \frac{1}{2}\sum_{k,l=1}^{n}\zeta^k\zeta^l\psi_{kl}(\zeta),$$

($\psi_{kl}(\zeta)$ is a smooth function on $|\zeta^1| \leq 1, \ldots, |\zeta^n| \leq 1$) by Taylor expansion, we obtain an estimate of the integral

$$\int_{|\zeta^1|\leq 1,\ldots,|\zeta^n|\leq 1} \exp\left(-N\sum_{k,l=1}^{n}v_{kl}\zeta^k\zeta^l\right)\psi(\zeta)d\zeta^1 \wedge \cdots \wedge d\zeta^n \tag{4.253}$$

$$= N^{-\frac{n}{2}}\int_{|\zeta^1|\leq\sqrt{N},\ldots,|\zeta^n|\leq\sqrt{N}} \exp\left(-\sum_{k,l=1}^{n}v_{kl}\zeta^k\zeta^l\right)\psi\left(\frac{\zeta}{\sqrt{N}}\right)d\zeta^1 \wedge \cdots \wedge d\zeta^n$$

$$= N^{-\frac{n}{2}}\int_{|\zeta^1|\leq\sqrt{N},\ldots,|\zeta^n|\leq\sqrt{N}} \exp\left(-\sum_{k,l=1}^{n}v_{kl}\zeta^k\zeta^l\right)$$

$$\left(\psi(0) + \frac{1}{\sqrt{N}}\sum_{k=1}^{n}\zeta^k\frac{\partial\psi(0)}{\partial\zeta^k} + \frac{1}{2N}\sum\zeta^k\zeta^l\psi_{kl}\left(\frac{\zeta}{\sqrt{N}}\right)\right)d\zeta^1 \wedge \cdots \wedge d\zeta^n$$

(here, we use the Gaussian integral (see, e.g., [AAR]))

$$= \pi^{\frac{n}{2}} \cdot N^{-\frac{n}{2}}\left(\frac{\psi(0)}{\sqrt{\det V_j}} + O\left(\frac{1}{N}\right)\right).$$

From this, (4.252) immediately follows.

To compute an asymptotic expansion of

$$\int_\sigma U(u; \alpha + N\eta) du_1 \wedge \cdots \wedge du_n \tag{4.254}$$

for any twisted cycle σ, we may proceed as follows. In $H_n(M, \mathcal{L}_\omega^\vee)$, σ is expressed as

$$\sigma = \sum_{j=1}^r g_j \sigma_j. \tag{4.255}$$

Here, since $g_j = g_j(\alpha)$ is a rational function of $e^{2\pi\sqrt{-1}\alpha_1}, \ldots, e^{2\pi\sqrt{-1}\alpha_m}$, $g_j(\alpha + N\eta)$ does not depend on N. Hence, we obtain

$$\int_\sigma U(u; \alpha + N\eta) du_1 \wedge \cdots \wedge du_n \tag{4.256}$$

$$= \sum_{j=1}^r g_j(\alpha) \int_{\sigma_j} U(u; \alpha + N\eta) du_1 \wedge \cdots \wedge du_n$$

$$= \sum_{j=1}^r g_j(\alpha) \exp(NF(c^{(j)})) \cdot U(c^{(j)}; \alpha) \frac{\pi^{\frac{n}{2}} N^{-\frac{n}{2}}}{[\det V_j]^{\frac{1}{2}}} \left(1 + O\left(\frac{1}{N}\right)\right).$$

When the basis $\varphi_1, \ldots, \varphi_r$ discussed in § 4.3.1 satisfies the inequality of the determinant

$$\det\left(\left(\frac{\varphi_\nu}{du_1 \wedge \cdots \wedge du_r}(c^{(j)})\right)\right)_{j,\nu=1}^n \neq 0, \tag{4.257}$$

(where $\frac{\varphi}{du_1 \wedge \cdots \wedge du_n}$ ($\varphi \in \Omega^n(*D)$) signifies the coefficient of φ in $du_1 \wedge \cdots \wedge du_n$) with respect to the critical points $c^{(1)}, \ldots, c^{(r)}$, we say that $\{\varphi_1, \ldots, \varphi_r\}$ is "non-degerenate". Then, similarly to Lemma 4.11, we obtain:

Proposition 4.4. *A square matrix* $((\int_{\sigma_j} U(u; \alpha)\varphi_\nu))_{j,\nu=1}^r$ *of order r satisfies the holonomic systems of difference equations (4.188)–(4.189) with respect to the variables* $\alpha = (\alpha_1, \ldots, \alpha_n)$. *When the direction* $\eta \in \mathbb{Z}^m \setminus \{0\}$ *satisfies Assumptions 3–4 and 8, η is a regular direction, and an asymptotic expansion in the direction of η is given by*

$$\int_{\sigma_j} U(u; \alpha + N\eta)\varphi_\nu(u) \sim U(c^{(j)}; \alpha + N\eta) \left[\frac{\varphi_\nu}{du_1 \wedge \cdots \wedge du_n}(c^{(j)})\right]$$

$$\frac{\pi^{\frac{n}{2}} N^{-\frac{n}{2}}}{[\det(V_j)]^{\frac{1}{2}}} \left(1 + O\left(\frac{1}{N}\right)\right).$$

In this case, the first characteristic index is equal to $F(c^{(1)}), F(c^{(2)}), \ldots, F(c^{(r)})$ and the second characteristic index is equal to $(-\frac{n}{2}, -\frac{n}{2}, \ldots, -\frac{n}{2})$ (see § 4.2.5, for definition).

The second characteristic index has the property that all of them coincide. With the notation (4.133), (4.134), we have the equality

$$(A_1^{*(0)})^{-1} A_1^{*(1)} = -\frac{n}{2} \cdot 1_r. \tag{4.258}$$

This formula really controls the difference equations (4.96) treated in § 4.2. See also [Or-Te2], [V4] for related topics.

4.4 Difference Equations Satisfied by the Hypergeometric Functions of Type $(n+1, m+1; \alpha)$

We consider the case when all of the $P_1, P_2, \cdots, P_m$ are inhomogeneous linear and each of their coefficients is real. We have already seen in § 3.2 (Theorem 3.1) that if the corresponding arrangement of hyperplanes is in general position, the dimension of $H^n(M, \mathcal{L}_\omega)$ is equal to $\binom{m-1}{n}$ and we naturally obtain a basis of $H_n(M, \mathcal{L}_\omega^\vee)$ from bounded chambers. In this section, by taking results obtained in the previous subsection into account, we again study this from a viewpoint of difference equations.

4.4.1 Bounded Chambers

First, we do not necessarily assume that an arrangement of hyperplanes is in general position.

Theorem 4.8 ([Ao4]). *For $\eta = (\eta_1, \cdots, \eta_m) \in \mathbb{Z}^m \setminus \{0\}$, if $\eta_j > 0$ $(1 \leq j \leq m)$, then we have:*

(1) Any zero c of $\omega_\eta = \sum_{j=1}^m \eta_j d \log P_j$ in M is real, and for each bounded connected component of $\mathbb{R}^n \setminus D \cap \mathbb{R}^n$, it uniquely exists. The level function $\Re F$ is non-degenerate at c, i.e., Assumption 3 is automatically satisfied. Moreover, a bounded chamber containing c is a contracting Lagrangian sub-manifold.

(2) A bounded chamber defines an n-dimensional locally finite twisted cycle, and the set of all such elements forms a basis of $H_n^{lf}(M, \mathcal{L}_\omega^\vee)$. In particular, we have

$$\dim H^n(M, \mathcal{L}_\omega) = \dim H_n(M, \mathcal{L}_\omega^\vee) \qquad (4.259)$$

$$= \; the \; number \; of \; the \; bounded \; chambers$$

Proof. The proof of (1).   Since P_j's are real, we can choose a real complete Kähler metric ds^2 on M. We consider the hight function $\Re F$ on a bounded chamber Δ. Since we have $\Re F = -\infty$ on the boundary of Δ by assumption, $\Re F$ is bounded above. Hence, there exists a point c where $\Re F$ attains a maximal. As $\mathrm{grad}\Re F = 0$ holds at c, c is a zero of ω_η. Since the symplectic form defined by ds^2 (cf. (4.205), (4.228))

$$\Omega = \sqrt{-1} \sum_{i=1}^{n} du_i \wedge d\overline{u}_i + \frac{\sqrt{-1}}{2} \sum_{j=1}^{m} d \log P_j \wedge d \log \overline{P}_j$$

vanishes identically on $M \cap \mathbb{R}^n$, it vanishes also on Δ, hence, Δ is a Lagrangian sub-manifold. We consider any line over $\mathbb{R}$ passing through c

$$\gamma : u_j = \lambda_j t + \mu_j \quad (\lambda_j, \mu_j \in \mathbb{R}) \qquad (4.260)$$

$(t \in \mathbb{C})$. Restricting ω_η to γ, we have

$$[\omega_\eta]_\gamma = \sum_{j=1}^{m'} \frac{\eta'_j}{t - a_j} dt \qquad (4.261)$$

$(a_1, \cdots, a_{m'} \in \mathbb{R}, \eta'_j > 0, m' \leqslant m)$, and we may assume that $a_1 < a_2 < \cdots < a_{m'}$. If the point on γ at $t = t_0$ coincides with c, we must have

$$\sum_{j=1}^{m'} \frac{\eta'_j}{t_0 - a_j} = 0. \qquad (4.262)$$

If c is a degenerated critical point, for an appropriate γ passing through c, its derivative further becomes 0, i.e., we must have

$$-\sum_{j=1}^{m'} \frac{\eta'_j}{(t_0 - a_j)^2} = 0. \qquad (4.263)$$

But, this is clearly a contradiction. Hence, c is a non-degenerate critical point. Next, we assume that Δ is an unbounded chamber and ω_η admits a zero c inside Δ. Then, there exists a real curve γ (4.260) passing through c satisfying {the points of γ corresponding to $t \geq t_0$} $\subset \Delta$. Restricting ω_η to γ, it can be expressed in the form (4.261) with $t_0 > a_1, \cdots, a_{m'}$. We should have $\omega_\eta = 0$ at $t = t_0$, but since

$$\sum_{j=1}^{m'} \frac{\eta_j'}{t_0 - a_j} > 0, \tag{4.264}$$

this is a contradiction. Hence, there is no zero of ω_η in Δ.

Next, we assume that there exists a non-real point c satisfying $\omega_\eta = 0$. Then, its complex conjugate $\bar{c}$ is also a point of M and $c \neq \bar{c}$. The line γ connecting c and $\bar{c}$ becomes a line over $\mathbb{R}$, hence it can be written in the form (4.260). And c corresponds to a non-real point t_0 ($t_0 \neq \overline{t_0}$) of t. But since all t's satisfying the equation

$$\sum_{j=1}^{m'} \frac{\eta_j'}{t - a_j} = 0 \tag{4.265}$$

are real, this is a contradiction. Namely, c has to be real.

(2) is a consequence of (1), (4.248) and Theorem 4.5.

Let us recall that, in § 3.2, we denote the twisted cycles corresponding to Theorem 4.7 by $\Delta_1(\omega)$, $\Delta_2(\omega)$, ..., $\Delta_r(\omega)$.

4.4.2 Derivation of Difference Equations

Regarding the hypergeometric functions of type $(n+1, m+1; \alpha)$ defined in § 3.4 as function of α, they satisfy the holonomic difference equations (4.188), (4.189). In this subsection, by applying some results from Chapter 3, we derive them concretely. We use the notation defined in § 3.4, 3.8. By the formula

$$dP_{j_1} \wedge \cdots \wedge dP_{j_n} = x\binom{N \setminus \{0\}}{J} du_1 \wedge \cdots \wedge du_n$$

for $J = \{j_1, \cdots, j_n\}$, setting $K = \{j_0\} \cup J$ for a $j_0 \notin J$, we have

$$T_{j_0}^{-1}\widehat{\varphi}\langle J;\alpha\rangle = \int_\sigma \frac{U}{P_{j_0}}\varphi\langle J\rangle \tag{4.266}$$

$$= x\binom{N\setminus\{0\}}{J}\int_\sigma \frac{U}{P_K}du_1 \wedge\cdots\wedge du_n$$

(by (3.100))

$$= \frac{x\binom{N\setminus\{0\}}{J}}{x\binom{N}{K}}\sum_{l=0}^n (-1)^l x\binom{N\setminus\{0\}}{K\setminus\{j_l\}}\int_\sigma U\frac{du_1\wedge\cdots\wedge du_n}{P_{K\setminus\{j_l\}}}$$

$$= \frac{x\binom{N\setminus\{0\}}{J}}{x\binom{N}{K}}\sum_{l=0}^n (-1)^l \widehat{\varphi}\langle K\setminus\{j_l\};\alpha\rangle.$$

On the other hand, since we have, for $1\le\mu\le n$,

$$T_{j_\mu}^{-1}\widehat{\varphi}\langle J;\alpha\rangle = x\binom{N\setminus\{0\}}{J}\int_\sigma U\frac{du_1\wedge\cdots\wedge du_n}{P_{j_\mu}P_J}, \tag{4.267}$$

by Stokes formula, we obtain

$$0 = \int_\sigma U\nabla_\omega\left((-1)^{\mu-1}\frac{\varphi\langle J\setminus\{j_\mu\}\rangle}{P_{j_\mu}}\right) \tag{4.268}$$

$$= (\alpha_{j_\mu}-1)\int_\sigma U\frac{\varphi\langle J\rangle}{P_{j_\mu}}$$

$$+ \sum_{\substack{j_0\notin J\\ K:=\{j_0\}^{\cup}J}}\alpha_{j_0}\int_\sigma U(-1)^{\mu-1}\frac{\varphi\langle K\setminus\{j_\mu\}\rangle}{P_{j_\mu}}$$

$$= (\alpha_{j_\mu}-1)T_{j_\mu}^{-1}\widehat{\varphi}\langle J;\alpha\rangle + \sum_{\substack{j_0\notin J\\ K:=\{j_0\}^{\cup}J}}\alpha_{j_0}(-1)^{\mu-1}\frac{x\binom{N\setminus\{0\}}{K\setminus\{j_\mu\}}}{x\binom{N}{K}}$$

$$\cdot\sum_{l=0}^n (-1)^l\widehat{\varphi}\langle K\setminus\{j_l\};\alpha\rangle$$

in a similar way to (4.266), which implies

$$(\alpha_{j_\mu}-1)T_{j_\mu}^{-1}\widehat{\varphi}\langle J;\alpha\rangle \tag{4.269}$$

$$= \sum_{\substack{j_0\notin J\\ K:=\{j_0\}^{\cup}J}}\alpha_{j_0}(-1)^\mu\frac{x\binom{N\setminus\{0\}}{K\setminus\{j_\mu\}}}{x\binom{N}{K}}\sum_{l=0}^n(-1)^l\widehat{\varphi}\langle K\setminus\{j_l\};\alpha\rangle.$$

We also have

$$T_{j_\mu} \widehat{\varphi} \langle J; \alpha \rangle = \frac{-1}{\sum_{k=1}^m \alpha_k + 1} \sum_{\substack{j_0 \notin J \\ K:=\{j_0\}^{\cup} J}} \alpha_{j_0} \frac{x\binom{N}{K}}{x\binom{N\setminus\{0\}}{K\setminus\{j_\mu\}}} \widehat{\varphi}\langle K \setminus \{j_\mu\}\rangle. \quad (4.270)$$

Below, we prove this, for simplicity, in the case when $\mu = 1$. By Stokes formula, we have

$$0 \sim \nabla_\omega \{P_{j_1} \varphi \langle J \setminus \{j_1\}\rangle\} \qquad\qquad (4.271)$$

$$= (\alpha_{j_1} + 1) P_{j_1} \varphi \langle J \rangle + \sum_{\substack{j_0 \notin J \\ K:=\{j_0\}^{\cup} J}} \alpha_{j_0} P_{j_1} \varphi \langle K \setminus \{j_1\}\rangle.$$

By (3.99) and $u_0 = 1$, we have the equality

$$x\binom{N\setminus\{0\}}{K\setminus\{j_1\}} P_{j_1} = \sum_{\substack{l=0 \\ l\neq 1}}^{n} (-1)^l x\binom{N\setminus\{0\}}{K\setminus\{j_l\}} P_{j_l} - x\binom{N}{K} \quad (4.272)$$

from which we obtain

$$0 = (\alpha_{j_1} + 1) P_{j_1} \varphi \langle J \rangle \qquad\qquad (4.273)$$

$$+ \sum_{j_0 \notin J} \alpha_{j_0} \left\{ \sum_{\substack{l=0 \\ l\neq 1}}^{n} (-1)^l P_{j_1} \varphi \langle K \setminus \{j_l\}\rangle - \frac{x\binom{N}{K}}{x\binom{N\setminus\{0\}}{K\setminus\{j_1\}}} \varphi \langle K \setminus \{j_1\}\rangle \right\}.$$

Now, we use the following lemma.

Lemma 4.12. *Suppose that $a_{i_1,\cdots,i_{n-1}}$ and $b_{i_1,\cdots,i_{n-1}}$ $(1 \leqslant i_1, \cdots, i_{n-1} \leqslant m)$ are skew-symmetric tensors with respect to $i_1, \cdots, i_{n-1}$. Given $b_{i_1,\cdots,i_{n-1}}$, the equations on $a_{i_1,\cdots,i_{n-1}}$*

$$\left(\rho + \sum_{j\notin I} \lambda_j\right) a_{i_1,\cdots,i_{n-1}} + \sum_{j\notin I} \sum_{l=1}^{n-1} (-1)^l \lambda_j a_{j i_1 \cdots i_{l-1} i_{l+1} \cdots i_{n-1}} \quad (4.274)$$

$$= b_{i_1,\cdots,i_{n-1}}, \qquad I = \{i_1, \cdots, i_{n-1}\}$$

have the unique solution given by

$$\rho(\rho + \lambda_1 + \cdots + \lambda_m) a_{i_1,\cdots,i_{n-1}} \qquad\qquad (4.275)$$

$$= (\rho + \lambda_{i_1} \cdots + \lambda_{i_{n-1}}) b_{i_1,\cdots,i_{n-1}} + \sum_{k\notin I} \sum_{l=1}^{n-1} \lambda_k (-1)^{l-1} b_{k i_1 \cdots i_{l-1} i_{l+1} \cdots i_{n-1}}.$$

Here, we assume that $\rho(\rho + \lambda_1 + \cdots + \lambda_m) \neq 0$.

Proof. Rewriting $b_{i_1,\cdots,i_{n-1}}$ in the right-hand side of (4.275) with the left-hand side of (4.274), we immediately obtain the left-hand side of (4.274). By the way, for $n = 2$, (4.274) gives us

$$\left(\rho + \sum_{j \neq i_1} \lambda_j\right) a_{i_1} + \sum_{j \neq i_1} (-\lambda_j) a_j = b_{i_1}.$$

Hence, rewriting the right-hand side of (4.275) with this formula, we obtain

$$(\rho + \lambda_{i_1}) b_{i_1} + \sum_{k \neq i_1} \lambda_k b_k = (\rho + \lambda_{i_1}) \left\{ \left(\rho + \sum_{j \neq i_1} \lambda_j\right) a_{i_1} - \sum_{j \neq i_1} \lambda_j a_j \right\}$$

$$+ \sum_{k \neq i_1} \lambda_k \left\{ \left(\rho + \sum_{j \neq k} \lambda_j\right) a_k - \sum_{j \neq k} \lambda_j a_j \right\} = \rho \left(\rho + \sum_{j=1}^{m} \lambda_j\right) a_{i_1}$$

and it coincides with the left-hand side of (4.275). For $n \geq 3$, a similar proof works.

In (4.273), fixing j_1 and setting

$$a_{j_2 \cdots j_n} = P_{j_1} \varphi \langle J \rangle,$$

$$b_{j_2 \cdots j_n} = - \sum_{\substack{j_0 \notin J \\ K := \{j_0\} \cup J}} \alpha_{j_0} \frac{x\binom{N}{K}}{x\binom{N \setminus \{0\}}{K \setminus \{j_1\}}} \varphi \langle K \setminus \{j_1\} \rangle$$

$\rho = 1 + \alpha_{j_1}$, $\lambda_j = \alpha_j$ $(j \neq j_1)$, $I = \{j_2, \cdots, j_n\}$, applying Lemma 4.12, we obtain

$$(1 + \alpha_{j_1})\left(1 + \sum_{j=1}^{m} \alpha_j\right) P_{j_1} \varphi \langle J \rangle \tag{4.276}$$

$$\sim - (1 + \sum_{j \in J} \alpha_j) \sum_{\substack{j_0 \notin J \\ K := \{j_0\} \cup J}} \alpha_{j_0} \frac{x\binom{N}{K}}{x\binom{N \setminus \{0\}}{K \setminus \{j_1\}}} \varphi \langle K \setminus \{j_1\} \rangle$$

$$+ \sum_{k \notin J} \alpha_k \sum_{\substack{l=2 \\ L := \{k\} \cup J}}^{n} \alpha_{j_l} \frac{x\binom{N}{L}}{x\binom{N \setminus \{0\}}{L \setminus \{j_1\}}} \varphi \langle L \setminus \{j_1\} \rangle$$

$$+ \sum_{k \notin K} \alpha_k \sum_{l=2}^{n} (-1)^l \left\{ \sum_{\substack{j_0 \notin J \\ L := \{k\} \cup K}} (-\alpha_{j_0}) \frac{x\binom{N}{L \setminus \{j_l\}}}{x\binom{N \setminus \{0\}}{L \setminus \{j_1, j_l\}}} \varphi \langle L \setminus \{j_1, j_l\} \rangle \right\}$$

$$= - (1 + \alpha_{j_1}) \sum_{j_0 \notin J} \alpha_{j_0} \frac{x\binom{N}{K}}{x\binom{N \setminus \{0\}}{K \setminus \{j_1\}}} \varphi \langle K \setminus \{j_1\} \rangle$$

$$+ \sum_{j_0, k \notin J} \alpha_k \alpha_{j_0} \sum_{l=2}^{n} (-1)^{l-1} \frac{x\binom{N}{L \setminus \{j_l\}}}{x\binom{N \setminus \{0\}}{L \setminus \{j_1, j_l\}}} \varphi \langle L \setminus \{j_1, j_l\} \rangle.$$

In the second term of the right-hand side, when $l, j_0 \notin J$, as the coefficient of $\alpha_l \alpha_{j_0}$ is skew-symmetric, its sum is 0. Hence, we obtain (4.270). One can prove similarly for $\mu \geq 2$. Moreover, since a shift operator $T_k (k \notin J)$ can be written as

$$T_k = \frac{\sum_{K := \{k\} \cup J, \nu=1}^{n} (-1)^{\nu-1} x\binom{N \setminus \{0\}}{K \setminus \{j_\nu\}} T_{j_\nu} + x\binom{N}{K}}{x\binom{N \setminus \{0\}}{J}}, \tag{4.277}$$

we obtain the identity ([Ao3])

$$T_k \widehat{\varphi} \langle J; \alpha \rangle = \frac{x\binom{N}{K}}{x\binom{N \setminus \{0\}}{J}} \widehat{\varphi} \langle J; \alpha \rangle \tag{4.278}$$

$$+ \sum_{\substack{j_0 \notin J \\ L : \{j_0\} \cup J}} \sum_{\nu=1}^{n} (-1)^\nu \alpha_{j_0} \frac{x\binom{N \setminus \{0\}}{K \setminus \{j_\nu\}} x\binom{N}{L}}{x\binom{N \setminus \{0\}}{J} x\binom{N \setminus \{0\}}{L \setminus \{j_\nu\}}} \cdot \frac{\widehat{\varphi} \langle L \setminus \{j_\nu\}; \alpha \rangle}{1 + \sum_{j=1}^{m} \alpha_j}, \quad 1 \leqslant k \leqslant m.$$

4.4.3 Asymptotic Expansion with a Fixed Direction

As was explained in § 2.9, under the assumption $\alpha_m \neq 0$, we can take $\varphi\langle J\rangle$, $1 \leqslant j_1 < \cdots < j_n \leqslant m-1$ as a basis of $H^n(\Omega^\bullet(*D), \nabla_\omega)$. The number of such elements is exactly equal to $r = \binom{m-1}{n}$. We denote them by $\varphi\langle J_1\rangle$, $\varphi\langle J_2\rangle, \cdots, \varphi\langle J_r\rangle$. We denote the period matrix obtained by integrating the basis $\{\varphi\langle J_\nu\rangle\}_{\nu=1}^r$ over the twisted cycles $\Delta_1(\omega), \cdots, \Delta_r(\omega)$ defined in § 3.5.1

$$\left(\left(\int_{\Delta_j(\omega)} U\varphi\langle J_\nu\rangle\right)\right)_{j,\nu=1}^r \tag{4.279}$$

by $\Phi(\alpha)$. $\Phi(\alpha)$ satisfies the holonomic system of difference equations (4.266), (4.269) or (4.270), (4.278) with respect to the variable α. Namely, it satisfies (4.96). The corresponding matrices $A_{\pm j}(\alpha)$ $(1 \leqslant j \leqslant m)$ are given by (4.266), (4.269) or (4.270), (4.278). Here, in these equations, we should replace $\widehat{\varphi}\langle i_1 \cdots i_{n-1}m; \alpha\rangle$ by

$$\widehat{\varphi}\langle i_1 \cdots i_{n-1}m; \alpha\rangle = -\sum_{j=1}^{m-1} \frac{\alpha_j}{\alpha_m}\widehat{\varphi}\langle i_1 \cdots i_{n-1}j; \alpha\rangle. \tag{4.280}$$

Now, suppose that $\eta \in \mathbb{Z}^m \setminus \{0\}$ satisfies the conditions

$$\eta_j \neq 0, \quad \eta_1 + \eta_2 + \cdots + \eta_m \neq 0, \quad 1 \leqslant j \leqslant m. \tag{4.281}$$

Each $A_{-j,\eta}^{(0)}, A_{j,\eta}^{(0)}$, as a linear operator acting on given skew-symmetric tensors $\{\widetilde{\varphi}_J\}$, is derived from (4.266), (4.269), (4.278) as a limit.

$$A_{-j_0,\eta}^{(0)} : \widetilde{\varphi}_J \mapsto \frac{x\binom{N\setminus\{0\}}{J}}{x\binom{N}{K}}\sum_{\nu=0}^n (-1)^\nu \widetilde{\varphi}_{K\setminus\{j_\nu\}}, \tag{4.282}$$

$$(K := \{j_0\}^\cup J, \ j_0 \notin J),$$

$$A_{-j_\mu,\eta}^{(0)} : \widetilde{\varphi}_J \mapsto \sum_{j_0 \notin J} \frac{\eta_{j_0}}{\eta_{j_\mu}}(-1)^\mu \frac{x\binom{N\setminus\{0\}}{K\setminus\{j_\mu\}}}{x\binom{N}{K}}\sum_{\nu=0}^n (-1)^\nu \widetilde{\varphi}_{K\setminus\{j_\nu\}}, \tag{4.283}$$

$$1 \leqslant \mu \leqslant m,$$

$$A_{k,\eta}^{(0)} : \quad \widetilde{\varphi}_J \mapsto \frac{x\binom{N}{K}}{x\binom{N\setminus\{0\}}{J}}\widetilde{\varphi}_J \qquad (4.284)$$

$$+ \sum_{j_0 \notin J}\sum_{\nu=1}^{n}(-1)^{\nu}\frac{x\binom{N\setminus\{0\}}{L\setminus\{j_\nu\}}x\binom{N}{K}}{x\binom{N\setminus\{0\}}{J}x\binom{N\setminus\{0\}}{K\setminus\{j_\nu\}}}\frac{\widetilde{\varphi}_{K\setminus\{j_\nu\}}}{\sum_{j=1}^{m}\eta_j}.$$

Here, we assume that the following equality is satisfied:

$$\sum_{j=1}^{m}\widetilde{\varphi}_{j_1\cdots j_{n-1}j}\eta_j = 0. \qquad (4.285)$$

4.4.4 Example

For the Appell–Lauricella type, i.e., hypergeometric functions of type $(2, m+1)$ (cf. § 3.6.3), setting

$$x = \begin{pmatrix} x_1 & x_2 & \cdots & x_m \\ 1 & 1 & \cdots & 1 \end{pmatrix},$$

(4.266), (4.269), (4.278) can be expressed as

$$T_j^{-1}\widehat{\varphi}\langle i; \alpha\rangle = \frac{1}{x_i - x_j}(\widehat{\varphi}\langle i; \alpha\rangle - \widehat{\varphi}\langle j; \alpha\rangle), \qquad (4.286)$$

$$(\alpha_i - 1)T_i^{-1}\widehat{\varphi}\langle i; \alpha\rangle = \sum_{j \neq i}\alpha_j\frac{\widehat{\varphi}\langle i; \alpha\rangle - \widehat{\varphi}\langle j; \alpha\rangle}{x_i - x_j}, \qquad (4.287)$$

$$T_j\widehat{\varphi}\langle i; \alpha\rangle = -\frac{\sum_{k=1}^{m}\alpha_k x_k\widehat{\varphi}\langle k; \alpha\rangle}{1 + \sum_{\nu=1}^{m}\alpha_\nu} + (x_i - x_j)\widehat{\varphi}\langle i; \alpha\rangle, \qquad (4.288)$$

respectively. Morover, if we set $m = 3$, then it is clear that $r = 2$ and the above formulas contain Gauss' contiguous relations explained in § 1.4 and § 4.2.8. The reader may verify it by himself (or herself) (cf. [Wa]).

4.4.5 Non-Degeneracy of Period Matrix

We have already seen in Chapter 3 that the period matrix (4.279) is non-degenerate, that is, the determinant of $\Phi(\alpha)$ is not zero. Here, we reconsider this important fact from the viewpoint of difference equations and an asymptotic expansion of their solutions.

Any polynomial $\psi \in \mathbb{C}[u_1, u_2, \cdots, u_n]$ is also a polynomial in $P_1, P_2, \cdots, P_n$. Hence, by

$$\psi(u)du_1 \wedge \cdots \wedge du_n \in \sum_{\nu_1, \cdots, \nu_n \geq 0} \mathbb{C}P_1^{\nu_1} \cdots P_n^{\nu_n} \varphi\langle 1, 2, \cdots, n\rangle, \quad (4.289)$$

using the recursion formula (4.278) repeatedly, we obtain, for an integral over the twisted cycle σ,

$$\psi \widehat{du_1 \wedge \cdots \wedge du_n} \quad (4.290)$$

$$= \sum_{1 \leq j_1 < \cdots < j_n \leq m-1} b_J \left(x; \frac{\alpha_1}{1 + \sum_{\nu=1}^m \alpha_\nu}, \cdots, \frac{\alpha_m}{1 + \sum_{\nu=1}^m \alpha_\nu} \right) \widehat{\varphi}\langle J\rangle.$$

Here, we have

$$b_J \left(x; \frac{\alpha_1}{1 + \sum_{\nu=1}^m \alpha_\nu}, \cdots, \frac{\alpha_m}{1 + \sum_{\nu=1}^m \alpha_\nu} \right)$$

$$\in \mathbb{C}(x) \otimes \mathbb{C} \left[\frac{\alpha_1}{1 + \sum_{\nu=1}^m \alpha_\nu}, \cdots, \frac{\alpha_m}{1 + \sum_{\nu=1}^m \alpha_\nu} \right].$$

For η satisfying the condition (4.281), taking an asymptotic expansion in the direction of η, by Proposition 4.4, for each integral over $\Delta_k(\omega)$, (4.290) becomes

$$\psi(c^{(k)}) = \sum_{l=1}^r b_{J_l} \left(x; \frac{\eta_1}{\sum_{\nu=1}^m \eta_\nu}, \cdots, \frac{\eta_m}{\sum_{\nu=1}^m \eta_\nu} \right) \quad (4.291)$$

$$\cdot \left[\frac{\varphi\langle J_l\rangle}{du_1 \wedge \cdots \wedge du_n} \right]_{u=c^{(k)}}.$$

In fact, we have

$$b_{J_l} \left(x; \frac{\eta_1}{\sum_{\nu=1}^m \eta_\nu}, \cdots, \frac{\eta_m}{\sum_{\nu=1}^m \eta_\nu} \right) \quad (4.292)$$

$$= \lim_{\substack{N \mapsto +\infty \\ \alpha=\alpha'+N\eta}} b_{J_l} \left(x; \frac{\alpha_1}{1 + \sum_{\nu=1}^m \alpha_\nu}, \cdots, \frac{\alpha_m}{1 + \sum_{\nu=1}^m \alpha_\nu} \right).$$

The principal term of an asymptotic expansion of $\det \Phi(\alpha)$ in the direction of η at $N \mapsto +\infty$ is written as

$$\det \Phi(\alpha) \qquad (4.293)$$

$$= \frac{\prod_{k=1}^{r} \left[U(c^{(k)}) \pi^{\frac{n}{2}} \right]}{\prod_{k=1}^{r} \left[(\det V_k)^{\frac{1}{2}} N^{\frac{n}{2}} \right]} \det \left(\left(\left[\frac{\varphi \langle J_l \rangle}{du_1 \wedge \cdots \wedge du_n} \right]_{u=c^{(k)}} \right) \right)_{k,l=1}^{r}.$$

Hence, if the matrix

$$\left(\left(\left[\frac{\varphi \langle J_l \rangle}{du_1 \wedge \cdots \wedge du_n} \right]_{u=c^{(k)}} \right) \right)_{k,l=1}^{r} \qquad (4.294)$$

is regular, then $\Phi(\alpha)$ is non-degenerate. It is quite cumbersome to prove this fact directly. So, we prove it as follows.

We take $\psi_1, \psi_2, \cdots, \psi_r \in \mathbf{C}[u_1, \cdots, u_n]$ in such a way that they satisfy $\psi_l(c^{(k)}) = \delta_{lk}$. For each ψ_j, we have (4.290). If we denote b_{J_l} corresponding to $\psi = \psi_j$ by b_{j,J_l}, by (4.291), we see that

$$\delta_{jk} = \sum_{l=1}^{r} b_{j,J_l} \left(x; \frac{\eta_1}{\sum_{\nu=1}^{m} \eta_\nu}, \cdots, \frac{\eta_m}{\sum_{\nu=1}^{m} \eta_\nu} \right) \qquad (4.295)$$

$$\cdot \left[\frac{\varphi \langle J_l \rangle}{du_1 \wedge \cdots \wedge du_n} \right]_{u=c^{(k)}},$$

hence, we obtain

$$\det \left(\left(\left[\frac{\varphi \langle J_l \rangle}{du_1 \wedge \cdots \wedge du_n} \right]_{u=c^{(k)}} \right) \right)_{l,k=1}^{r} \neq 0. \qquad (4.296)$$

As a consequence, we obtain the following theorem.

Theorem 4.9. *Let* $\{\varphi \langle J_1 \rangle, \cdots, \varphi \langle J_r \rangle\}$ *be the basis of* $H^n(\Omega^\bullet(*D), \nabla_\omega)$ *formed by* $d \log P_{j_1} \wedge \cdots \wedge d \log P_{j_n}$, $1 \leqslant j_1 < \cdots < j_n \leqslant m-1$. *The period matrix (4.279) is non-degenerate.* $\eta \in \mathbb{Z}^m \setminus \{0\}$ *satisfying (4.281) provides a regular direction of the holonomic systems of difference equations (4.266), (4.269), (4.278). At* $\alpha = \alpha' + N\eta$ (α' *fixed) and* $N \mapsto +\infty$, *an asymptotic expansion of* $\Phi(\alpha)$ *is given by*

$$\int_{\Delta_k(\omega)} U(u) \varphi \langle J_l \rangle \qquad (4.297)$$

$$= \frac{\pi^{\frac{n}{2}}}{N^{\frac{n}{2}}} \frac{U(c^{(k)})}{\sqrt{\det V_k}} \cdot \left[\frac{\varphi \langle J_l \rangle}{du_1 \wedge \cdots \wedge du_n} \right]_{u=c^{(k)}} \left(1 + O\left(\frac{1}{N} \right) \right), \quad 1 \leqslant l, k \leqslant r,$$

hence, $\Phi(\alpha)$ *is a regular matrix.*

As we have stated in § 3.5.8, the explicit formula of the determinant of (4.297) is known as the Varchenko formula (Lemma 3.7). He derived it in a completely different manner from what we have explained here.

Exercise 4.1. By the above consideration and Theorem 3.10, compute directly the determinant of the period matrix (4.279) and clarify a relation to the Varchenko formula.

From (4.282) and (4.283), we can derive an important relation between the critical points $c^{(k)}$ and the eigenvalues of the mutually commutative matrices $A_{\pm j,\eta}^{(0)}$ in the following way.

Theorem 4.10. *For a fixed k, j $(1 \le k \le r), (1 \le j \le m)$, the r-dimensional vector $(\tilde{\varphi}_{J_l}) = \left(\left[\frac{\varphi(J_l)}{du_1 \wedge \cdots \wedge du_n} \right]_{u=c^{(k)}}, 1 \le l \le r \right)$ is a simultaneous eigenvector for the operators $A_{-j,\eta}^{(0)}, A_{j,\eta}^{(0)}$ and $A_1^{*(0)}$ defined by (4.135) with the eigenvalues $P_j(c^{(k)})^{-1}, P_j(c^{(k)})$ and $P_1(c^{(k)})^{\eta_1} \cdots P_m(c^{(k)})^{\eta_m}$ respectively.*

In fact, one has only to take the limit $N \to \infty$ in (4.266), (4.269) and (4.270) for $\alpha = \alpha' + N\eta$, and apply the formula (4.297).

4.5 Connection Problem of System of Difference Equations

As was stated in Chapter 1, Gauss' formula

$$\Gamma(z)\Gamma(1-z) = \frac{\pi}{\sin \pi z}$$

can be regarded as a connection relation of the difference equation (4.1). In this section, we generalize the connection problem to a holonomic system of difference equations for the case of several variables, and show some examples in a simple case.

4.5.1 Formulation

Suppose that a holonomic system of difference equations (4.96) is given. For two regular directions $\eta, \eta' \in \mathbb{Z}^m \setminus \{0\}$, let $\Phi_\eta(z), \Phi_{\eta'}(z)$ be solutions of (4.96) which have an asymptotic expansion in the direction of η, η', respectively; we may set

$$\Phi_{\eta'}(z) = \Phi_\eta(z) P_{\eta\eta'}(z). \tag{4.298}$$

$P_{\eta\eta'}(z)$ is a periodic matrix. That is, for a standard basis $\{e_j\}_{j=1}^m$ of $\mathbb{Z}^m$, since it satisfies

$$P_{\eta\eta'}(z + e_j) = P_{\eta\eta'}(z), \quad 1 \le j \le m. \tag{4.299}$$

$P_{\eta\eta'}(z)$ is a meromorphic function of $e^{2\pi\sqrt{-1}z_1}, \cdots, e^{2\pi\sqrt{-1}z_m}$. Fixing η, η', the problem to find the matrix $P_{\eta\eta'}(z)$ is called a connection problem, $P_{\eta\eta'}(z)$ is called a connection matrix, and each of its components is called a connection function. Now, if three regular directions $\eta, \eta', \eta'' \in \mathbb{Z}^m \setminus \{0\}$ are given, for each pair (η, η'), (η', η''), (η, η''), three connection matrices $P_{\eta\eta'}$, $P_{\eta'\eta''}$, $P_{\eta\eta''}$ are defined and they satisfy the following relations:

$$P_{\eta\eta'}(z)P_{\eta'\eta''}(z) = P_{\eta\eta''}(z), \qquad (4.300)$$

$$P_{\eta\eta'}(z)P_{\eta'\eta}(z) = 1. \qquad (4.301)$$

By specializing equations further, we consider (4.188), (4.189). (Notice that we consider them with respect to α and not z.) For $\eta = (\eta_1, \cdots, \eta_m)$, $\eta' = (\eta'_1, \cdots, \eta'_m)$, we consider the gradient vector fields $\mathbf{V}_\eta = \text{grad } \Re F_\eta$, $\mathbf{V}_{\eta'} = \text{grad } \Re F_{\eta'}$ associated to the level functions $\Re F_\eta(u) = \sum_{j=1}^m \eta_j \log|P_j|$, $\Re F_{\eta'}(u) = \sum_{j=1}^m \eta'_j \log|P_j|$, respectively. If both $\Re F_\eta$ and $\Re F_{\eta'}$ are non-degenerate, the contracting cycles $\{\sigma_1(\eta), \cdots, \sigma_r(\eta)\}$, $\{\sigma_1(\eta'), \cdots, \sigma_r(\eta')\}$ constructed by $\mathbf{V}_\eta$, $\mathbf{V}'_\eta$, respectively, provide bases of $H_n(M, \mathcal{L}_\omega^\vee)$. Hence, the one should be a linear combination of the other:

$$\sigma_i(\eta') = \sum_{j=1}^r P_{\eta\eta', ji}(\alpha)\sigma_j(\eta). \qquad (4.302)$$

Since the base field defining the local system $\mathcal{L}_\omega$ belongs to the function field $\mathbb{C}(e^{2\pi\sqrt{-1}\alpha_1}, \cdots, e^{2\pi\sqrt{-1}\alpha_m})$, all $P_{\eta\eta', ji}(\alpha)$'s are rational functions of $e^{2\pi\sqrt{-1}\alpha_1}, \cdots, e^{2\pi\sqrt{-1}\alpha_m}$. Translating this relation to integral representations, it gives us the relation (4.298). Here, we have

$$\Phi_\eta(\alpha) = \left(\left(\int_{\sigma_j(\eta)} U\varphi_k\right)\right)^r_{j,\,k=1}, \qquad (4.303)$$

$$\Phi_{\eta'}(\alpha) = \left(\left(\int_{\sigma_j(\eta')} U\varphi_k\right)\right)^r_{j,\,k=1}, \qquad (4.304)$$

namely, $P_{\eta\eta'}(\alpha) = ((P_{\eta\eta',ij}(\alpha)))^r_{i,j=1}$ is a connection matrix of equations (4.188), (4.189). In this way, the connection problem is reduced to finding the linear relations among two bases of $H_n(M, \mathcal{L}_\omega^\vee)$.

But, in general, fixing two directions η, η', it is very hard to compute $P_{\eta\eta'}(\alpha)$ concretely, and the calculation has been done only for some simple cases (cf. see, for example, [Ao9]). Next, we provide an example in the case of hypergeometric functions of Appell–Lauricella type.

4.5.2 The Case of Appell–Lauricella Hypergeometric Functions

In this case, as was considered in § 2.2.1, the functions are given by integrals of Jordan-Pochhammer type:

$$F(\alpha_1, \cdots, \alpha_m; x) = \int_\sigma \prod_{j=1}^m (u - x_j)^{\alpha_j} du. \tag{4.305}$$

(σ is an appropriate twisted cycle). Now, we assume that x_j's are all real and they satisfy

$$x_1 < x_2 < \cdots < x_m. \tag{4.306}$$

$M = \mathbb{C} \setminus \{x_1, \cdots, x_m\}$ and the bounded chambers of $M \cap \mathbb{R}$ are given by the intervals $\Delta_1^+ = [x_1, x_2], \cdots, \Delta_{m-1}^+ = [x_{m-1}, x_m]$. We assume that $\eta_j > 0$ for any j. By using the flat Kähler metric $ds^2 = |du|^2$ on M, we define the gradient vector field $\mathbf{V}$ by

$$\mathbf{V} = \mathrm{grad}(\Re F), \tag{4.307}$$

$$F(u) = \sum_{j=1}^m \eta_j \log(u - x_j). \tag{4.308}$$

On M, there is exactly one point in each open interval (x_j, x_{j+1}) at which $\mathbf{V}$ vanishes. We denote them by $c^{(1)}, c^{(2)}, \cdots, c^{(m-1)}$. Namely, $c^{(1)}, c^{(2)}, \cdots, c^{(m-1)}$ are the points u of M that satisfy the equation

$$\sum_{j=1}^m \frac{\eta_j}{u - x_j} = 0. \tag{4.309}$$

Setting $\alpha_j = \alpha_j' + \eta_j N$ (α_j' fixed), an asymptotic expansion at $N \mapsto +\infty$ is given by (4.252). Setting $\alpha_j = \alpha_j' - \eta_j N$, we can also consider an asymptotic expansion at $N \mapsto +\infty$. What can be the contracting twisted cycles in this case? Since the points where $\mathbf{V}$ vanishes are given by $c^{(1)}, c^{(2)}, \cdots, c^{(m-1)}$, the trajectory of $\mathbf{V}$ passing through $c^{(k)}$ is the contracting twisted cycle Δ_k^- associated to the direction $-\eta$. They are given by the equation

$$\Im F(u) = \Im F(c^{(k)}),$$

i.e.,

$$\sum_{j=1}^m \eta_j \arg(u - x_j) = \sum_{j=1}^m \eta_j \arg(c^{(k)} - x_j). \tag{4.310}$$

As all η_j's are positive, $\Delta_1^{(-)}, \cdots, \Delta_{m-1}^-$ do not mutually intersect. We choose the branch of $\Delta_j^{\pm}$ with respect to $U = \prod_{j=1}^m (u - x_j)^{\alpha_j}$ in such a way that U becomes a real positive number at $c^{(j)}$. The phase diagram of Δ_j^- is given in Figure 4.4. $\{\Delta_1^+, \cdots, \Delta_{m-1}^+\}$ and $\{\Delta_1^-, \cdots, \Delta_{m-1}^-\}$ provide bases of $H_1^{lf}(M, \mathcal{L}_\omega^\vee)$, respectively, and its regularization $\{\Delta_1^+(\omega), \cdots, \Delta_{m-1}^+(\omega)\}$, $\{\Delta_1^-(\omega), \cdots, \Delta_{m-1}^-(\omega)\}$ provide bases of $H_1(M, \mathcal{L}_\omega^\vee)$, respectively (cf. § 2.3). As is clear from the phase diagram, we have

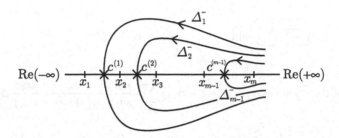

Fig. 4.4

$$\Delta_j^- = (e^{\pi\sqrt{-1}\alpha_{j+1}} - e^{-\pi\sqrt{-1}\alpha_{j+1}})\Delta_{j+1}^+ \tag{4.311}$$

$$+ \cdots + \left(e^{\pi\sqrt{-1}(\alpha_{j+1}+\cdots+\alpha_{m-1})} - e^{-\pi\sqrt{-1}(\alpha_{j+1}+\cdots+\alpha_{m-1})}\right)\Delta_{m-1}^+$$

$$+ (e^{\pi\sqrt{-1}(\alpha_{j+1}+\cdots+\alpha_m)} - e^{-\pi\sqrt{-1}(\alpha_{j+1}+\cdots+\alpha_m)})\mathrm{reg}[x_m, +\infty].$$

Here, $\mathrm{reg}[x_m, +\infty] \in H_1(M, \mathcal{L}_\omega^\vee)$ signifies the regularization of $[x_m, +\infty]$. Since, we also have

$$0 = \partial_\omega(\Im u \geq 0) \tag{4.312}$$

$$= \mathrm{reg}[-\infty, x_1] + e^{-\pi\sqrt{-1}\alpha_1}\Delta_1^+ + \cdots + e^{-\pi\sqrt{-1}(\alpha_1+\cdots+\alpha_{m-1})}\Delta_{m-1}^+$$

$$+ e^{-\pi\sqrt{-1}(\alpha_1+\cdots+\alpha_m)}\mathrm{reg}[x_m, +\infty],$$

$$0 = -\partial_\omega(\Im u \leqslant 0) \tag{4 313}$$

$$= \mathrm{reg}[-\infty, x_1] + e^{\pi\sqrt{-1}\alpha_1}\Delta_1^+ + \cdots + e^{\pi\sqrt{-1}(\alpha_1+\cdots+\alpha_{m-1})}\Delta_{m-1}^+$$

$$+ e^{\pi\sqrt{-1}(\alpha_1+\cdots+\alpha_m)}\mathrm{reg}[x_m, +\infty],$$

we obtain the formulas

$$\operatorname{reg}[x_m, +\infty] = -\sum_{j=1}^{m-1} \frac{\sin \pi(\alpha_1 + \cdots + \alpha_j)}{\sin \pi(\alpha_1 + \cdots + \alpha_m)} \Delta_j^+, \qquad (4.314)$$

$$\operatorname{reg}[-\infty, x_1] = -\sum_{j=1}^{m-1} \frac{\sin \pi(\alpha_{j+1} + \cdots + \alpha_m)}{\sin \pi(\alpha_1 + \cdots + \alpha_m)} \Delta_j^+. \qquad (4.315)$$

Rewriting (4.311) with (4.314), (4.315), we have

$$\Delta_j^- = 2\sqrt{-1} \sum_{k=j+1}^{m-1} \sin \pi(\alpha_{j+1} + \cdots + \alpha_k) \Delta_k^+ \qquad (4.316)$$

$$- 2\sqrt{-1} \sin \pi(\alpha_{j+1} + \cdots + \alpha_m) \sum_{k=1}^{m-1} \frac{\sin \pi(\alpha_1 + \cdots + \alpha_k)}{\sin \pi(\alpha_1 + \cdots + \alpha_m)} \Delta_k^+$$

$$= - 2\sqrt{-1} \sum_{k=1}^{j} \frac{\sin \pi(\alpha_1 + \cdots + \alpha_k) \sin \pi(\alpha_{j+1} + \cdots + \alpha_m)}{\sin \pi(\alpha_1 + \cdots + \alpha_m)} \Delta_k^+$$

$$- 2\sqrt{-1} \sum_{k=j+1}^{m-1} \frac{\sin \pi(\alpha_1 + \cdots + \alpha_j) \sin \pi(\alpha_{k+1} + \cdots + \alpha_m)}{\sin \pi(\alpha_1 + \cdots + \alpha_m)} \Delta_k^+.$$

This is nothing but the connection relations of the contracting twisted cycles in the directions of η and of $-\eta$.

The linear relations expressing Δ_k^+ by Δ_j^- can be obtained by solving (4.316). Conversely, this can be also obtained using the intersection numbers of the twisted cycles explained in § 2.3. Indeed, since the geometric intersection number of Δ_k^+ and Δ_j^- is given by

$$I_c(\Delta_k^+, \Delta_j^-) = -\delta_{kj}.$$

Δ_k^+ and Δ_j^- are given a relation between elements of the homology $H_1(M, \mathcal{L}_\omega^\vee)$

$$\Delta_k^+ = \sum_{j=1}^{m-1} I_c(\Delta_k^+, \Delta_j^\vee) \cdot \Delta_j^-.$$

Here, $\Delta_j^\vee = \operatorname{reg}^{-1} \Delta_j^+ \in H_1^{lf}(M, \mathcal{L}_\omega)$ and

$$I_c(\Delta_k^+, \Delta_j^\vee) = \begin{cases} \dfrac{1}{2\sqrt{-1}} \dfrac{1}{\sin \pi \alpha_{j+1}} & k = j + 1, \\[2mm] -\dfrac{1}{2\sqrt{-1}} \dfrac{\sin \pi(\alpha_j + \alpha_{j+1})}{\sin \pi \alpha_j \cdot \sin \pi \alpha_{j+1}} & k + j, \\[2mm] \dfrac{1}{2\sqrt{-1}} \dfrac{1}{\sin \pi \alpha_j} & k = j - 1, \\[2mm] 0 & \text{otherwise.} \end{cases}$$

(cf. § 2.3.3. For a higher-dimensional generalization, see [Ao12]).

For $m = 3$, i.e., Gauss' hypergeometric functions, the complete list of the connection relations is given in [Wa] (cf. [Ao11]).

Appendix A
Mellin's Generalized Hypergeometric Functions

Here, we will introduce a generalized hypergeometric series generalizing the $\nu!$ appearing in the denominator of the coefficient of hypergeometric series (3.1) defined in § 3.1 to $\prod_{i=1}^{n} b_i(\nu)!$, where $b_i(\nu)$'s are integer-valued linear forms on the lattice $L = \mathbb{Z}^n$, and will derive a system of partial differential equations and integral representation of Euler type. We will also explain its relation to the Bernstein−Mellin−Sato b-function. We use the notation defined in § 3.1.

A.1 Definition

An element a of the dual lattice $L^\vee$ of the lattice $L = \mathbb{Z}^n$ naturally extends to a linear form on $L \otimes \mathbb{C}$. Here and after, we denote this linear form on $\mathbb{C}$ by the same symbol a. We denote a point of L as ν, a point of $L \otimes \mathbb{C}$ as s, and if necessary, we denote them by $a(\nu)$, $a(s)$ to distinguish them.

Suppose that $m + n$ integer-valued linear forms $a_j^+ \in L^\vee$, $1 \leq j \leq m$, $a_i^- \in L^\vee$, $1 \leq i \leq n$ on L and

$$\alpha = (\alpha_1, \cdots, \alpha_m) \in \mathbb{C}^m$$

are given satisfying the following two assumptions:

$$\det(a_i^-(e_k)) \neq 0, \tag{A.1}$$

(namely, $a_1^-(s), \cdots, a_n^-(s)$ regarded as elements of $L^\vee \otimes \mathbb{C}$ are linearly independent) and the condition corresponding to (3.2)

$$\sum_{j=1}^{m} a_j^+(e_k) = \sum_{i=1}^{n} a_i^-(e_k), \quad 1 \leq k \leq n. \tag{A.2}$$

Then, we consider the following Laurent series

K. Aomoto et al., *Theory of Hypergeometric Functions*, Springer Monographs in Mathematics, DOI 10.1007/978-4-431-53938-4, © Springer 2011

$$F(\alpha, x) = \sum_{\nu \in \mathcal{C} \cap L} \frac{\prod_{j=1}^{m} \Gamma(a_j^+(\nu) + \alpha_j)}{\prod_{i=1}^{n} \Gamma(a_i^-(\nu) + 1)} x^\nu \qquad (A.3)$$

as a function of $x \in \mathbb{C}^n$, which is called Mellin's hypergeometric series of general type. Here, $\mathcal{C}$ is the cone in $L \otimes \mathbb{R} = \mathbb{R}^n$

$$\mathcal{C} := \{s \in \mathbb{R}^n | a_i^-(s) \geq 0, \quad 1 \leq i \leq n\} \qquad (A.4)$$

defined by linearly independent elements $a_1^-(s), \cdots, a_n^-(s)$ as elements of $L^\vee \otimes \mathbb{R}$, and the sum is taken over all of the lattice points in $\mathcal{C}$. Moreover, for each coefficient of (A.3) to be finite, we impose the condition: for any $\nu \in \mathcal{C} \cap L$,

$$a_j^+(\nu) + \alpha_j \notin \mathbb{Z}_{\leq 0}, \quad 1 \leq j \leq m. \qquad (A.5)$$

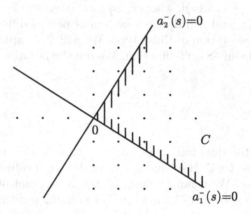

Fig. A.1

Integrals related to Mellin's hypergeometric series have been treated in many references, for example in [Ad-Sp], [A-K], [Ao9], [Bai], [Bel], [G-G-R], [Kim], [Kit2], [Ok], [Sl].

A.2 Kummer's Method

Following § 3.3, we would like to show the convergence of this series and elementary integral representation of Euler type at the same time by Kummer's method. For this purpose, we first prepare three lemmata.

By $\sum_{j=1}^{m} a_j^+(\nu) = \sum_{i=1}^{n} a_i^-(\nu)$ which follows easily from (A.2), using (3.18), we obtain the following lemma:

Lemma A.1. *Set*

$$U = (u_1, \cdots, u_{m-1}) \in \mathbb{C}^{m-1}, \quad u_m := 1 - \sum_{j=1}^{m-1} u_j,$$

$$U(u) := \prod_{j=1}^{m} u_j^{a_j^+(\nu)+\alpha_j-1}, \quad \omega = dU/U,$$

$\Delta^{m-1}(\omega)$: *the twisted cycle defined from the $(m-1)$-simplex*

$$\Delta^{m-1} = \{u \in \mathbb{R}^{m-1} | u_j \geq 0,\ 1 \leq j \leq m\}, \quad |\alpha| = \sum_{j=1}^{m} \alpha_j.$$

Under the assumption $\alpha_j \in \mathbb{C} \setminus \mathbb{Z}$, $1 \leq j \leq m$, we have

$$\frac{\prod_{j=1}^{m} \Gamma(a_j^+(\nu)+\alpha_j)}{\Gamma(\sum_{j=1}^{m} a_j^+(\nu)+|\alpha|)} = \int_{\Delta^{m-1}(\omega)} U(u) du_1 \cdots du_{m-1}. \tag{A.6}$$

Next, for later use, we generalize the multinomial theorem which played an essential role in the proof of Theorem 3.2. For this purpose, we first prepare the following fact about the zero-dimensional toric varieties (see [Od] for toric varieties).

Lemma A.2. *For a given $y \in (\mathbb{C}^*)^n$, we denote the set of finite points defined by the equations (a zero-dimensional toric variety) by $P(y)$:*

$$P(y) := \left\{ w \in (\mathbb{C}^*)^n | w_1^{a_1^-(e_i)} \cdots w_n^{a_n^-(e_i)} = y_i,\ 1 \leq i \leq n \right\}.$$

Then, if there exists $s \in L$ such that $\nu_i = a_i^-(s)$, $1 \leq i \leq n$ for a $\nu \in L$, we have

$$\sum_{w \in P(y)} w^\nu = |P(y)| y^s, \tag{A.7}$$

and for a $\nu \in L$ that does not admit such expression, we have

$$\sum_{w \in P(y)} w^\nu = 0. \tag{A.8}$$

Here, $|P(y)|$ signifies the cardinality of $P(y)$.

Proof. Since any of y_i, $1 \leq i \leq n$ are not zero, we can define its argument and we fix one among them, and we denote by $\log y_i$ the branch of logarithms defined by it. A condition to be $w \in P(y)$ can be expressed by using a solution of

$$a_1^-(e_i)X_1 + \cdots + a_n^-(e_i)X_n = \log y_i + 2r_i\pi\sqrt{-1},$$

$$1 \leq i \leq n, \quad r_i \in \mathbb{Z},$$

as $w_i = e^{X_i}$, $1 \leq i \leq n$. By the condition (A.1) and the fact that $a_i^-(e_k)$ is an integer, we have $\det(a_i^-(e_k)) \in \mathbb{Z} \setminus \{0\}$ and the number of such $X_1, \cdots, X_n$ mod $2\pi\sqrt{-1}$ is finite by the Cramer formula. Hence, $P(y)$ is a finite set. Next, notice that $w^\nu = \exp(\sum \nu_k X_k) = \Pi_i \exp(s_i \sum_k a_k^-(e_i)X_k)$. When each s_i is an integer, this concludes (A.7). When ν cannot be expressed as $\nu_i = a_i^-(s)$, $s \in L$, by the above consideration, at least one of s_i's is a rational number which is not an integer, from which we have

$$\sum_{k=0}^{r-1} e^{2\pi\sqrt{-1}k/r} = 0,$$

and (A.8) follows.

A.3 Toric Multinomial Theorem

As the third lemma, we state a fact that should be called a toric version of the ordinary multinomial theorem.

Lemma A.3. *Set*

$$\Psi_{|\alpha|}(y) := \sum_{w \in P(y)} \frac{(1 - w_1 \cdots - w_n)^{-|\alpha|}}{|P(y)|}.$$

If any $w \in P(y)$ satisfies

$$|w_1| + \cdots + |w_n| < 1,$$

then the multinomial expansion

$$\Psi_{|\alpha|}(y) = \sum_{s \in \mathcal{C} \cap L} \frac{\Gamma(\sum_{i=1}^n a_i^-(s) + |\alpha|)}{\Gamma(|\alpha|) \prod_{i=1}^n \Gamma(a_i^-(s) + 1)} y^s \qquad (A.9)$$

holds and it converges uniformly on every compact subset.

Proof. Here, we abbreviate $(a_1^-(s), \cdots, a_n^-(s))$ as $a^-(s)$. Under the condition $\sum_{i=1}^n |w_i| < 1$, the series arising in the following process converge uniformly on every compact subset:

$$\sum_{w \in P(y)} \left(1 - \sum_{i=1}^{n} w_i\right)^{-|\alpha|} /|P(y)| = \sum_{w \in P(y)} \sum_{\nu \in \mathbb{Z}_{\geq 0}^n} \frac{(|\alpha|; |\nu|)}{\nu!} w^\nu /|P(y)|$$

$$= \sum_{\nu \in \mathbb{Z}_{\geq 0}^n} \frac{(|\alpha|; |\nu|)}{\nu!} \sum_{w \in P(y)} \frac{w^\nu}{|P(y)|}$$

$$= \sum_{\substack{\nu \in \mathbb{Z}_{\geq 0}^n \\ \nu = a^-(s) \\ s \in L}} \frac{(|\alpha|; |\nu|)}{\nu!} y^s \quad \text{(by (A.7) and (A.8))}$$

$$= \sum_{s \in C_\cap L} \frac{\Gamma(\sum a_i^-(s) + |\alpha|)}{\Gamma(|\alpha|) \prod_{i=1}^{n} \Gamma(a_i^-(s) + 1)} y^s \quad \text{(by (A.4))}.$$

Remark A.1. The sum in the right-hand side of (A.9) is taken over the lattice points of $C_\cap L$, and as we see from Figure A.1, in general, negative integers appear as components of $s \in C_\cap L$. Hence, the (A.9) converges not on an appropriate neighborhood of the origin as in the ordinary multinomial theorem but on the image of $(\mathbb{C}^*)^n \cap \{\sum_{i=1}^{n} |w_1| < 1\}$ by

$$y_i = w_i^{a_1^-(e_i)} \cdots w_n^{a_n^-(e_i)}, \qquad 1 \leq i \leq n. \tag{A.10}$$

To show the convergence of the hypergeometric series (A.3), we state a little bit on the convergent domain of a Laurent series in several variables. To simplify the notation, we set

$$\exp \Gamma := \{y \in (\mathbb{C}^*)^n | |y_i| = e^{\eta_i}, (\eta_1, \cdots, \eta_n) \in \Gamma\},$$

for a domain Γ in $\mathbb{R}^n$. Then, setting

$$\Gamma_\zeta := \left\{\zeta \in \mathbb{R}^n | \zeta_i < \log \frac{1}{2n}, 1 \leq i \leq n\right\},$$

we clearly have

$$\exp \Gamma_\zeta \subset (\mathbb{C}^*)^n \cap \left\{\sum_{i=1}^{n} |w_i| < 1\right\}.$$

Now, we consider the map induced from (A.10):

$$\Phi: \ \mathbb{R}^n \longrightarrow \mathbb{R}^n,$$

$$\zeta \longmapsto \eta,$$

$$\eta_i = a_1^-(e_i)\zeta_1 + \cdots\cdots + a_n^-(e_i)\zeta_n,$$

$$\zeta_i = \log|w_i|, \quad \eta_i = \log|y_i|, \quad 1 \le i \le n.$$

By Assumption (A.1), Φ is a non-singular linear map and $\Phi(\Gamma_\zeta) = \Gamma_\eta$ becomes a infinite convex domain surrounded by n hyperplanes. Clearly, (A.9) converges absolutely and uniformly on $\exp\Gamma_\eta \subset (\mathbb{C}^*)^n$.

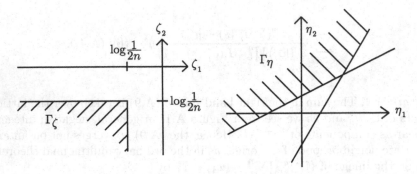

Fig. A.2

A.4 Elementary Integral Representations

Now, we have prepared to apply Kummer's method. The rest of the arguments can be formally developed as follows. First, rewriting (A.3), we have

$$F(\alpha, x) = \sum_{\nu \in \mathcal{C} \cap L} \frac{\Gamma(\sum a_i^-(\nu) + |\alpha|)}{\prod \Gamma(a_i^-(\nu) + 1)} x^\nu \frac{\prod \Gamma(a_j^+(\nu) + \alpha_j)}{\Gamma(\sum a_i^-(\nu) + |\alpha|)}$$

$$= \sum_{\nu \in \mathcal{C} \cap L} \frac{\Gamma(\sum a_i^-(\nu) + |\alpha|)}{\prod \Gamma(a_i^-(\nu) + 1)} x^\nu \int_{\Delta^{m-1}(\omega)} U(u) du_1 \cdots du_{m-1}$$

(by (A.6))

$$= \sum_{\nu \in \mathcal{C} \cap L} \int_{\Delta^{m-1}(\omega)} \prod_{j=1}^m u_j^{\alpha_j - 1} \cdot \frac{\Gamma(\sum a_i^-(\nu) + |\alpha|)}{\prod \Gamma(a_i^-(\nu) + 1)} y^\nu du_1 \cdots du_{m-1}.$$

Here, we set

$$y_i = x_i \prod_{j=1}^{m} u_j^{a_j^+(e_i)}, \quad 1 \le i \le n.$$

By the construction of the twisted cycle $\Delta^{m-1}(\omega)$ (cf. § 3.2), independent of the sign of $a_j^+(e_i)$, for $u \in \Delta^{m-1}(\omega)$, there exists positive constants $c_1 < c_2$ such that

$$c_1 \le \left| \prod_{j=1}^{m} u_j^{a_j^+(e_i)} \right| \le c_2, \quad 1 \le i \le n.$$

Hence, we have

$$\log|y_i| = \log|x_i| + \log \left| \prod_{j=1}^{m} u_j^{a_j^+(e_i)} \right|, \quad 1 \le i \le n,$$

and as Γ_η is an infinite domain surrounded by n hyperplanes, there exists an appropriate vector $c \in \mathbb{R}^n$ such that if we always have $(\log|x_1|, \cdots, \log|x_n|) \in c + \Gamma_\eta$ for any $u \in \Delta^{m-1}(\omega)$, then we can let $(\log|y_1|, \cdots, \log|y_n|) \in \Gamma_\eta$. Hence, by Remark A.1, the series

$$\sum_{\nu \in C \cap L} \frac{\Gamma(\sum a_i^-(\nu) + |\alpha|)}{\prod \Gamma(a_i^-(\nu) + 1)} y^\nu,$$

for $u \in \Delta^{m-1}(\omega)$, converges uniformly to $\Gamma(|\alpha|)\Psi_{|\alpha|}(y)$, and we can apply the Kummer method.

Remark A.2. The domain $c + \Gamma_\eta$ contains an n-dimensional cube parallel to the coordinate axes, hence, $\exp(c + \Gamma_n)$ contains the direct product of the annulus

$$\{x \in (\mathbb{C}^*)^n \mid r_i < |x_i| < R_i, \quad 1 \le i \le n\}.$$

Summarizing, we obtain the following theorem.

Theorem A.1. *The Laurent series (A.3) converges absolutely and uniformly on an appropriate product of annulus* $\{x \in (\mathbb{C}^*)^n \mid r_i \le |x_i| \le R_i, (1 \le i \le n)\}$, *and it possesses an elementary integral representation of Euler type:*

$$F(\alpha, x) = \Gamma(|\alpha|) \int_{\Lambda^{m-1}(\omega)} \prod_{j=1}^{m} u^{\alpha_j - 1} \Psi_{|\alpha|} \left(r_1 \prod_{j=1}^{m} u_j^{a_j^+(e_1)}, \cdots, x_n \prod_{j=1}^{m} u_j^{a_j^+(e_n)} \right)$$

$$\cdot du_1 \wedge \cdots \wedge du_{m-1}.$$

There are lots of hypergeometric functions whose integral representations can be expressed by an arrangement of hyperquadrics or hypersurfaces of higher-degree. For this aspect, see for example, [Ao6], [Ao7], [Ao13], [AKOT], [Dw], [G-K-Z], [G-R-S], [Hat-Kim], [Ka1], [Kimu], [No], [Ter3], [Yos1] etc.

A.5 Differential Equations of Mellin Type

As in Chapter 4, for $\nu = (\nu_1, \cdots, \nu_n) \in L$, we set the shift operator $T^\nu = T_1^{\nu_1} \cdots T_n^{\nu_n}$ as

$$T^\nu \varphi(s) = \varphi(s + \nu), \quad s \in L \otimes \mathbb{C} = \mathbb{C}^n.$$

For the meromorphic function of s

$$\Phi(s) := \frac{\prod_{j=1}^m \Gamma(a_j^+(s) + \alpha_j)}{\prod_{i=1}^n \Gamma(a_i^-(s) + 1))} x^s,$$

we set

$$b_\nu(s) := \Phi(s + \nu)/(\Phi(s)x^\nu). \tag{A.11}$$

Using the notation $(\alpha; k) = \Gamma(\alpha + k)/\Gamma(\alpha)$, $b_\nu(s)$ can be written as

$$b_\nu(s) = \prod_{j=1}^m (a_j^+(s) + \alpha_j; a_j^+(\nu))/ \prod_{i=1}^n (a_i^-(s) + 1; a_i^-(\nu)),$$

hence, it is a rational function of s. Expressing this as the quotient of coprime polynomials

$$b_\nu(s) = b_\nu^+(s)/b_\nu^-(s), \quad b_\nu^\pm(s) \in \mathbb{C}[s],$$

by (A.11), we have

$$b_\nu^-(s)\Phi(s + \nu) = x^\nu b_\nu^+(s)\Phi(s). \tag{A.12}$$

Here, using the notation

$$\langle \varphi(s) \rangle = \sum_{\mu \in L} \varphi(\mu),$$

by the fact that $1/\Gamma(a_i^-(\mu) + 1)$ is zero for $\mu \in L$ such that $a_i^-(\mu) < 0$, Theorem A.1 implies

$$F(\alpha, x) = \langle \Phi(s) \rangle.$$

Notice that, we have

$$\vartheta_k F(\alpha, x) = \sum_{\mu \in C \cap L} \frac{\prod \Gamma(a_j^+(\mu) + \alpha_j)}{\prod \Gamma(a_i^-(\mu) + 1)} \mu_k x^\mu \qquad (A.13)$$

$$= \langle s_k \Phi(s) \rangle,$$

where we set $\vartheta_k := x_k \frac{\partial}{\partial x_k}$. Since we clearly have

$$\langle T^\nu \varphi(s) \rangle = \langle \varphi(s) \rangle$$

for any $\nu \in L$, together with (A.12), we obtain the identity:

$$\langle (b_\nu^-(s-\nu)\Phi(s)) \rangle = \langle T^\nu (b_\nu^-(s-\nu)\Phi(s)) \rangle$$

$$= \langle b_\nu^-(s)\Phi(s+\nu) \rangle$$

$$= \langle x^\nu b_\nu^+(s)\Phi(s) \rangle$$

$$= x^\nu \langle b_\nu^+(s)\Phi(s) \rangle.$$

By (A.13), the above formula transforms to the following countable partial differential equations satisfied by $F(\alpha, x)$:

$$b_\nu^-(\vartheta_1 - \nu_1, \cdots, \vartheta_n - \nu_n)F(\alpha, x) \qquad (A.14)$$

$$= x^\nu b_\nu^+(\vartheta_1, \cdots, \vartheta_n)F(\alpha, x), \qquad \nu \in L.$$

These are called the hypergeometric differential equations of Mellin type . As a simple calculation shows

$$\vartheta_k(x^{-\nu}F) = x^{-\nu}(\vartheta_k - \nu_k)F,$$

(A.14) can be rewritten in the following form:

$$b_\nu^+(\vartheta_1, \cdots, \vartheta_n)F(\alpha, x) = b_\nu^-(\vartheta_1, \cdots, \vartheta_n)x^{-\nu}F(\alpha, x), \quad \nu \in L. \quad (A.15)$$

These are the systems of partial differential equations defined by G. Mellin in 1907 (cf. [Bel]). This system of equations was independently proposed by M. Sato in 1971 as a holonomic $\mathcal{D}$-module (cf. [Hot]).

A.6 *b*-Functions

More generally, for a lattice $L = \mathbb{Z}^n$, we consider a 1-cocyle $\{b_\nu(s) | \nu \in L\}$ on $L \otimes \mathbb{C} = \mathbb{C}^n$ with values in the multiplicative group $\mathbb{C}(s)^\times := \mathbb{C}(s) \setminus \{0\}$ of the rational function field $\mathbb{C}(s)$. Namely, for any $\nu, \nu' \in L$, we have

$$b_{\nu+\nu'}(s) = b_\nu(s)b_{\nu'}(s+\nu),$$

$$b_0(s) = 1.$$

It can be easily verified that $\{b_\nu(s) | \nu \in L\}$ defined in (A.11) satisfy these cocycle conditions. Then, as in § A.5, we have an expression

$$b_\nu(s) = b_\nu^+(s)/b_\nu^-(s), \quad b_\nu^\pm(s) \text{ are coprime polynomials.}$$

In the cohomology class of $\{b_\nu(s)\}$, one can choose its representative in such a way that $b_\nu^\pm(s)$ are finite products of appropriate linear functions of s. To make the statement precise, we introduce the symbol:
   For a function f on $\mathbb{Z}$, we set

$$\prod_{l=0}^{k-1} f(l) = \begin{cases} \displaystyle\prod_{l=0}^{k-1} f(l), & k \ge 1, \\[2ex] \displaystyle 1/\prod_{l=k}^{-1} f(l), & k < 0, \\[2ex] 1, & k = 0. \end{cases}$$

Then, the following Sato theorem is known.

Theorem A.2 ([S-S-M]). *We define the action of an n-dimensional lattice $L = \mathbb{Z}^n$ on $\mathbb{C}(s)^\times$ by $R(s) \longmapsto R(s+\nu)$, $R \in \mathbb{C}(s)^\times$, $\nu \in L$ and regard $\mathbb{C}(s)^\times$ as an L-module. Then, each cohomology class of the one-dimensional cohomology $H^1(L, \mathbb{C}(s)^\times)$ of the group L with coefficients in $\mathbb{C}(s)^\times$ is represented by rational functions $b_\nu(s)$, $\nu \in L$ of the following form: there exists k integervalued linear forms $a_\kappa \in L^\vee$ on L, complex numbers $\alpha_\kappa \in \mathbb{C}$, $1 \le \kappa \le k$, and $\beta = (\beta_1, \cdots, \beta_n) \in \mathbb{C}^n$ such that*

$$b_\nu(s) = e^{(\nu,\beta)} \prod_{\kappa=1}^{k} \left\{ \prod_{l=0}^{a_\kappa(\nu)-1} (a_\kappa(s) + \alpha_\kappa + l) \right\},$$

where we set $(\nu, \beta) = \sum_{i=1}^{n} \nu_i \beta_i$.

This $\{b_\nu(s) | \nu \in L\}$ is called the b-function, à la Bernstein, Mellin and Sato (see, also [Lo-Sab]).
   When an analytic function $F(x)$ of $x = (x_1, \cdots, x_n) \in \mathbb{C}^n$ satisfies the system of partial differential equations (A.14) or (A.15) determined by countable $b_\nu(s)$, $s \in L$, $F(x)$ is called a Mellin generalized hypergeometric function.

A.7 Action of Algebraic Torus

Mellin's system of hypergeometric functions (A.14) is, in comparison with the system of hypergeometric functions $E(n + 1, m + 1; \alpha)$ introduced in § 3.4, a system of differential equations obtained by killing the actions of $GL_{n+1}(\mathbb{C})$ and H_{m+1}. The system of differential equations which keeps the action of H_{m+1} (cf. (3.31)), which is essentially the same as Mellin's system, was derived in [G-G-Z] from a Lie group theoretical point of view, and was described more systematically in [G-Z-K] from the viewpoint of a $\mathcal{D}$-module in its relation to the theory of toric varieties. A similar result is also obtained in [Hra] (cf. see also [SaiM]). Here, we summarize the results of [G-G-Z], and explain a relation to Mellin's hypergeometric functions.

Let V be an l-dimensional complex vector space, $GL(V)$ be the group of non-singular linear endomorphisms on V, and H a subgroup of $GL(V)$ isomorphic to an m-dimensional complex torus: $H \simeq (\mathbb{C}^*)^m$. We impose the following condition on H:

$$\mathbb{C}^* \cdot 1_V \subset H, \tag{A.16}$$

where 1_V is the identity map on V. As it is known that H is simultaneously diagonalizable, there exists a basis $e_1, \cdots, e_l$ of V such that, for any $h \in H$, we have

$$he_k = \chi_k(h)e_k, \quad 1 \le k \le l. \tag{A.17}$$

Evidently, $\chi_k : H \longrightarrow \mathbb{C}^*$ is a homomorphism. We denote the group of rational homomorphisms from H to $\mathbb{C}^*$ by $X(H)$ and it is called the character group of H. An element of $X(H)$ is called a character of H. We have $\chi_k \in X(H)$, $1 \le k \le l$. Now, writing the isomorphism $H \simeq (\mathbb{C}^*)^m$

$$H \xrightarrow{\sim} (\mathbb{C}^*)^m,$$
$$h \longmapsto (z_1, \cdots, z_m)$$

each z_j, $1 \le j \le m$, becomes a character of H and $X(H)$ is a free abelian group generated by $z_1, \cdots, z_m$:

$$X(H) \xleftarrow{\sim} \mathbb{Z}^m, \tag{A.18}$$

$$\chi = z_1^{p_1} \cdots z_m^{p_m} \longleftrightarrow (p_1, \cdots, p_m).$$

A.8 Vector Fields of Torus Action

The Lie algebra $\mathfrak{h}$ of the left invariant vector fields on the Lie group H is an m-dimensional complex vector space generated by the vector fields $z_1 \frac{\partial}{\partial z_1}, \cdots, z_m \frac{\partial}{\partial z_m}$, and the dual space $\mathfrak{h}^\vee$ of $\mathfrak{h}$ is a vector space generated by the left invariant holomorphic 1-forms $\frac{dz_1}{z_1}, \cdots, \frac{dz_m}{z_m}$. Then, a natural map

$$\delta : X(H) \longrightarrow \mathfrak{h}^\vee \qquad\qquad (A.19)$$

$$z_j \longmapsto dz_j/z_j$$

is defined as follows. For any $A \in \mathfrak{h}$, we set

$$(\delta\chi)(A) = \frac{d}{dt}\chi(\exp(tA))\Big|_{t=0} \qquad\qquad (A.20)$$

Here, $t \in \mathbb{R}$ is a parameter. By (A.19), we obtain an isomorphism

$$X(H) \otimes_\mathbb{Z} \mathbb{C} \simeq \mathfrak{h}^\vee.$$

On the other hand, a one-parameter subgroup $\exp tA$, $A \in \mathfrak{h}$ of H defines a vector field $\mathbb{A}$ on V as follows:

$$V \ni v \longmapsto \frac{d}{dt}(\exp tA) \cdot v\Big|_{t=0}$$

$$= (\delta\chi_1(A)v_1, \cdots, \delta\chi_l(A)v_l).$$

Here, we identified V with $\mathbb{C}^l$ via the basis $e_1, \cdots, e_l$ and used (A.17) and (A.20). We have

$$\mathbb{A} = \sum_{k=1}^l \delta\chi_k(A)v_k \frac{\partial}{\partial v_k}. \qquad\qquad (A.21)$$

A.9 Lattice Defined by the Characters

Via the identification (A.18), l characters χ_k, $1 \le k \le l$ define l points of the lattice $\mathbb{Z}^m$. To make the story precise, under this identification, we write $\chi_k = {}^t(\chi_{1k}, \cdots, \chi_{mk}) \in \mathbb{Z}^m$ and the $m \times l$ matrix obtained by arranging l column vectors $\chi_1, \cdots, \chi_l$ is denoted by (χ_{jk}). Here, we impose the condition:

$$\chi_1, \cdots, \chi_l \text{ generate the lattice } \mathbb{Z}^m. \qquad\qquad (A.22)$$

This assumption is equivalent to:

there is an $m \times m$ submatrix of $(\chi_{jk}) \in M_{m,l}(\mathbb{Z})$ (A.23)

whose determinant is ± 1.

Under this assumption, let us rewrite the condition (A.16). The condition that a one-parameter subgroup $\exp(tA)$, $A \in \mathfrak{h}$ is contained in $\mathbb{C}^* \cdot 1_V$ is

$$\chi_1(\exp tA) = \cdots = \chi_l(\exp tA),$$

by (A.17). Hence, by (A.19), (A.20), setting $A = \sum_{j=1}^{m} a_j z_j \frac{\partial}{\partial z_j}$, we obtain

$$\delta\chi_1(A) = \cdots = \delta\chi_l(A),$$

$$\sum_{j=1}^{m} a_j \chi_{j1} = \cdots = \sum_{j=1}^{l} a_j \chi_{jl}.$$

Hence, the system of linear equations

$$(a_1, \cdots, a_m)(\chi_{jk}) = (1, \cdots, 1)$$

admits a non-trivial solution and since (χ_{jk}) is a matrix with integral coefficients, there exists $c_0 \in \mathbb{Z}$ and m integers $c_1, \cdots, c_m$ which are relatively prime to each other such that

$$(c_1, \cdots, c_m)(\chi_{jk}) = c_0(1, \cdots, 1).$$

The condition (A.23) implies $c_0 = \pm 1$.

Lemma A.4. *There exists m integers $c_1, \cdots, c_m$ such that*

$$(c_1, \cdots, c_m)(\chi_{jk}) = (1, \cdots, 1).$$

Let L be the lattice defined by the linear relations of l points $\chi_1, \cdots, \chi_l$ of $\mathbb{Z}^m$ as follows.

$$L := \{a = \begin{pmatrix} a_1 \\ \vdots \\ a_l \end{pmatrix} \in \mathbb{Z}^l | (\chi_{jk})a = 0\}.$$

This is the same condition as $\chi_1^{a_1} \cdots \chi_l^{a_l} = 1$ in $X(H)$. By Lemma A.4, $a \in L$ implies

$$\sum_{k=1}^{l} a_k = 0. \qquad\qquad\qquad (A.24)$$

By the condition (A.23), there exists a matrix with integral coefficients $Q \in GL_m(\mathbb{Z})$ and $R \in GL_l(\mathbb{Z})$ with determinant ± 1 such that

$$Q(\chi_{jk})R = \begin{pmatrix} 1 & & 0 \cdots 0 \\ & \ddots & \vdots & \vdots \\ & 1 & 0 \cdots 0 \end{pmatrix} \begin{matrix} \uparrow \\ m \\ \downarrow \end{matrix} \quad .$$

$$\longleftarrow l \longrightarrow$$

Hence, we have the following lemma.

Lemma A.5. *There exists a basis* $w_1, \cdots, w_l$ *of* $\mathbb{Z}^l$ *such that* L *can be expressed as*

$$L = \mathbb{Z}w_1 \oplus \cdots \oplus \mathbb{Z}w_n, \quad n = l - m.$$

Hence, we have rank $L = l - m$.

A.10 G-G-Z Equation

With these preliminaries, we define a system of hypergeometric differential equations with the parameter $\beta \in \mathfrak{h}^\vee$, which we call the G-G-Z equation after the paper [G-G-Z]: here, we abbreviate as $\partial_k = \partial/\partial v_k$.

1. For any $A \in \mathfrak{h}$,

$$\mathbb{A}\Phi = \beta(A)\Phi \qquad (\text{cf. (A.21)}). \tag{A.25}$$

2. For any $a \in L$, we set

$$\square_a = \prod_{a_k > 0} \partial_k^{a_k} - \prod_{a_k < 0} \partial_k^{-a_k}$$

and

$$\square_a \Phi = 0. \tag{A.26}$$

Notice that this is homogeneous by (A.24).

As we set $\delta\chi_k = \sum_{j=1}^m \chi_{jk}\delta z_j$ in § A.9, by (A.21), we have

$$\mathbb{A} = \sum_{j=1}^m \delta z_j(A) \sum_{k=1}^l \chi_{jk}v_k\partial_k.$$

Since $\mathfrak{h}$ is generated by $z_j \frac{\partial}{\partial z_j}$, $1 \le j \le m$, equation (A.25) is equivalent to the following m equations: setting $\beta_j = \beta(z_j \frac{\partial}{\partial z_j})$ and defining m differential operators $Z_1, \cdots, Z_m$ by

$$
\begin{pmatrix} Z_1 \\ \vdots \\ Z_m \end{pmatrix} := (\chi_{jk}) \begin{pmatrix} v_1 \partial_1 \\ \vdots \\ v_l \partial_l \end{pmatrix} - \begin{pmatrix} \beta_1 \\ \vdots \\ \beta_m \end{pmatrix},
$$

we have

$$
Z_1 \Phi = 0, \cdots, Z_m \Phi = 0. \tag{A.27}
$$

A formal Laurent solution of this system of differential equations is given as follows.

Fix $\alpha = {}^t(\alpha_1, \cdots \alpha_l) \in \mathbb{C}^l$ in a way that satisfies

$$
(\chi_{jk}) \begin{pmatrix} \alpha_1 \\ \vdots \\ \alpha_l \end{pmatrix} = \begin{pmatrix} \beta_1 \\ \vdots \\ \beta_m \end{pmatrix},
$$

and we consider the formal Laurent series:

$$
\Phi(\alpha, v) := v^\alpha \sum_{a \in L} \frac{1}{\prod_{k=1}^l \Gamma(\alpha_k + a_k + 1)} v^a. \tag{A.28}
$$

Here, we set $a = {}^t(a_1, \cdots, a_l) \in L$, $v^a = \prod v_k^{a_k}$. Then, in completely the same way as in § A.5, one can show that $\Phi(\alpha, v)$ satisfies the system of partial differential equations (A.26), (A.27).

The sum in the series (A.28) is taken over the sublattice of $\mathbb{Z}^l$ of rank $n = l - m$. We would like to rewrite this sum as the sum over $\mathbb{Z}^n$ by using the identification $L \simeq \mathbb{Z}^n$ asserted in Lemma A.5. For an element a of L, the map $a \in L \longrightarrow a_k \in \mathbb{Z}$ associating a to its kth coordinate a_k can be regarded as an integer-valued linear form on L. Below, we use the following notation. Under the identification

$$
\nu = (\nu_1, \cdots, \nu_n) \in \mathbb{Z}^n \longleftrightarrow a = \sum \nu_i w_i \in L,
$$

we denote an element of $\mathbb{Z}^n$ by ν and a standard basis of $\mathbb{Z}^n$ by $e_1 = (1, 0, \cdots, 0), \cdots, e_n = (0, \cdots, 0, 1)$. Under this identification, regarding a_k as a linear form on $\mathbb{Z}^n$, we denote it by $a_k(\nu)$. Noting that

$$
v^a = \prod_{k=1}^l v_k^{a_k(\nu)} = \prod_{i=1}^n \left(\prod_{k=1}^l v_k^{a_k(e_i)} \right)^{\nu_i},
$$

we set

$$x_i = \prod_{k=1}^{l} v_k^{a_k(e_i)}, \quad 1 \le i \le n.$$

Let $\gamma \in \mathbb{C}^n = L \otimes \mathbb{C}$ be a vector satisfying $a_k(\gamma) = \alpha_k$, $1 \le k \le l$. Then, we have

$$x^\gamma = \prod_{i=1}^{n} x_i^{\gamma_i} = \prod_{k=1}^{l} v_k^{a_k(\gamma)} = v^\alpha.$$

Hence, (A.28) is rewritten as a formal Laurent series in x

$$F(\alpha, x) = x^\gamma \sum_{\nu \in \mathbb{Z}^n} \frac{1}{\prod_{k=1}^{l} \Gamma(\alpha_k + a_k(\nu) + 1)} x^\nu. \tag{A.29}$$

A.11 Convergence

Apparently, this series may not converge in this form. Hence, we take α_k as follows. First, L is a sub-lattice of $\mathbb{Z}^l$ of rank n and a_k is the kth coordinate of $a \in L$ regarded as an element of $\mathbb{Z}^l$. Hence, there are n indices $k_1, \cdots, k_n$ such that $a_{k_1}(\nu), \cdots, a_{k_n}(\nu)$ are linearly independent as elements of $L^\vee \otimes \mathbb{C}$. To be consistent with the notation in § A.1, we set

$$a_1^-(\nu) = a_{k_1}(\nu), \cdots, a_n^-(\nu) = a_{k_n}(\nu).$$

Arranging the remaining $l - n = m$ $a_k(\nu)$ appropriately, we set those with the opposite sign as $a_1^+(\nu), \cdots, a_m^+(\nu)$, and correspondingly, α_k with the opposite sign by $\alpha_1^+, \cdots, \alpha_m^+$. As a convergent condition, we impose

$$\alpha_{k_1} = \cdots = \alpha_{k_n} = 0. \tag{A.30}$$

Then, by the formula $\Gamma(z)\Gamma(1 - z) = \pi/\sin \pi z$, we obtain

$$1/\Gamma(\alpha_k + a_k(\nu) + 1) = \frac{(-1)^{a_k(\nu)+1} \sin \alpha_k}{\pi} \Gamma(-\alpha_k - a_k(\nu)).$$

Hence, (A.29) is rewritten, up to the sign of x and to a constant factor, as follows:

$$F(\alpha, x) = x^\gamma \sum_{\nu \in \mathbb{Z}^n} \frac{\prod_{j=1}^{m} \Gamma(\alpha_j^+ + a_j^+(\nu))}{\prod_{i=1}^{n} \Gamma(a_i^-(\nu) + 1)} x^\nu.$$

Here, recalling that $1/\Gamma(z)$ is zero for $z \in \mathbb{Z}_{\le 0}$, the above sum is taken over the subset of $\mathbb{Z}^n$

$$\{\nu \in \mathbb{Z}^n \,|\, a_i^-(\nu) \geq 0, \quad 1 \leq i \leq n\}.$$

Finally, by (A.24) and the definition of $a_j^{\pm}(\nu)$, we have

$$\sum_{j=1}^{m} a_j^+(\nu) = \sum_{i=1}^{n} a_i^-(\nu), \quad \nu \in \mathbb{Z}^n.$$

Therefore, it is shown that $F(\alpha, x)$ is nothing but Mellin's hypergeometric series of general type introduced in § A.1.

Appendix B
The Selberg Integral and Hypergeometric Function of BC Type

B.1 Selberg's Integral

For $\alpha, \beta > -1, \gamma > 0$, the integral

$$S(\alpha, \beta, \gamma) = \int_{[0,1]^n} \prod_{i=1}^n u_i{}^\alpha (1 - u_i)^\beta \prod_{1 \leq i < j \leq n} |u_i - u_j|^{2\gamma} du_1 \wedge \cdots \wedge du_n \quad (B.1)$$

is called the Selberg integral. In [Sel], A. Selberg has proved the following proposition.

Proposition B.1.

$$S(\alpha, \beta, \gamma) = \prod_{i=1}^n \frac{\Gamma(1 + i\gamma)\Gamma(\alpha + 1 + (i - 1)\gamma)\Gamma(\beta + 1 + (i - 1)\gamma)}{\Gamma(1 + \gamma)\Gamma(\alpha + \beta + 2 + (n + i - 2)\gamma)}. \quad (B.2)$$

This formula had been forgotten for a long time, but recently, it has commanded attention unexpectedly. One of the reasons is that it gives a correlation function of a random matrix ([Me]). Another reason is that it is effective to provide an explicit expression of vertex operators in two-dimensional conformal field theory ([Tsu-Ka1]) and a further reason is that it has a deep relation to orthogonal polynomials that are the spherical functions of type A ([B-O]) etc.

As we have discussed in Chapters 3 and 4, (B.1) is an integral associated to the arrangement of hyperplanes

$$D = \bigcup_{i=1}^n (u_i = 0) \bigcup_{i=1}^n (u_i - 1 = 0) \bigcup_{1 \leq i < j \leq n} (u_i - u_j = 0). \quad (B.3)$$

The relatively compact chambers are given by

$$\Delta : 0 \leq u_1 \leq u_2 \leq \cdots \leq u_n \leq 1, \quad (B.4)$$

and its transforms by the elements of the symmetric group $\mathfrak{S}_n$ of degree n. That is,

$$\dim H^n(\Omega^\bullet(*D), \nabla_\omega) = n!. \tag{B.5}$$

Hence, the invariant subspace of the cohomology with respect to $\mathfrak{S}_n$ is of dimension

$$\dim H^n(\Omega^\bullet(*D), \nabla_\omega)^{\mathfrak{S}_n} = 1. \tag{B.6}$$

From results obtained in Chapter 4, by the shift operators $\alpha \mapsto \alpha \pm 1$, $\beta \mapsto \beta \pm 1$, $\gamma \mapsto \gamma \pm \frac{1}{2}$, $S(\alpha, \beta, \gamma)$ satisfies the difference relations

$$S(\alpha + 1, \beta, \gamma) = S(\alpha, \beta, \gamma) A_1(\alpha, \beta, \gamma),$$

$$S(\alpha, \beta + 1, \gamma) = S(\alpha, \beta, \gamma) A_2(\alpha, \beta, \gamma), \tag{B.7}$$

$$S\left(\alpha, \beta, \gamma + \frac{1}{2}\right) = S(\alpha, \beta, \gamma) A_3(\alpha, \beta, \gamma).$$

Here, $A_i(\alpha, \beta, \gamma)$ are rational functions of α, β, γ. The formula (B.2) can be obtained by computing $A_i(\alpha, \beta, \gamma)$'s concretely ([AAR], [Ao10], [BarCar]).

This formula had been generalized by I. Macdonald to an integral formula associated to any root system ([Mac]), and E. Opdam finally proved it ([Op]).

B.2 Generalization to Correlation Functions

We consider a more general integral than (B.1)

$$S_{n,m}(x; \alpha, \beta, \gamma) \tag{B.8}$$

$$= \int_{[0,1]^n} \prod_{k=1}^m \prod_{i=1}^n (u_i - x_k) \prod_{i=1}^n u_i^\alpha (1 - u_i)^\beta \prod_{1 \le i < j \le n} |u_i - u_j|^{2\gamma} du_1 \wedge \cdots \wedge du_n.$$

$S_{n,m}(x; \alpha, \beta, \gamma)$ is a polynomial in $x = (x_1, \cdots, x_m)$ of degree nm. J. Kaneko showed the following theorem ([Ka2]).

Theorem B.1. $F(x) = S_{n,m}(x; \alpha, \beta, \gamma)$ *satisfies the holonomic system of partial differential equations*

$$x_i(1 - x_i)\frac{\partial^2 F}{\partial x_i^2} + \left\{c - \frac{(m-1)}{\gamma} - (a+b+1-\frac{m-1}{\gamma})x_i\right\}\frac{\partial F}{\partial x_i} \quad \text{(B.9)}$$

$$+ \frac{1}{\gamma}\left\{\sum_{\substack{j=1\\j\neq i}}^{m}\frac{x_i(1-x_i)}{x_i - x_j}\frac{\partial F}{\partial x_i} - \sum_{\substack{j=1\\j\neq i}}^{m}\frac{x_j(1-x_j)}{x_i - x_j}\frac{\partial F}{\partial x_j}\right\} = 0$$

$$a = -n, \quad b = \frac{\alpha + \beta + m + 1}{\gamma} + n - 1, \quad c = \frac{\alpha + m}{\gamma} \quad (1 \leq i \leq m).$$

$F(x)$ is uniquely determined as a symmetric (in fact, a polynomial) solution of (B.9) that is holomorphic at the origin.

(B.9) is a special case of the hypergeometric differential equations of type BC defined by Heckman and Opdam. For $m = 3$, (B.9) is equivalent to Gauss' hypergeometric differential equation and $F(x)$ becomes a Jacobi polynomial. For $m = 4$, (B.9) coincides with Appell's differential equation of type IV. For Appell's hypergeometric function of type IV, see Example 3.1.

For a general m, $F(x)$ is known to be expanded using Jack polynomials. For this, see, e.g., [Ka2], [B-O].

Appendix C
Monodromy Representation of Hypergeometric Functions of Type $(2, m+1; \alpha)$

C.1 Isotopic Deformation and Monodromy

Let us give a formula of the monodromy of the hypergeometric integral of type $(2, m+1; \alpha)$ discussed in § 2.3. The integral over the twisted cycle $\Delta_\nu(\omega)$ (cf. (2.47))

$$\varphi_\nu \langle j \rangle = \int_{\Delta_\nu(\omega)} U(u) d\log(u - x_j), \qquad (\text{C.1})$$

as we calculated at the end of § 3.8, satisfies the Gauss–Manin connection independent of ν

$$d\varphi_\nu \langle j \rangle = \sum_{k \neq j} \alpha_k d\log(x_j - x_k)(\varphi_\nu \langle j \rangle - \varphi_\nu \langle k \rangle), \qquad (\text{C.2})$$

and the identity

$$\sum_{j=1}^{m} \alpha_j \varphi_\nu \langle j \rangle = 0. \qquad (\text{C.3})$$

(C.1) is a multi-valued analytic function defined on the affine variety

$$X_0 = \{(x_1, \cdots, x_m) \in \mathbb{C}^m; x_i \neq x_j (i \neq j)\}.$$

To show explicitly the continuous deformation of M by $(x) = (x_1, \cdots, x_m)$, we set $M = M(x_1, \cdots, x_m)$.

Now, we assume that $x_1, \cdots, x_m$ are real and satisfy $x_1 < \cdots < x_m$. Let (x) move continuously in X_0 in such a way that $(x_{j+1} - x_j)/|x_{j+1} - x_j|$ turns half in the positive direction and the other vectors $(x_{k+1} - x_k)/|x_{k+1} - x_k|$ $(k \neq j)$ stay as before (Figure C.1).

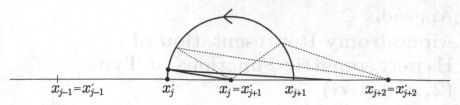

$$x_{j-1}=x'_{j-1} \qquad x'_j \qquad x_j=x'_{j+1} \quad x_{j+1} \qquad x_{j+2}=x'_{j+2}$$

Fig. C.1

For a given set of points, we set in the increasing order $(x') : x'_1 < \cdots < x'_m$. $M(x_1, \cdots, x_m)$ is deformed isotopically to $M(x'_1, \cdots, x'_m)$ (for the definition of "isotopic", see [Birm]). According to that, a linear map

$$\tau_j : H_1^{lf}(M(x); \mathcal{L}_\omega^\vee) \longrightarrow H_1^{lf}(M(x'); \mathcal{L}_\omega^\vee) \tag{C.4}$$

is induced. For each basis $\Delta_\nu(\omega) = \text{reg}[x_\nu, x_{\nu+1}]$, $\Delta'_\nu(\omega) = \text{reg}[x'_\nu, x'_{\nu+1}]$ (reg signifies the regularization), as one sees from Figure C.1, looking carefully the branch of functions, we obtain the formulas

$$\Delta_k(\omega) \longrightarrow \Delta'_k(\omega) \quad (|k - j| \geq 2),$$

$$\Delta_{j-1}(\omega) \longrightarrow \Delta'_{j-1}(\omega) + \Delta'_j(\omega), \tag{C.5}$$

$$\Delta_j(\omega) \longrightarrow -e^{2\pi\sqrt{-1}\alpha_j}\Delta'_j(\omega),$$

$$\Delta_{j+1}(\omega) \longrightarrow \Delta'_{j+1}(\omega) + e^{2\pi\sqrt{-1}\alpha_j}\Delta'_j(\omega).$$

Moreover, as in the previous case, deforming continuously $\{x'_1, \cdots, x'_m\}$ in X_0 in such a way that $(x'_{j+1} - x'_j)/|x'_{j+1} - x'_j|$ turns half and obtaining the pairs $(x'') : x''_1 < \cdots < x''_m$, we obtain a similar expression to (C.5) of the map

$$\tau_j : H_1^{lf}(M(x'); \mathcal{L}_\omega^\vee) \longrightarrow H_1^{lf}(M(x''); \mathcal{L}_\omega^\vee) \tag{C.6}$$

in the basis $\Delta'_\nu(\omega)$, $\Delta''_\nu(\omega) = \text{reg}[x''_\nu, x''_{\nu+1}]$.

Deforming τ_j to be $x''=x$, we obtain an automorphism τ_j^2 of $H_1^{lf}(M(x); \mathcal{L}_\omega^\vee)$

$$\Delta_k(\omega) \longrightarrow \Delta_k(\omega), \quad (|k - j| \geq 2),$$

$$\Delta_{j-1}(\omega) \longrightarrow \Delta_{j-1}(\omega) + (1 - e^{2\pi\sqrt{-1}\alpha_{j+1}})\Delta_j(\omega), \tag{C.7}$$

$$\Delta_j(\omega) \longrightarrow e^{2\pi\sqrt{-1}(\alpha_j + \alpha_{j+1})}\Delta_j(\omega),$$

$$\Delta_{j+1}(\omega) \longrightarrow \Delta_{j+1}(\omega) + e^{2\pi\sqrt{-1}\alpha_{j+1}}(1 - e^{2\pi\sqrt{-1}\alpha_j})\Delta_j(\omega).$$

This is the monodromy representation corresponding to τ_j^2. $\{\tau_1^2, \cdots, \tau_m^2\}$ generate the pure braid group P_m on m strands, and hence, (C.7) provides an $(m-1)$-dimensional representation of P_m. This is called the Gassner representation. Moreover, setting $\alpha_1 = \cdots = \alpha_m$, τ_j induces an automorphism of $H_1^{lf}(M(x), \mathcal{L}_\omega^\vee)$. As $\{\tau_1, \cdots, \tau_m\}$ generate the braid group B_m, (C.5) provides an $(m-1)$-dimensional representation of B_m. This is called the Burau representation (see, e.g., [Birm], [Koh3], and Appendix D).

The monodromy representation for a general type $(n+1, m+1; \alpha)$ gives one of the representations of a higher braid group defined by Manin and Schechtman. For these topics, see [Ao4], [M-S-T-Y1], [Koh4], [Man-Sch], [Ter2] etc.

Appendix D
KZ Equation
Structures of Monodromy Representations and Their Applications to Invariants of Knots

Toshitake Kohno

In [K-Z], the Knizhnik–Zamolodchikov (KZ) equation was obtained as the differential equation satisfied by the n-point functions of the conformal field theory on the Riemann sphere derived from the Wess–Zumino–Witten model. Significant algebraic structures on the monodromy representations of the KZ equation have since been studied in relation to quantum groups and Iwahori–Hecke algebra etc. ([Tsu-Ka2], [Koh2], [Dr]). Tsuchiya and Kanie ([Tsu-Ka2]) clearly formulated the concept of vertex operators, and the monodromy representations of the n-point functions were studied in detail from the viewpoint of the connection matrix of hypergeometric functions. Moreover, the theory of quasi-Hopf algebras introduced by Drinfel'd ([Dr]) through his research on the KZ equation had a great impact on several domains.

The aim of this appendix is to present the monodromy representations of the KZ equation and their application focusing on their relation to hypergeometric functions. First, in § D.1, we will formulate the KZ equation and in § D.2, we will explain how the KZ equation has been derived in conformal field theory. Second, in § D.3, the connection problem of the KZ equations will be discussed from the viewpoint of a compactification of the configuration spaces of points. In particular, for $\mathfrak{sl}_m(\mathbb{C})$, in § D.4, we will show how representations of the Iwahori–Hecke algebra appear as monodromy representations and how the connection problem of four-point functions can be reduced to the classical connection formulas of Gauss' hypergeometric function. Moreover, we will also explain a background of the formulation of quasi-Hopf algebras due to Drinfel'd from the viewpoint of a connection matrix. In § D.5, we will explain the Kontsevich integral, which is a generalization of iterated integral representations of solutions of the KZ equation to a construction of invariants of knots. And in § D.6, we will discuss integral representations of solutions of the KZ equation as generalized hypergeometric functions. In particular, following Varchenko ([V3]), we will explain the Gauss–Manin connection, which provides the Burau–Gassner representation of the braid group explained in § C.1; it appears as the KZ connection which acts on the space of null vectors in the tensor product of Verma modules.

D.1 Knizhnik−Zamolodchikov Equation

Let H_{ij} be the hyperplane in the complex vector space $\mathbb{C}^n$ defined by $z_i = z_j$ with respect to the coordinates $(z_1, \cdots, z_n)$. The space

$$X_n = \mathbb{C}^n \setminus \cup_{i<j} H_{ij}$$

is called the configuration space of ordered, distinct n points in $\mathbb{C}$. The fundamental group of X_n is called the pure braid group and is denoted by P_n. The fundamental group of its quotient space by the action of the symmetric group of degree n, via the permutations of coordinates, is called the braid group and is denoted by B_n. It is known (cf. [Birm]) that the braid group B_n is generated by the elements σ_i, $i = 1, \cdots, n-1$ shown in Figure D.1 with the relations

$$\sigma_i \sigma_{i+1} \sigma_i = \sigma_{i+1} \sigma_i \sigma_{i+1}, \quad i = 1, \cdots, n-2, \tag{D.1}$$

$$\sigma_i \sigma_j = \sigma_j \sigma_i, \quad |i-j| > 1. \tag{D.2}$$

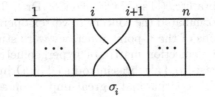

Fig. D.1

We consider the logarithmic differential form

$$\omega_{ij} = d \log(z_i - z_j) = \frac{dz_i - dz_j}{z_i - z_j}, \quad i \neq j,$$

on the configuration space X_n. They generate the cohomology ring of X_n and satisfy the quadratic relations

$$\omega_{ij} \wedge \omega_{jk} + \omega_{jk} \wedge \omega_{ki} + \omega_{ki} \wedge \omega_{ij} = 0, \quad i < j < k. \tag{D.3}$$

Fix a complex finite-dimensional simple Lie algebra $\mathfrak{g}$ and its representations $\rho_i : \mathfrak{g} \to End(V_i), 1 \leq i \leq n$. Let $\{I_\mu\}$ be an orthonormal basis of the Lie algebra $\mathfrak{g}$ with respect to the Cartan−Killing form. Set $\Omega = \sum_\mu I_\mu \otimes I_\mu$ and we denote the action of Ω on the ith and jth components of $V_1 \otimes V_2 \otimes \cdots \otimes V_n$ by $\Omega_{ij}, i \neq j$. Namely, by using the representations $\rho_i : \mathfrak{g} \to End(V_i)$, it can be expressed as

$$\Omega_{ij} = \sum_\mu 1 \otimes \cdots \otimes \rho_i(I_\mu) \otimes \cdots \otimes \rho_j(I_\mu) \otimes \cdots \otimes 1. \qquad (D.4)$$

The KZ equation is the differential equation

$$\frac{\partial W}{\partial z_i} = \frac{1}{\kappa} \sum_{j \neq i} \frac{\Omega_{ij}}{z_i - z_j} W \qquad (D.5)$$

satisfied by a function $W(z_1, \cdots, z_n)$ on X_n with values in $V_1 \otimes V_2 \otimes \cdots \otimes V_n$. Here, κ is a non-zero parameter. This can be, by setting

$$\omega = \frac{1}{\kappa} \sum_{i<j} \Omega_{ij} \omega_{ij},$$

expressed as the total differential equation

$$dW = \omega W. \qquad (D.6)$$

One can check that Ω_{ij}'s satisfy the relations

$$[\Omega_{ij} + \Omega_{jk}, \Omega_{ik}] = 0, \quad (i, j, k \text{ are mutually distinct}), \qquad (D.7)$$

$$[\Omega_{ij}, \Omega_{kl}] = 0, \quad (i, j, k, l \text{ are mutually distinct}). \qquad (D.8)$$

(D.8) is evident, and (D.7) follows from the fact that $\sum_\mu I_\mu \cdot I_\mu$ is the so-called Casimir element and lies in the center of the universal enveloping algebra $U\mathfrak{g}$. Combining these with (D.3), we obtain

$$d\omega = \omega \wedge \omega = 0. \qquad (D.9)$$

Namely, ω defines an integrable connection on the trivial vector bundle over the base space X_n whose fiber is $V_1 \otimes V_2 \otimes \cdots \otimes V_n$, which is called the KZ connection.

D.2 Review of Conformal Field Theory

The conformal field theory was introduced by Belavin, Polyakov, and Zamolodchikov in [B-P-Z] as a field theory to treat the phenomena at critical points in two-dimensional statistical mechanical models. Moreover, in [K-Z], they constructed a conformal field theory with the symmetry of an affine Lie algebra from the Wess−Zumino−Witten model and derived the KZ equation as differential equations satisfied by n-point functions. Here, an important role is played by the concept of vertex operators, or primary fields. Following [Tsu-Ka2], we formulate it with the aid of representation theory of affine Lie algebras and explain a process to derive the KZ equation. For details, we refer the reader to [Tsu-Ka2], [Koh6] etc. The arguments given in this section are not used after § D.3. A basic reference for affine Lie algebras is [Kac].

We define the affine Lie algebra $\widehat{\mathfrak{g}}$ associated to a complex finite-dimensional simple Lie algebra $\mathfrak{g}$ as the central extension

$$\widehat{\mathfrak{g}} = (\mathfrak{g} \otimes \mathbb{C}[t, t^{-1}]) \oplus \mathbb{C}c \tag{D.10}$$

of the loop algebra $\mathfrak{g} \otimes \mathbb{C}[t, t^{-1}]$. Here, $\mathbb{C}[t, t^{-1}]$ signifies the ring of Laurent polynomials. Denoting the Cartan-Killing form on the Lie algebra $\mathfrak{g}$ by $\langle \ , \ \rangle$, the commutation relation among the elements of $\mathfrak{g} \otimes \mathbb{C}[t, t^{-1}]$ is given by

$$[X \otimes f, Y \otimes g] = [X, Y] \otimes fg + \langle X, Y \rangle \mathrm{Res}_{t=0}(g\,df) \cdot c. \tag{D.11}$$

We also set $\widehat{\mathfrak{g}}_{\pm} = \oplus_{n>0}\mathfrak{g} \otimes t^{\pm n}$.

Here and after, we fix a positive integer K, called the level. We would like to construct irreducible representations of $\widehat{\mathfrak{g}}$ of level K, i.e., irreducible representations where the central element c acts as the multiplication by K. Let P_K be the set of all dominant integral weights λ of $\mathfrak{g}$ satisfying $\langle \lambda, \theta \rangle \leq K$. Here, θ is the longest root and we normalize the Cartan–Killing form as $\langle \theta, \theta \rangle = 2$. Then, it is known that there uniquely exists, up to an isomorphism, the left $\widehat{\mathfrak{g}}$-module of level K whose subspace annihilated by the action of $\widehat{\mathfrak{g}}_{+}$ coincides with the highest weight $\mathfrak{g}$-module V_λ with highest weight λ, which is denoted by $\mathcal{H}_\lambda$. This is constructed as the quotient of the Verma module M_λ by the submodule generated by $(X_\theta \otimes t^{-1})^{K-\langle \lambda, \theta \rangle + 1}v$, where v is a highest weight vector of V_λ. Similarly, we define the irreducible right $\widehat{\mathfrak{g}}$-module $\mathcal{H}_\lambda^\dagger$ by replacing, in the above construction, $\widehat{\mathfrak{g}}_{\pm}$ with $\widehat{\mathfrak{g}}_{\mp}$. It is known that the Virasoro algebra acts on the representation space $\mathcal{H}_\lambda$. Concretely, this is constructed by the following Sugawara form. For the Casimir element $C = \sum_\mu I_\mu \cdot I_\mu$, the adjoint action of $\frac{1}{2}C$ on $\mathfrak{g}$ is given by a scalar which we denote by h. We define the Sugawara form by

$$L_n = \frac{1}{2(K+h)} \sum_\mu \sum_{m \in \mathbb{Z}} : I_\mu \otimes t^{n-m} \cdot I_\mu \otimes t^m : . \tag{D.12}$$

Here, $:\ :$ signifies the normal order, i.e., $: X \otimes t^m \cdot Y \otimes t^n :$ means $(X \otimes t^m)(Y \otimes t^n)$ if $m < n$, $\frac{1}{2}\{(X \otimes t^m)(Y \otimes t^n) + (Y \otimes t^n)(X \otimes t^m)\}$ if $m = n$, and $(Y \otimes t^n)(X \otimes t^m)$ if $m > n$. We can check

$$[L_n, X \otimes t^m] = -mX \otimes t^{m+n}. \tag{D.13}$$

That is, the operator L_n corresponds to the action of the vector field $-t^{n+1}\frac{d}{dt}$ for Laurent polynomials with values in the Lie algebra $\mathfrak{g}$. Moreover, L_n, as operators on $\mathcal{H}_\lambda$, satisfy the commutation relations

$$[L_m, L_n] = (m-n)L_{m+n} + \frac{m^3 - m}{12} \frac{K \dim \mathfrak{g}}{K+h} \delta_{m+n,0}. \tag{D.14}$$

That is, via the Sugawara form, a representation of the Virasoro algebra with central charge $K \dim \mathfrak{g}/(K + h)$ is constructed.

Fixing a coordinate function t on the Riemann sphere $\mathbb{P}^1(\mathbb{C})$, we take a point Q different from 0 and ∞ and set $t(Q) = z$. We take $\lambda, \lambda_0, \lambda_\infty \in P_K$ and attach the representation spaces $V_\lambda, \mathcal{H}_{\lambda_0}, \mathcal{H}^\dagger_{\lambda_\infty}$ to Q, 0 and ∞, respectively. We let $\mathfrak{g} \otimes \mathbb{C}[t, t^{-1}]$ act on the tensor product $\mathcal{H}^\dagger_{\lambda_\infty} \otimes V_\lambda \otimes \mathcal{H}_{\lambda_0}$ by

$$X \otimes t^n (v \otimes u \otimes w) \tag{D.15}$$
$$= -(v(X \otimes t^n)) \otimes u \otimes w + v \otimes (z^n X u) \otimes w + v \otimes u \otimes ((X \otimes t^n) w).$$

We denote the set of linear maps

$$\varphi : \mathcal{H}^\dagger_{\lambda_\infty} \otimes V_\lambda \otimes \mathcal{H}_{\lambda_0} \to \mathbb{C} \tag{D.16}$$

that are invariant under the action (D.15) by

$$V(\mathbb{P}^1(\mathbb{C}); 0, z, \infty; \lambda_0, \lambda, \lambda_\infty).$$

This is what is called the space of conformal blocks associated to $\mathbb{P}^1(\mathbb{C})$ with three marked points. A linear map φ satisfying the above condition defines a bilinear map

$$\phi(u, z) : \mathcal{H}^\dagger_{\lambda_\infty} \otimes \mathcal{H}_{\lambda_0} \to \mathbb{C} \tag{D.17}$$

depending on $z \in \mathbb{C} \setminus \{0\}$ which is linear in $u \in V_\lambda$. Regarding this map as an operator from $\mathcal{H}_{\lambda_0}$ to $\mathcal{H}_{\lambda_\infty}$, the invariance with respect to the action (D.15) can be described by the commutation relation

$$[X \otimes t^n, \phi(u, z)] = z^n \phi(X u, z). \tag{D.18}$$

This is called the gauge invariance. We call $\phi(u, z)$ a vertex operator if it has the above gauge invariance and if $\phi(u, z)$ is multi-valued holomorphic in z and satisfies the commutation relation

$$[L_n, \phi(u, z)] = z^n \{ z \frac{d}{dz} + (n + 1) \Delta_\lambda \} \phi(u, z). \tag{D.19}$$

Here, Δ_λ signifies the action of L_0 on the representation V_λ which is given by

$$\Delta_\lambda - \frac{\langle \lambda, \lambda + 2\rho \rangle}{2(K + h)}, \tag{D.20}$$

where ρ is the half sum of the positive roots of $\mathfrak{g}$. As we explain below, the condition (D.19) can be interpreted as the invariance of $\phi(u, z)(dz)^{\Delta_\lambda}$ by a conformal transformation. Assuming the invariance with respect to the conformal transformation $f_\epsilon(z) = z - \epsilon z^{n+1}$, the variation $\delta_\epsilon \phi(u, z)$ of $\phi(u, z)$

by the transformation f_ϵ corresponds to the right-hand side of the condition (D.19). At the same time, as the transformation f_ϵ generates the vector field $-z^{n+1}\frac{d}{dz}$, this can be considered as in the left-hand side of the condition (D.19) as an infinitesimal transformation.

We define operators $X(z)$ and $T(z)$, $X \in \mathfrak{g}$, $z \in \mathbb{C} \setminus \{0\}$ on the representation space $\mathcal{H}_\lambda$ by

$$X(z) = \sum_{n \in \mathbf{Z}} (X \otimes t^n) z^{-n-1}, \tag{D.21}$$

$$T(z) = \sum_{n \in \mathbf{Z}} L_n z^{-n-2}. \tag{D.22}$$

$T(z)$ is called the energy momentum tensor.

On the domain $|w| > |z| > 0$, there is an operator product expansion (OPE) of the form

$$X(w)\phi(u,z) \tag{D.23}$$
$$= \frac{1}{w-z}\phi(Xu,z) + \text{(holomorphic terms with respect to } w - z),$$

which continues analytically to $\phi(u,z)X(w)$ on $|z| > |w| > 0$. For $T(z)$, there is a similar operator product expansion:

$$T(w)\phi(u,z) \tag{D.24}$$
$$= \left(\frac{\Delta_\lambda}{(w-z)^2} + \frac{1}{w-z}\frac{\partial}{\partial z} \right) \phi(u,z)$$
$$+ \text{(holomorphic terms with respect to } w - z).$$

Next, let us define the n-point function. Let $\lambda_1, \cdots, \lambda_n \in P_K$. For a sequence of weights

$$\mu_0, \mu_1, \cdots, \mu_n \in P_K, \tag{D.25}$$

we take vertex operators

$$\phi_j(u_j, z_j) : \mathcal{H}^\dagger_{\mu_{j-1}} \otimes \mathcal{H}_{\mu_j} \to \mathbb{C}, \quad u_j \in V_{\lambda_j}, \quad j = 1, \cdots, n, \tag{D.26}$$

and on the domain $|z_1| > \cdots > |z_n| > 0$, we consider their composition

$$\phi_1(u_1, z_1) \cdots \phi_n(u_n, z_n) : \mathcal{H}^\dagger_{\mu_0} \otimes \mathcal{H}_{\mu_n} \to \mathbb{C}. \tag{D.27}$$

In particular, for $\mu_0 = \mu_n = 0$, the image of the tensor product of the highest weight vectors $v_0^\dagger \otimes v_0$ is denoted by

$$\langle \phi_1(u_1, z_1) \cdots \phi_n(u_n, z_n) \rangle$$

and is called the n-point function.

The commutation relations of $X \otimes t^n$ and L_n with $\phi(u, z)$ can be expressed as

$$[X \otimes t^n, \phi(u, z)] = \frac{1}{2\pi\sqrt{-1}} \int_\Gamma w^n X(w)\phi(u, z)dw, \qquad (D.28)$$

$$[L_n, \phi(u, z)] = \frac{1}{2\pi\sqrt{-1}} \int_\Gamma w^{n+1}T(w)\phi(u, z)dw. \qquad (D.29)$$

Here, the cycle Γ is a small circle in the w-plane turning around z once in the positive direction. In fact, letting Γ_1 be a contour in the w-plane turning around the origin once satisfying $|w| > |z|$ in the positive direction, $L_n\phi(u, z)$ can be written as

$$\frac{1}{2\pi\sqrt{-1}} \int_{\Gamma_1} w^{n+1}T(w)\phi(u, z)dw,$$

and letting Γ_2 be a contour in the w-plane turning around the origin once satisfying $|z| > |w|$ in the positive direction, $\phi(u, z)L_n$ can also be written as

$$\frac{1}{2\pi\sqrt{-1}} \int_{\Gamma_2} w^{n+1}\phi(u, z)T(w)dw.$$

Hence, as $T(w)\phi(u, z)$ ($|w| > |z| > 0$) continues analytically to $\phi(u, z)T(w)$ ($|z| > |w| > 0$), we obtain the result. Comparing this with the operator product expansion (D.24), we recover the commutation relation (D.19). In this way, we see that the commutation relations among operators are dominated by the operator product expansions. A similar result holds for $X \otimes t^n$.

By operator product expansions, we can derive the following identities:

$$\langle X(z)\phi_1(u_1, z_1) \cdots \phi_1(u_n, z_n)\rangle \qquad (D.30)$$

$$= \sum_{i=1}^{n} \frac{1}{z - z_i} \langle \phi_1(u_1, z_1) \cdots \phi_i(Xu_i, z_i) \cdots \phi_n(u_n, z_n)\rangle,$$

$$\langle T(z)\phi_1(u_1, z_1) \cdots \phi_1(u_n, z_n)\rangle \qquad (D.31)$$

$$= \sum_{i=1}^{n} \left(\frac{\Delta_{\lambda_i}}{(z - z_i)^2} + \frac{1}{z - z_i} \frac{\partial}{\partial z_i} \right) \langle \phi_1(u_1, z_1) \cdots \phi_n(u_n, z_n)\rangle.$$

As L_{-1} is given by the Sugawara form and $[L_{-1}, \phi(u, z)] = \frac{\partial}{\partial z}\phi(u, z)$, we obtain

$$(K + h)\frac{\partial}{\partial z}\phi(u, z) \qquad (D.32)$$

$$= \lim_{w \to z} \left\{ \sum_\mu I_\mu(w)\phi(I_\mu u, z) - \frac{1}{w - z}\phi\left(\sum_\mu (I_\mu \cdot I_\mu)u, z\right) \right\}.$$

From this and the identity (D.30), it can be shown that the n-point function satisfies the KZ equation by setting $\kappa = K + h$.

D.3 Connection Matrices of KZ Equation

Here, regarding κ as a complex parameter, we study algebraic structures of the monodromy representation of the KZ equation (D.5). First, for $n = 3$, let us see that the KZ equation reduces to a Fuchsian differential equation of one variable with regular singularities at $0, 1, \infty$. This corresponds to a four-point function in conformal field theory by regarding infinity as the fourth point. The KZ equation (D.5) for $n = 3$ has a solution of the form

$$W(z_1, z_2, z_3) = (z_3 - z_1)^{\frac{1}{\kappa}(\Omega_{12}+\Omega_{13}+\Omega_{23})} G\left(\frac{z_2 - z_1}{z_3 - z_1}\right), \qquad (D.33)$$

and it can be checked that G satisfies the differential equation

$$G'(x) = \frac{1}{\kappa}\left(\frac{\Omega_{12}}{x} + \frac{\Omega_{23}}{x-1}\right)G(x). \qquad (D.34)$$

Let us consider the fundamental system of solutions $G_1(x)$, $G_2(x)$ of the differential equation (D.34) having asymptotic behaviors

$$G_1(x) \sim x^{\frac{1}{\kappa}\Omega_{12}}, \quad x \to 0$$
$$G_2(x) \sim (1-x)^{\frac{1}{\kappa}\Omega_{23}}, \quad x \to 1$$

on the open real interval $0 < x < 1$. That is, $G_1(x)\exp(-\frac{1}{\kappa}\Omega_{12}\log x)$ is analytic around $x = 0$ and its value at $x = 0$ is the unit matrix. We have a similar interpretation for $G_2(x)$ at $x = 1$. By analytic continuation, there exists a matrix F independent of x satisfying

$$G_1(x) = G_2(x)F. \qquad (D.35)$$

This matrix F is the connection matrix of the solutions $G_1(x)$ and $G_2(x)$.

Returning to the KZ equation, let us consider the corresponding solutions. By the coordinate transforms $u_1 u_2 = z_2 - z_1$, $u_2 = z_3 - z_1$, G_1 provides a solution which admits an expression around $u_1 = u_2 = 0$

$$W_1(u_1, u_2) = u_1^{\frac{1}{\kappa}\Omega_{12}} u_2^{\frac{1}{\kappa}(\Omega_{12}+\Omega_{13}+\Omega_{23})} H_1(u_1, u_2), \qquad (D.36)$$

where H_1 is a single-valued holomorphic function. And for G_2, by the coordinate transforms $v_1 v_2 = z_3 - z_2$, $v_2 = z_3 - z_1$, G_2 provides a solution which admits an expression around $v_1 = v_2 = 0$

$$W_2(v_1, v_2) = v_1^{\frac{1}{\kappa}\Omega_{23}} v_2^{\frac{1}{\kappa}(\Omega_{12}+\Omega_{13}+\Omega_{23})} H_2(v_1, v_2), \qquad (D.37)$$

where H_2 is a single-valued holomorphic function. The coordinate transformations used here are the blow-ups along $z_1 = z_2 = z_3$, and we can consider them as partial compactifications of the configuration space X_3. The connection matrix F can be interpreted as a relation between solutions W_1 and W_2 around different intersection points of the normal crossing divisors obtained by the blow-up. We express the solutions obtained in such a way as the graphs in Figure D.2.

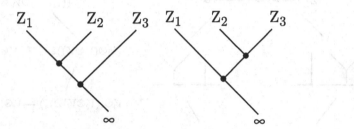

Fig. D.2

Furthermore, in the case of a general n, for a tree with $n+1$ outer vertices, one can define a partial compactification of the configuration space X_n and a normalized solutions of the KZ equation. For example, by the coordinate transform $u_k u_{k+1} \cdots u_{n-1} = z_{k+1} - z_1$, $k = 1, \cdots, n-1$, we can consider a solution having an asymptotic behavior around $u_1 = \cdots = u_{n-1} = 0$

$$W_1 \sim u_1^{\frac{1}{\kappa}\Omega_{12}} u_2^{\frac{1}{\kappa}(\Omega_{12}+\Omega_{13}+\Omega_{23})} \cdots u_{n-1}^{\frac{1}{\kappa}\sum_{1\leq i<j\leq n}\Omega_{ij}}. \qquad (D.38)$$

Figure D.3 corresponds to the case $n = 4$, and the solution (D.38) corresponds to the topmost tree. For the tree in the lower right part, setting $v_1 v_2 v_3 = z_2 - z_1$, $v_2 v_3 = z_4 - z_3$, $v_3 = z_4 - z_1$, the corresponding solution has an asymptotic behavior

$$W_2 \sim v_1^{\frac{1}{\kappa}\Omega_{12}} v_2^{\frac{1}{\kappa}(\Omega_{12}+\Omega_{34})} v_3^{\frac{1}{\kappa}\sum_{1\leq i<j\leq 4}\Omega_{ij}}. \qquad (D.39)$$

Notice that, on the domain $z_1 < z_2 < z_3 < z_4$, by considering

$$\lim_{z_2 \to z_1} W_i \cdot (z_2 - z_1)^{-\frac{1}{\kappa}\Omega_{12}}, \quad i = 1, 2, \qquad (D.40)$$

the connection matrix that relates these two solutions is given by the connection matrix that we have already considered for $n = 3$.

As in this case, one can naturally derive the fact that the connection problem of solutions reduces to four-point functions from the viewpoint of vertex operators explained in the previous section ([Tsu-Ka2]). The connection matrix of the conformal field theory associated to minimal unitary series of the Virasoro algebra was considered in [D-F].

As is shown in Figure D.3, in the case of $n = 4$, for each pair of two trees, there are two paths to connect them. Since the KZ equation is defined by an integrable connection, the connection matrices corresponding to these two paths coincide. This is called the pentagonal relation.

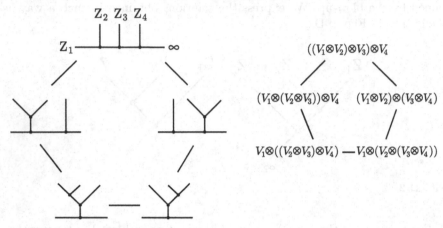

Fig. D.3

D.4 Iwahori−Hecke Algebra and Quasi-Hopf Algebras

Let us consider the representations of the braid group obtained as the monodromy representations of the KZ equation. First, we start from the normalized solution (D.38) on the domain $z_1 < z_2 < \cdots < z_i < z_{i+1} < \cdots < z_n$ and continue it analytically along a path corresponding to the generator σ_i of the braid group. By looking at its relation to the normalized solution on the domain $z_1 < z_2 < \cdots < z_{i+1} < z_i < \cdots < z_n$, we obtain an action of the braid group

$$\rho(\sigma_i) : V_1 \otimes V_2 \otimes \cdots \otimes V_i \otimes V_{i+1} \otimes \cdots \otimes V_n \qquad (D.41)$$
$$\to V_1 \otimes V_2 \otimes \cdots \otimes V_{i+1} \otimes V_i \otimes \cdots \otimes V_n.$$

First, consider the case $n = 3$. For the solution W_1 defined in (D.36), it can be seen that the action of the braid σ_1 is given by $P_{12} \exp \frac{\pi\sqrt{-1}}{\kappa} \Omega_{12}$. The action of the braid σ_2, for the solution W_2 defined in (D.37) is given by $P_{23} \exp \frac{\pi\sqrt{-1}}{\kappa} \Omega_{23}$. Here, P_{ij} signifies the permutation of the ith and the jth components of the tensor product. Hence, we see that the computation of the global action of the braid group is managed by the connection matrix F. Moreover, as in the consideration on the connection matrix of solutions in the previous section, for a general n, the computation of the action of the braid σ_i is reduced to the case $n = 3$.

Second, let us consider the case when our Lie algebra $\mathfrak{g}$ is $\mathfrak{sl}_m(\mathbb{C})$ and all of the representations $V_1, \cdots, V_n$ are its natural action on $V = \mathbb{C}^m$ (the vector representation). By blowing up as was explained in § D.3 and looking at the multi-valuedness of the part $(z_{i+1} - z_i)^{\frac{1}{\kappa}\Omega_{i,i+1}}$, one can compute the eigenvalues of the monodromy matrix $\rho(\sigma_i)$. Setting $q = \exp(\pi\sqrt{-1}/\kappa)$ and $t = \exp(\pi\sqrt{-1}/m\kappa)$, it can be verified that $g_i = -t^{-1}\rho(\sigma_i)$ satisfies the relation

$$(g_i - q)(g_i + q^{-1}) = 0. \tag{D.42}$$

We denote the representation of the braid group, obtained by associating the generator σ_i to the above g_i, by

$$\tilde{\rho} : B_n \to GL(V^{\otimes n}). \tag{D.43}$$

In general, for a non-zero complex parameter q, the algebra over $\mathbb{C}$ generated by 1, T_i ($1 \leq i \leq n - 1$) satisfying the relations

$$T_i T_{i+1} T_i = T_{i+1} T_i T_{i+1}, \quad i = 1, \cdots, n - 2 \tag{D.44}$$

$$T_i T_j = T_j T_i, \quad |i - j| > 1 \tag{D.45}$$

$$(T_i - q)(T_i + q^{-1}) = 0 \tag{D.46}$$

is called the Iwahori—Hecke algebra and is denoted by $H_n(q)$. Summarizing the above consideration, the following theorem is proved.

Theorem D.1. *The representation of the braid group $\tilde{\rho} : B_n \to GL(V^{\otimes n})$ obtained from the monodromy representation of the KZ equation defined by the Lie algebra $\mathfrak{sl}_m(\mathbb{C})$ and its vector representation provides us a representation of the Iwahori—Hecke algebra.*

As was stated in [Koh2], for a generic parameter q, this representation is equivalent to the representation used by Jones [Jo] to construct polynomial invariants of links. As there are several references, containing [Jo], which explain a recipe to obtain polynomial invariants from a representation of the braid group, we only remark that the invariant of links that is naturally constructed from the monodromy representation of the above KZ equation is obtained from the Jones polynomial $P_L(x, y)$ in two variables satisfying the skein relation

$$x^{-1}P_{L_+} - xP_{L_-} = yP_{L_0} \tag{D.47}$$

by the specializations

$$x = q^m, \quad y = q^{-1} - q. \tag{D.48}$$

Third, let us explain a relation between the connection problem of the KZ equation and the connection formulas of the classical Gauss hypergeometric function. We represent the KZ equation considered up to now, by noting the invariant subspace of the diagonal action of the Lie algebra $\mathfrak{sl}_m(\mathbb{C})$, by the so-called path basis. We fix a highest weight λ of an irreducible finite-dimensional representation of the Lie algebra $\mathfrak{sl}_m(\mathbb{C})$ and consider the set of all sequences of dominant integral weights $(\lambda_0, \lambda_1, \cdots, \lambda_n)$, $\lambda_0 = 0, \lambda_n = \lambda$ satisfying $V_{\lambda_{i+1}} \subset V_{\lambda_i} \otimes V, i = 0, 1, \cdots n - 1$, as representation spaces of $\mathfrak{sl}_m(\mathbb{C})$. Here, V_λ signifies the representation with highest weight λ. Describing the KZ equation with this path basis and using the method to reduce to four-point functions as was explained in § D.3, we obtain an equation of the form (D.34) where Ω_{ij} is a square matrix of degree at most 2. Hence, this can be solved by using Gauss' hypergeometric function, and the monodromy representation of the KZ equation can be completely described by using the connection formula of Gauss' hypergeometric function explained as in [AAR] and by the action of the symmetric group on the path basis. For a detailed result such as its relation to the action of $H_q(n)$ on the path basis due to Hoefsmit and Wenzl, see [Tsu-Ka2]. In [Tsu-Ka2], a unitary representation on the restricted path basis, for q a root of unity, is discussed from the viewpoint of the conformal field theory. Moreover, for a generalization of this representation to a representation of the mapping class group of a Riemann surface, or, for its application to the topological invariant of 3-manifolds proposed by Witten [Wi], see [Re-Tu], [Koh4].

Drinfel'd, in [Dr], defined the concept of quasi-Hopf algebras in the process of describing a relation between the monodromy representation of the KZ equation and the quantum group. Let us explain an outline of its idea. Here, a quasi-Hopf algebra is an algebra, instead of assuming the co-associativity of the coproduct of a Hopf algebra $\Delta : A \to A \otimes A$, which assumes the existence of an intertwiner $\Phi \in A \otimes A \otimes A$ between two coproducts, i.e., satisfying

$$(id \otimes \Delta)(\Delta(a)) = \Phi \cdot (\Delta \otimes id)(\Delta(a)) \cdot \Phi^{-1}, \ a \in A. \tag{D.49}$$

For a detailed definition, see [Dr]. By this, we obtain an isomorphism of representations

$$F : (V_1 \otimes V_2) \otimes V_3 \cong V_1 \otimes (V_2 \otimes V_3), \tag{D.50}$$

and we impose a universal condition to be able to obtain the pentagon relation among the tensor product of four representation spaces.

For a complex semi-simple Lie algebra $\mathfrak{g}$ and a formal parameter h, we set $A = U\mathfrak{g}[[h]]$. Setting $\frac{1}{\kappa} = \frac{h}{2\pi\sqrt{-1}}$ and regarding the KZ equation (D.34) as an equation with values in $A \otimes A \otimes A$, similarly to (D.35), we obtain $\Phi_{\mathrm{KZ}} \in A \otimes A \otimes A$. This is called the Drinfel'd associator. The connection matrix introduced in (D.35) represents this and we can consider that it provides us the isomorphism of tensor representations (D.50). Recall that the action of the braid σ_1 on $(V_1 \otimes V_2) \otimes V_3$ is given by $P_{12} \exp(\frac{h}{2}\Omega_{12})$. By using the integrability of the KZ equation, we obtain a commutative diagram as is shown in Figure D.4. In this way, in addition to Φ, an isomorphism $R = \exp(\frac{h}{2}\Omega)$ between $V_1 \otimes V_2$ and $V_2 \otimes V_1$ is added to the algebra A, and such an algebra is called a quasi-triangular quasi-Hopf algebra. Drinfel'd ([Dr]) proved an equivalence between the monodromy representation of the KZ equation and the representation of the braid group defined by the R-matrix of the quantum group $U_h\mathfrak{g}$, by comparing the structures of this algebra A and the quantum group $U_h\mathfrak{g}$.

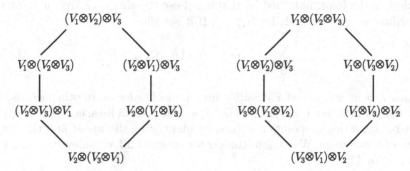

Fig. D.4

D.5 Kontsevich Integral and Its Application

One of the expressions of the monodromy representation of the KZ equation is given by the sum of Chen's iterated integrals ([Ch]) generalizing those mentioned in Chapter 1. Indeed, by rewriting the total differential equation $d\omega = W\omega$ as an integral equation, and solving it along $\gamma(t), 0 \le t \le 1$, by Picard's iteration, a solution can be expressed as

$$W(\gamma(1)) = I + \sum_{m=1}^{\infty} J_m(\gamma), \tag{D.51}$$

where

$$J_m(\gamma) = \int_\gamma \underbrace{\omega \cdots \omega}_{m}. \tag{D.52}$$

Here, $J_m(\gamma)$ is the iterated integral defined as follows. In general, for 1-forms $\omega_1, \cdots, \omega_m$, if we denote the pull-back of them by the path γ by $\alpha_1(t)dt_1, \cdots, \alpha_m(t)dt_m, 0 \le t \le 1$, then the iterated integral is defined by

$$\int_\gamma \omega_1 \cdots \omega_m = \int_{0 \leq t_1 \leq \cdots \leq t_m \leq 1} \alpha_1(t_1) \cdots \alpha_m(t_m) \, dt_1 \cdots dt_m. \qquad (D.53)$$

Kontsevich ([Kon]) generalized this expression to the case of knots and obtained a universal expression of the Vassiliev invaraints of knots. The Vassiliev invariants were introduced in the process of the research on the cohomology of the space of all knots ([Va]). For its combinatorial approach, see, e.g., [Birm-Lin], [Bar].

An invariant v of oriented knots in S^3 with values in complex numbers called a Vassiliev invariant of order m is defined as follows. Noting the k crossings of the knot diagram of a knot K, for each $\epsilon_j = \pm 1, j = 1, \cdots, k$, we denote the knot, obtained by making these crossings positive or negative according as $\epsilon_j = 1$ or -1, by $K_{\epsilon_1 \cdots \epsilon_k}$. If it satisfies

$$\sum_{\epsilon_j = \pm 1, j = 1, \cdots, k} \epsilon_1 \cdots \epsilon_k \; v(K_{\epsilon_1 \cdots \epsilon_k}) = 0 \qquad (D.54)$$

for any $k > m$, v is called a Vassiliev invariant of order m. In other words, we may think of the left-hand side of (D.54) as defining an invariant of a singular knot having k double points obtained by identifying the upper and the lower parts of k crossings. We denote the vector space of all Vassiliev invariants of order m by V_m.

Taking $2m$ points on an oriented circle, consider a diagram connecting two of these by a dotted line (cf. Figure D.6). We call such a diagram a chord diagram with m chords. When two such chord diagrams are mapped to each other by an orientation preserving homeomorphism, we identify them. In general, if v is a Vassiliev invariant of order m, we define its weight on a chord diagram with m chords as the value of v for the singular knot with m double points obtained by identifying two points connected by a dotted line. The weight of v defined in such a way satisfies the relations RI, RII indicated in Figure D.5. RI has its origin in the braid relation and is called the four-term relation. And RII can be derived from the fact that v does not depend on the framing of a knot.

Let $\mathcal{A}_m$ be the quotient of the complex vector space formally generated by all chord diagrams with m chords by the relations RI, RII. By the above construction, we obtain an injection

$$V_m/V_{m-1} \to Hom_{\mathbb{C}}(A_m, \mathbb{C}). \qquad (D.55)$$

Kontsevich ([Kon]) constructed its inverse map via iterated integrals and showed that (D.55) is an isomorphism. Below, we explain about the Kontsevich integral.

As in Figure D.6, we consider an oriented knot K in $\mathbb{R} \times \mathbb{C}$. Let t be a coordinate function in the direction of $\mathbb{R}$ and $z_i(t)$ be a curve connecting a maximal and a minimal point with respect to this parameter. Here, we assume that t has only non-degenerate critical points on K. In Figure D.6,

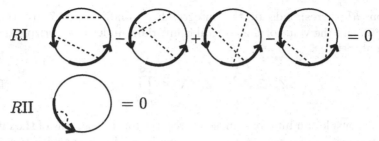

RI

RII

Fig. D.5 The parts attached by an arrow represent a local change. The first relation is called the four-term relation.

we have $i = 1, \cdots, 4$. Consider the iterated integral

$$Z_m(K) = \frac{1}{(2\pi\sqrt{-1})^m} \int_{t_1 < t_2 < \cdots < t_m} \sum_P (-1)^{\epsilon(P)} D_P \bigwedge_{k=1}^{m} \omega_{i_k j_k}(t_k). \quad (\text{D.56})$$

Here, $P = (i_1 j_1, \cdots, i_m j_m)$ signifies a way to choose two curves $z_i(t)$ for each level in $t_1 < \cdots < t_m$, and the sum is taken over all such choices. Each P corresponds to a chord diagram D_P with m chords as in Figure D.6; we regard it as an element of $\mathcal{A}_m$ by taking modulo the relations RI and RII. Moreover, we set

$$\omega_{ij}(t) = \frac{dz_i(t) - dz_j(t)}{z_i(t) - z_j(t)}, \quad (\text{D.57})$$

and let $\epsilon(P)$ be the number of the endpoints of the dotted line in D_P whose orientation in K is downward. The integral is regarded as an element of $\mathcal{A}_m$.

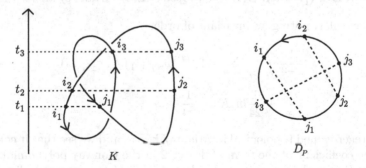

Fig. D.6

By the relation RII, this integral does not diverge at the maximal and the minimal points and always has a finite value. In addition, the four-term relation RI corresponds to the integrability condition (D.7) of the KZ equation, and the value of the integral is invariant under any horizontal move of a knot. Set

$$Z(K) = \sum_{m=0}^{\infty} Z_m(K) \in \prod_{m=0}^{\infty} \mathcal{A}_m. \tag{D.58}$$

Here, we consider an infinite sum, so we regard it as an element of the infinite direct product of $\mathcal{A}_m, m = 0, 1, \cdots$. As this is not invariant with respect to a deformation as in Figure D.7, we modify it as follows. Let K_0 be a trivial knot as in Figure D.7 that has four maximal and minimal points, and set

$$\tilde{Z}(K) = Z(K) \cdot Z(K_0)^{-m(k)+1}. \tag{D.59}$$

Here, $m(K)$ is the number of the maximal points of K.

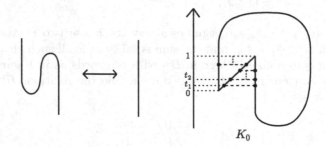

Fig. D.7

Theorem D.2 ([Kon]). $\tilde{Z}(K)$ *is a topological invariant of knots.*

In particular, noting an invariant of order 2,

$$\frac{1}{4\pi^2} \int_{t_1 < t_2} \sum_{P} (-1)^{\epsilon(P)} \omega_{i_1 j_1}(t_1) \wedge \omega_{i_2 j_2}(t_2) \tag{D.60}$$

$$+ \frac{1}{24} m(K) - \frac{1}{24}$$

is an integer-valued topological invariant of knots, and we see that it coincides with the coefficient of the term of degree 2 in the Conway polynomial. Here, we take P in such a way that the dotted lines of the corresponding chord diagram D_P cross each other.

As an application of the Kontsevich integral, in [L-M1], some relations among special values of the multiple zeta functions

$$\zeta(s_1, s_2, \cdots, s_k) = \sum_{0 < m_1 < m_2 < \cdots < m_k} \frac{1}{m_1^{s_1} \cdots m_k^{s_k}} \qquad (D.61)$$

were obtained. We consider the case when $\mathfrak{g} = sl_m(\mathbb{C})$ and all of $V_1, \cdots, V_n$ are the vector representations. First, since the monodromy representation of the braid group becomes a representation of the Iwahori–Hecke algebra, as was seen in Theorem D.1, it can be shown that the Kontsevich integral can be obtained by specializing the Jones polynomial in two variables (D.47) as in (D.48). From this, we obtain

$$Z(K_0) = \frac{m(q - q^{-1})}{q^m - q^{-m}}. \qquad (D.62)$$

On the other hand, in the situation of Figure D.7, if we directly compute the iterated integrals of $d\log t$ and $d\log(1 - t)$ as was explained in § 1.2.5, they are expressed in terms of higher logarithmic functions and they become special values of the multiple zeta functions. By comparing the above two computations, we obtain, for example,

$$\left(\frac{1}{2^{2n-2}} - 1\right)\zeta(2n) - \zeta(1, 2n - 1) + \cdots \qquad (D.63)$$

$$\cdots + \zeta(\underbrace{1, \cdots, 1}_{2n-2}, 2) = 0$$

etc. ([L-M1]).

As was shown in [L-M2] etc., by the Kontsevich integral, one obtains an expression of the Drinfel'd associator by special values of the multiple zeta functions. Setting $\frac{h}{2\pi\sqrt{-1}}\Omega_{12} = X$, $\frac{h}{2\pi\sqrt{-1}}\Omega_{23} = Y$ and computing the Drinfel'd associator Φ_{KZ} associated to $A = U\mathfrak{g}[[h]]$, we see that it has an expression of the form

$$\Phi_{\mathrm{KZ}} = 1 - \zeta(2)[X, Y] - \zeta(3)[X, [X, Y]] - \zeta(3)[Y, [X, Y]] \qquad (D.64)$$

$$-\zeta(4)[X, [X, [X, Y]]] - \zeta(4)[Y, [Y, [X, Y]]]$$

$$-\zeta(1, 3)[X, [Y, [X, Y]]] + \frac{1}{2}\zeta(2)^2[X, Y]^2 + \cdots.$$

D.6 Integral Representation of Solutions of the KZ Equation

Here, we set $\mathfrak{g} = sl_2(\mathbb{C})$ and for a complex number m, we denote the Verma module with highest weight m by M_m. Namely, with a basis H, E, F of $\mathfrak{sl}_2(\mathbb{C})$ satisfying $[H, E] = 2E$, $[H, F] = -2F$, $[E, F] = H$, M_m is generated by a vector v satisfying $Hv = mv$, $Ev = 0$ and is an infinite-dimensional representation space with the basis $F^j v, j = 0, 1, \cdots$.

For solutions of the KZ equation, applying a method explained in § 3.8 to a difference product, their integral representations were calculated by [D-J-M-M], [Sch-Va1] etc. See [Tsu] etc., for treating them from the point of view of conformal field theory.

Theorem D.3 ([D-J-M-M], [Sch-Va1]). *The solutions of the KZ equation (D.6) with coefficients in the tensor product of Verma modules $M_{m_1} \otimes \cdots \otimes M_{m_n}$ have integral representations of the form*

$$I(z) = \int_\Delta \prod_{1 \le i < j \le n} (z_i - z_j)^{\frac{m_i m_j}{2\kappa}} \prod_{l=1}^{n} \prod_{i=1}^{k} (t_i - z_l)^{-\frac{m_l}{\kappa}}$$
$$\times \prod_{1 \le i < j \le k} (t_i - t_j)^{\frac{2}{\kappa}} R(t, z) dt_1 \wedge \cdots \wedge dt_k.$$

Here, $R(t, z)$ is an appropriate rational function and Δ is a k-dimensional twisted cycle.

Following a method of Varchenko [V3], let us state how the Gassner representation of the pure braid group which has appeared in Appendix C relates to the KZ equation. Set

$$U(t, z) = \prod_{1 \le i < j \le n} (z_i - z_j)^{\frac{m_i m_j}{2\kappa}} \prod_{j=1}^{n} (t - z_j)^{-\frac{m_j}{\kappa}}. \tag{D.65}$$

Take a twisted cycle $\Delta \in H_1(X(z); \mathcal{L}_\omega^\vee)$ on $X(z) = \mathbb{C} \setminus \{z_1, \cdots, z_n\}$ and consider the integral

$$\varphi_j = \int_\Delta U(t, z) \frac{dt}{t - z_j}, \quad j = 1, \cdots, n. \tag{D.66}$$

They satisfy

$$\sum_{j=1}^{n} m_j \varphi_j = 0, \tag{D.67}$$

and we obtain the Gassner representation as the action of the pure braid group P_n on $H_1(X(z); \mathcal{L}_\omega^\vee)$.

Its relation to the KZ equation can be stated as follows. We consider the space of null vectors

$$N_{m-2} = \{x \in M \mid Ex = 0, \ Hx = (m-2)x\} \tag{D.68}$$

in the tensor product of Verma modules

$$M = M_{m_1} \otimes \cdots \otimes M_{m_n},$$

where we set $m = \sum_{j=1}^n m_j$. As the KZ connection commutes with the action of $\mathfrak{sl}_2(\mathbb{C})$, we can consider the KZ equation with values in N_{m-2}. One can check that this coincides with the Gauss–Manin connection satisfied by φ_j, $j = 1, \cdots, n$. In this way, we can naturally realize the Gauss–Manin connection, which provides the Gassner representation, as the KZ connection with values in the space of the null vectors.

References

[Ad-Sp]   A. Adolphson and S. Sperber, *On twisted de Rham cohomology*, Nagoya J. Math. **146**, (1997), 55–81.

[AAR]   G.E. Andrews, R. Askey and R. Roy. *Special Functions*, Cambridge Univ. Press, 1999.

[An]   E. Andronikof, *Intégrale de Nilsson et faisceaux constructibles*, Bull. Soc. Math. France **120**, (1992), 51–85.

[Ao1]   K. Aomoto, *Un théorème du type de Matsushima-Murakami concernant l'intégrale des fonctions multiformes*, J. Math. Pures Appl. **52**, (1973), 1–11.

[Ao2]   K. Aomoto, *On vanishing of cohomology attached to certain many valued meromorphic functions*, J. Math. Soc. Japan **27**, (1975), 248–255.

[Ao3]   K. Aomoto, *Les équations aux différences linéaires et les intégrales des fonctions multiformes*, J. Fac. Sci. Univ. Tokyo, Sec. IA, **22**, (1975), 271–297 and: *Une correction et un complément à l'article "Les équations aux différences linéaires et les intégrales des fonctions multiformes"*, ibid. **26**, (1979), 519–523.

[Ao4]   K. Aomoto, *On the structure of integrals of power products of linear functions*, Sci., Papers, Coll. Gen. Education, Univ. Tokyo **27**, (1977) 49–61.

[Ao5]   K. Aomoto, *Fonctions hyperlogarithmiques et groupe de monodromie unipotente*, Jour. Fac. Sci., Univ. Tokyo **25**, (1978), 149–156.

[Ao6]   K. Aomoto, *Configurations and invariant Gauss-Manin connections of integrals I*, Tokyo J. Math. **5**, (1982), 249–287.

[Ao7]   K. Aomoto, *Configurations and invariant Gauss-Manin connections for integrals II*, Tokyo J. Math. **5**, (1982), 1–24.

[Ao8]   K. Aomoto, *Addition theorem of Abel type for hyperlogarithms*, Nagoya Math. J. **88**, (1982), 55–71.

[Ao9]   K. Aomoto, *Special value of the hypergeometric function $_3F_2$ and connection formulae among asymptotic expansions*, J. Indian Math. Soc. **51**, (1987), 161–221.

[Ao10]   K. Aomoto, *Jacobi polynomials associated with Selberg integrals*, SIAM J. Math. Anal. **18**, (1987), 545–549.

[Ao11]   K. Aomoto, *Analytic difference equations and connection problem*, in *Open problems and conjectures*, J. Difference Eqs. Appl. **4**, (1998), 597–603.

[Ao12]   K. Aomoto, *Connection Problem for difference equations associated with real hyperplane arrangements*, Proc. Int. Conf. on the Works of Ramanujan, Eds. Chandrashekhar Adiga and D. D. Somashekhara, (2002), 35–53.

[Ao13]   K. Aomoto, *Gauss-Manin connections of Schläfli type for hypersphere arrangements*, Ann. Inst. Fourier **53**, (2003), 977–995.

307

[AKOT]   K. Aomoto, M. Kita, P. Orlik and H. Terao, *Twisted de Rham cohomology groups of logarithmic forms*, Adv. Math. **128**, (1997), 119–152.

[A-K]   P. Appell et J. Kampé de Fériet, *Fonctions hypergéometriques et hypersphériques, polynômes d'Hermite*, Gauthier-Villars, Paris, 1926.

[Ar]   V.I. Arnold, *Geometrical Methods in the Theory of Ordinary Differential Equations*, 2nd ed., Springer-Verlag, 1988.

[Ar-Av]   V.I. Arnold and A. Avez, *Ergodic Problems of Classical Mechanics*, Benjamin, 1968.

[Ar-GZ-V]   V.I. Arnold, S.M. Gussein-Zade and A.N. Varchenko, *Singularities of Differentiable Maps II, Chap 11, Complex Oscillatory Integrals*, Birkhäuser, 1988, 296–315.

[A-B-G]   M.F. Atiyah, R. Bott and L. Gårding, *Lacunas for hyperbolic differential operators with constant coefficients II*, Acta Math. **131**, (1973), 145–206.

[Bai]   W.N. Bailey, *Generalized Hypergeometric Functions*, Cambridge Univ. Press, 1935.

[Bar]   D. Bar-Natan, *On the Vassiliev knot invariants*, Topology **34**, (1995), 503–547

[BarCar]   D. Barsky and M. Carpentier, *Polynômes de Jacobi généralisés et intégrales de Selberg*, Electron. J. Combinatorics **3**, (1996), no. 2.

[B-O]   R.J. Beerends and E.M. Opdam, *Certain hypergeometric series related to the root system BC*, Trans. Amer. Math. Soc. **339**, (1993), 581–610.

[B-P-Z]   A.A. Belavin, A.N. Polyakov and A.B. Zamolodchikov, *Infinite dimensional symmetries in two dimensional quantum field theory*, Nucl. Phys. **B241**, (1984), 333–380.

[Bel]   G. Bellardinelli, *Fonctions Hypergéométriques de Plusieurs Variables et Resolutions Analytiques des Équations Algébriques Générales*, Paris, Gauthiers Villars, 1960.

[Bir1]   G.D. Birkhoff, *Dynamical Systems*, Amer. Math. Soc. Colloq. Publ., 1927.

[Bir2]   G.D. Birkhoff, *General theory of linear difference equations*, Collected Math. Paper I, (1950), 476–517.

[Birm]   J.S. Birman, *Braids, links, and mapping class groups*, Annals of Math. Studies **82**, Princeton Univ. Press, 1975.

[Birm-Lin] J. S. Birman and X.-S. Lin, *Knot polynomials and Vassiliev's invariants*, Invent. Math. **111**, (1993), 225–270.

[Br]   G. E. Bredon, *Sheaf Theory*, Second ed., Grad. Texts in Math. **170**, Springer-Verlag, 1997.

[Bru]   N.G. De Bruijin, *Asymptotic Methods in Analysis*, Dover, 1951.

[B-T]   R. Bott and L.W. Tu, *Differential Forms in Algebraic Topology*, Springer-Verlag, 1982.

[Ca]   B.C. Carlson, *Special Functions of Applied Mathematics*, Acad. Press, 1977.

[Ch]   K.T. Chen, *Iterated integrals, fundamental groups and covering spaces*, Trans. Amer. Math. Soc. **206**, (1975), 83–98.

[Cho]   K. Cho, *A generalization of Kita and Noumi's vanishing theorems of cohomology groups of local system*, Nagoya J. Math. **147**, (1997), 63–69.

[Cho-Ma]   K. Cho and K. Matumoto, *Intersection theory for twisted cohomologies and twisted Riemann's period relations I*, Nagoya J. Math. **139**, (1995), 67–86.

[Co]   D. Cohen, *Morse inequalities for arrangements*, Adv. Math. **134**, (1998), 43–45.

[C-D-O]   D. C. Cohen, A. Dimca and P. Orlik, *Nonresonance conditions for arrangements*, Ann. Inst. Fourier **53**, (2003), 1883–1896.

[C-O1]   D. C. Cohen and P. Orlik, *Gauss-Manin connections for arrangements, I, Eigenvalues*, Compositio Math. **136**, (2003), 299–316.

[C-O2]   D. C. Cohen and P. Orlik, *Gauss-Manin connections for arrangements, II, Nonresonant weights*, Amer. J. Math. **127**, (2005), 569–594.

[D-J-M-M] E. Date, M. Jimbo, A. Matsuo and T. Miwa, *Hypergeometric type integrals and the* $\mathfrak{sl}(2,\mathbb{C})$ *Knizhnik-Zamolodchikov equations*, Intern. J. Modern Phys. **B4**, (1990), 1049–1057.

[De]   P. Deligne, *Équations différentielles à points singuliers réguliers*, Lecture Notes in Math. **163**, Springer-Verlag, 1970.

[De-Mo]   P. Deligne and G.D. Mostow, *Monodromy of hypergeometric functions and non-lattice integral monodromy*, Publ. Math. I.H.E.S. **63**, (1986), 5–89.

[deR1]   G. de Rham, *Sur la division de formes et de courants par une forme linéaire*, Comment. Math. Helv. **28**, (1954), 346–352.

[deR2]   G. de Rham, *Variétés différentiables*, Hermann, Paris, 1960.

[D-F]   V. Dotsenko and V. Fateev, *Conformal algebra and multipoint correlation functions in 2D statistical models*, Nucl. Phys. **B240**, (1984), 312–348.

[Do-Te]   A. Douai and H. Terao, *The determinant of a hypergeometric period matrix*, Invent. Math. **128**, (1997), 417–436.

[Dr]   V.G. Drinfel'd, *On quasi-Hopf algebras*, Leningrad Math. J. **1**, (1990), 1419–1457.

[Dw]   B. Dwork, *Generalized Hypergeometric Functions*, Oxford Univ. Press, 1990.

[Dw-Lo]   B. Dwork and F. Loeser, *Hypergeometric series*, Japanese J. Math. **19**, (1993), 81–129.

[Er1]   A. Erdélyi, *Higher Transcendental Functions, vol. I*, McGraw-Hill, 1953.

[Er2]   A. Erdélyi, *Asymptotic Expansions*, Dover, 1956.

[E-S-V]   H. Esnault, V. Schechtman and E. Viehweg, *Cohomology of local systems of the complement of hyperplanes*, Invent. Math. **109**, (1992), 557–561.

[Fo]   O. Forster, *Lectures on Riemann Surfaces*, Grad. Texts Math. **81**, Springer-Verlag, 1981.

[Gan]   F. Gantmacher, *Theory of Matrices I, Chap 2*, Chelsea, 1959.

[Gau]   G.F. Gauss, *Disquisitiones generales circa seriem infinitam*, $1 + \frac{\alpha\beta}{\gamma}x + \frac{\alpha(\alpha+1)\beta(\beta+1)}{\gamma(\gamma+1)1\cdot2}x^2 + \cdots\cdots$, (1812), C.F. Gauss Werke 3, 123–162.

[Ge]   I.M. Gelfand, *General theory of hypergeometric functions*, Soviet Math. Dokl. **33**, (1986), 573–577.

[Ge-Ge]   I.M. Gelfand and S.I. Gelfand, *Generalized hypergeometric equations*, Soviet Math. Dokl. **33**, (1986), 643–646.

[Ge-Gr]   I.M. Gelfand and M.I. Graev, *A duality theorem for general hypergeometric functions*, Soviet Math. Dokl. **34**, (1987), 9–13.

[G-G-R]   I.M. Gelfand, M.I. Graev and V.S. Retakh, *General hypergeometric systems of equations and series of hypergeometric type*, Russian Math. Surveys **47**, (1992), 1–88.

[G-G-Z]   I.M. Gelfand, M.I. Graev and A.V. Zelevinskii, *Holonomic systems and series of hypergeometric type*, Soviet Math. Dokl. **36**, (1988), 5–10.

[G-K-Z]   I.M. Gelfand, M.M. Kapranov and A.V. Zelevinskii, *Discriminants, Resultants and Multidimensional Determinants*, Birkhäuser, 1994.

[G-R-S]   I.M. Gelfand, V.S. Retakh and V.V. Serganova, *Generalized Airy functions, Schubert cells and Jordan groups*, Dokl. Akad. Nauk SSSR **298**, (1988), 17–21.

[G-Z-K]   I.M. Gelfand, A.V. Zelevinskii and M.M. Kapranov, *Hypergeometric functions and toral manifolds*, Functional Anal. Appl. **23**, (1989), 94–106.

[Gelfo]   A.O. Gelfond, *Calculus of Finite Differences*, Hindustan, 1971.

[God]   R. Godement, *Topologie algébrique et théorie des faisceaux*, 1958, Hermann, Paris.

[Gon]   A.V. Goncharov, *Geometry of configurations, polylogarithms and motivic cohomology*, Adv. Math. **114**, (1995), 197–318.

[Gra-Rem] H. Grauert and R. Remmert, *Theory of Stein Spaces*, Springer-Verlag, 1979.

[Gro] A. Grothendieck, *On the de Rham cohomology of algebraic varieties*, Publ. Math. I.H.E.S. **29**, (1966), 95–103.

[Gui-St] V. Guillemin and S. Sternberg, *Geometric Asymptotics*, 2nd ed., Amer. Math. Soc., 1990.

[Had] J. Hadamard, *Lectures on Cauchy's Problem in Linear Partial Differential Equations*, reprinted by Dover, New York, 1952.

[Ha-M] M. Hanamura and R. Macpherson, *Geometric construction of polylogarithms*, Duke Math. J. **70**, (1993), 481–516.

[Ha-Y] M. Hanamura and M.Yoshida, *Hodge structure on twisted cohomologies and twisted Riemann inequalities, I.*, Nagoya J. Math. **154**, (1999), 123–139.

[Har] Y. Haraoka, *Chyokika Kansuu (Hypergeometric functions)*, in Japanese, Asakura Pub., 2002.

[Hat] A. Hattori, *Topology of $\mathbb{C}^n$ minus a finite number of affine hyperplanes in general position*, J. Fac. Univ. Tokyo, Sci. **IA 22**, (1975), 205–219.

[Hat-Kim] A. Hattori and T. Kimura, *On the Euler integral representations of hypergeometric functions in several variables*, J. Math. Soc. Japan **26**, (1974), 1–16.

[Hir] F. Hirzebruch, *Topological Methods in Algebraic Geometry*, 3rd enlarged ed., Springer-Verlag, 1966.

[Hor] E. Horikawa, *Transformations and contiguity relations for Gelfand's hypergeometric functions*, J. Math. Sci., Univ. of Tokyo **1** (1994), 181–203.

[Hot] R. Hotta, *Equivariant D-modules, preprint*, to appear in Proc. of Wuhan Spring School by CIMPA.

[H-T-T] R. Hotta, K. Takeuchi and T. Tanisaki, *D-modules, Perverse Sheaves, and Representation Theory*, Progr. Math. **236**, Birkhäuser, 2008.

[Hra] J. Hrabowskii, *Multiple hypergeometric functions and simple Lie groups SL and Sp.*, SIAM J. Math. Anal. **16**, (1985), 876–886.

[In] E. L. Ince, *Ordinary Differential Equations*, Dover Publ., New York, 1956.

[It] H. Ito, *Convergence of Birkhoff normal forms for integrable systems*, Comment. Math. Helv. **64**, (1989), 413–461.

[I-K-S-Y] K. Iwasaki, H. Kimura, S. Shimomura and M. Yoshida, *From Gauss to Painlevé*, 1991, Vieweg, Wiesbaden.

[Iw-Kit1] K. Iwasaki and M. Kita, *Exterior power structure of the twisted de Rham cohomology associated with hypergeometric function of type $(n+1, m+1)$*, J. Math. Pure. Appl. **75**, (1996), 69–84.

[Iw-Kit2] K. Iwasaki and M. Kita, *Twisted homology associated with hypergeometric functions*, Kumamoto J. Math. **12**, (1999), 9–72.

[J-T-W] L. Jacobsen, W. J. Thron and H. Waadeland, *Julius Worpitzky, his contributions to the analytic theory of continued fractions and his times*, in Analytic theory of continued fractions III (Redstone, CO, 1988), 25–47, Lect. Notes in Math. **1406**, Springer, Berlin, 1989.

[Jo] V. Jones, *Hecke algebra representations of braid groups and link polynomials*, Ann. Math. **126**, (1987), 335–388.

[Kac] V. Kac, *Infinite Dimensional Lie Algebras*, 3rd ed., Cambridge University Press, 1990.

[Ka] A. Kaneko, *Introduction to Hyperfunctions*, translated from the Japanese by Y. Yamamoto, Math. and its Appl. (Japanese Ser.) **3**, Kluwer Academic Press, Dodrecht, 1988.

[Ka1] J. Kaneko, *Monodromy group of Appell's system (F_4)*, Tokyo J. Math. **4**, (1981), 35–54.

[Ka2] J. Kaneko, *Selberg integrals and hypergeometric functions associated with Jack polynomials*, SIAM. J. Math. Anal. **24**, (1993), 1086–1110.

[Kas1]   M. Kashiwara, *Index theorem for a maximally overdetermined system of linear differential equations*, Proc. Japan Acad. **49**, (1973), 803–804.

[Kas2]   M. Kashiwara, *Systems of Microdifferential Equations*, Notes and Translations by T.M. Fernandes, Birkhäuser, 1983.

[Kim]   T. Kimura, *Hypergeometric functions of two variables*, Lecture Notes, Univ. of Minnesota (1971–1972).

[Kimu]   H. Kimura, *On rational de Rham cohomology associated with the general Airy functions*, Ann. Scuola Norm. Sup. Pisa **24**, (1997), 351–366.

[Kit1]   M. Kita, *On hypergeometric functions in several variables 1. New integral representations of Euler type*, Japan J. Math. **18**, (1992), 25–74.

[Kit2]   M. Kita, *On hypergeometric functions in several variables 2. The Wronskian of the hypergeometric functions of type $(n+1, m+1)$*, J. Math. Soc. Japan **45**, (1993), 645–669.

[Kit3]   M. Kita, *On vanishing of the twisted rational de Rham cohomology associated with hypergeometric functions*, Nagoya Math. J. **135**, (1994), 55–85.

[Kit-It]   M. Kita and M. Ito, *On the rank of hypergeometric system $E(n + 1, m + 1; \alpha)$*, Kyushu J. Math. **50**, (1996), 285–295.

[Kit-Ma]   M. Kita and K. Matumoto, *On the duality of the hypergeometric functions of type $(n + 1, m + 1)$*, Compositio Math. **108**, (1997), 77–106.

[Kit-No]   M. Kita and M. Noumi, *On the structure of cohomology groups attached to the integral of certain many-valued analytic functions*, Japan J. Math. **9**, (1983), 113–157.

[Kit-Yos1] M. Kita and M. Yoshida, *Intersection theory for twisted cycles*, Math. Nachr. **166**, (1994), 287–304.

[Kit-Yos2] M. Kita and M. Yoshida, *Intersection theory for twisted cycles II*, Math. Nachr. **168**, (1994), 171–190.

[K-Z]   V.G. Knizhnik and A.B. Zamolodchikov, Current algebra and Wess-Zumino models in two dimensions, Nucl. Phys. B247 (1984), 83–103.

[Koh1]   T. Kohno, *Homology of a local system on the complement of hyperplanes*, Proc. Japan Acad., Ser. **A. 62**, (1986), 144–147.

[Koh2]   T. Kohno, *Monodromy representations of braid groups and Yang-Baxter equations*, Ann. Inst. Fourier **37-4**, (1987), 139–160.

[Koh3]   T. Kohno, *Linear representations of braid groups and classical Yang-Baxter equations*, Contemp. Math. **78**, (1988), 339–363.

[Koh4]   T. Kohno, *Topological invariants for 3-manifolds using representations of mapping class groups I*, Topology **31-2**, (1992), 203–230.

[Koh5]   T. Kohno, *Integrable connections related to Manin and Schechtman's higher braid groups*, Illinois J. Math. **34**, (1990), 476–484.

[Koh6]   T. Kohno, *Conformal Field Theory and Topology*, Translation of Mathematical Monographs **210**, Amer. Math. Soc., 2002.

[Koh7]   T. Kohno, *Geometry of Iterated Integrals* (in Japanese), Springer, Japan, 2009.

[Kon]   M. Kontsevich, *Vassiliev's knot invariants*, Adv. Sov. Math. **16**, (1993), 137–150.

[Ku]   E. Kummer, *De integralibus definitis et seriebus infinitis*, J. Reine Angew. Math. **17**, (1837), 210–227.

[La]   J.A. Lappo-Danilevsky, *Mémoires sur la Théorie des Systèmes des Équations Différèntielles Linéaires*, Chelsea, 1953.

[L-M1]   T.Q.T. Le and J. Murakami, *On Kontsevich's integral for the Homfly polynomial and relations of mixed Euler numbers*, Topology Appl. **62**, (1995), 535–562.

[L-M2]   T.Q.T. Le and J. Murakami, *The universal Vassiliev-Kontsevich invariant for framed links*, Compositio Math. **102**, (1996), 41–64.

[Le]   J. Leray, *Un complément au théorème de Nilsson sur les intégrales de
            formes différèntielles à support singulier algébrique*, Bull. Soc. Math.
            France **95**, (1967), 313–374.

[Lew]   L. Lewin, *Polylogarithms and Associated Functions*, North Holland, 1981.

[Lin]   X.-S. Lin, *Knot invariants and iterated integrals*, preprint, 1994.

[Lo-Sab]   F. Loeser et C. Sabbah, *Équations aux différences finies et déterminants
            d'intégrales de fonctions multiformes*, Comment. Math. Helv. **66**, (1991),
            458–503.

[Mac]   I.G. Macdonald, *Some conjectures for root systems*, SIAM J. Math. Anal.
            13, (1982), 988–1007.

[Mag]   W. Magnus, F. Oberhettinger and R.P. Sonin, *Formulas and Theorems
            of Mathematical Physics*, Springer-Verlag, 1966.

[Man-Sch]   Yu.I. Manin and V.V. Schechtman, *Arrangement of hyperplanes, higher
            braid groups and higher Bruhat orders*, Adv. Stud. Pure Math. **17**, (1989),
            289–308.

[Mat]   J. Mather, *Notes on Topological Stability*, Mimeographed Notes, Harvard
            University 1970.

[Matsu]   H. Matsumura, *Commutative Algebra*, The Benjamin/Cummings Pub-
            lishing Company, 1980.

[Matu]   K. Matumoto, *Intersection numbers for logarithmic k-forms*, Osaka J.
            Math. **35**, (1998), 873–893.

[M-M]   Y. Matsushima and S. Murakami, *On certain cohomology groups attached
            to Hermitian symmetric spaces* (I) , Osaka J. Math. **2** (1965), 1–35, (II),
            ibid, **5** (1968), 223–241.

[M-S-Y]   K. Matumoto, T. Sasaki and M. Yoshida, *The monodromy of the pe-
            riod map of a 4-parameter family of K3 surfaces and the hypergeometric
            function of type* $(3,6)$, Internat. J. Math. **3**, (1992), 1–164.

[M-S-T-Y1] K. Matumoto, T. Sasaki, N. Takayama and M. Yoshida, *Monodromy of
            the hypergeometric differential equation of type* $(3,6)$ (I), Duke Math. J.
            71, (1993), 403–426.

[M-S-T-Y2] K. Matumoto, T. Sasaki, N. Takayama and M. Yoshida, *Monodromy of
            the hypergeometric differential equation of type* $(3,6)$ (II), *The unitary
            reflection group of order* $2^9 \cdot 3^7 \cdot 5 \cdot 7$, Ann. Scuola Norm. Sup. Pisa **20**,
            (1993), 617–631.

[Me]   M.L. Mehta, *Random Matrices*, Acad. Press 1990.

[Mi]   J. Milnor, *Morse Theory*, Annals of Math. Studies **51**, Princeton Univ.
            Press, 1963.

[M-O-Y]   K. Mimachi, H. Ochiai and M. Yoshida, *Intersection theory for loaded
            cycles, IV, Resonant cases*, Math. Nachr. **260**, (2003), 67–77.

[Mo]   J. Moser, *A rapidly convergent iteration method and non-linear differen-
            tial equations, I*, Ann. Scuola Norm. Pisa **20**, (1966), 265–315.

[N]   N.E. Nörlund, *Differenzenrechnung*, Springer-Verlag, 1924.

[No]   M. Noumi, *Expansion of the solutions of a Gauss-Manin system at a
            point of infinity*, Tokyo J. Math. **7**, (1984), 1–60.

[Od]   T. Oda, *Convex Bodies and Algebraic Geometry, An Introduction to the
            Theory of Toric Varieties*, translated from the Japanese, Ergebnisse der
            Mathematik und ihrer Grenzengebiete (3) **15**, Springer-Verlag, Berlin,
            1988.

[Ok]   K. Okubo, *On the group of Fuchsian equations*, Seminary reports, Tokyo
            metropolitan Univ. , 1987.

[Ol]   F.W.J. Olver, *Asymptotics and Special Functions*, A.K. Peters, Mas-
            sachusetts, 1997.

[Op]   E.M. Opdam, *Some applications of hypergeometric shift operators*, In-
            vent. Math. **98**, (1989), 1–18.

[Or-Te1]   P. Orlik and H. Terao, *Arrangement of Hyperplanes*, Springer-Verlag, 1991.
[Or-Te2]   P. Orlik and H. Terao, *The number of critical points of a product of power of linear functions*, Invent. Math. **120**, (1995), 1–14.
[Or-Te3]   P. Orlik and H. Terao, *Arrangement and hypergeometric integrals*, MSJ Memoirs **9**, 2001.
[Or-Te4]   P. Orlik and H. Terao, *Moduli space of combinatorially arrangements of hyperplanes and logarithmic Gauss-Manin connections*, Topology and its Appl. **118**, (2002), 549–558.
[Pe]   O. Perron, *Die Lehre von den Kettenbrüchen II*, Teubner, 1957.
[Pha1]   F. Pham, *Introduction à l'étude topologique des singularités de Landau*, Gauthiers Villars, 1967.
[Pha2]   F. Pham, *Singularités des Systèmes Différentielles de Gauss-Manin*, Birkhäuser, 1979.
[Pr]   C. Praagman, *Formal decomposition of n commuting linear difference operators*, Duke Math. J. **51**, (1984), 331–353.
[Re-Tu]   N. Y. Reshetikhin and V. G. Turaev, *Invariants of 3-manifolds via link polynomials and quantum groups*, Invent. Math. **103**, (1991), 547–597.
[Ri1]   B. Riemann, *Beiträge zur Theorie der durch die Gauss'sche Reihe $F(\alpha, \beta, \gamma, x)$ darstellbaren Funktionen*, Collected Papers, Springer-Verlag, 1990, 99–115.
[Ri2]   B. Riemann, *Vorlesungen über die hypergeometrische Reihe*, ibid, 667–691.
[R-T]   L. Rose and H. Terao, *A free resolution of the module of logarithmic forms of a generic arrangement*, J. Algebra **136**, (1991), 376–400.
[Sab1]   C. Sabbah, *Lieu des pôles d'un système holonome d'équations aux différences finies*, Bull. Soc. Math. France **120**, (1992), 371–396.
[Sab2]   C. Sabbah, *On the comparison theorem for elementary irregular D-modules*, Nagoya J. Math. **141**, (1996), 107–124.
[Sai1]   K. Saito, *On a generalization of de Rham lemma*, Ann. Inst. Fourier, Grenoble **26**, (1976), 165–170.
[Sai2]   K. Saito, *Theory of logarithmic differential forms and logarithmic vector fields*, J. Fac. Sci. Univ. Tokyo, Sec. **IA 27**, (1980), 265–291.
[SaiM]   M. Saito, *Symmetry algebras of normal generalized hypergeometric systems*, Hokkaido Math. J. **25**, (1996), 591–619.
[S-S-T]   M. Saito, B. Sturmfels, and N. Takayama, *Gröbner deformation of hypergeometric differential equations*, Algorithms and Computations in Mathematics **6**, Springer 2000.
[Sas]   T. Sasaki, *Contiguity relations of Aomoto-Gelfand hypergeometric functions and applications to Appell's system F_3 and Goursat's system $_3F_2$*, SIAM J. Math. Anal. **22**, (1991), 821–846.
[Sata]   I. Satake, *Linear Algebra*, translated from the Japanese by S. Koh, T. Akiba and S. Ihara. Pure and Appl. Math. **29**, Marcel Dekker Inc, New York, 1975.
[S-S-M]   M. Sato, T. Shintani and M. Muro, *Theory of prehomogeneous vector spaces (Algebraic part) – The English translation of Sato's lecture from Shintani's note*, Nagoya Math. J. **120**, (1990), 1–34.
[Sch-Va1]   V.V. Schechtman and A.N. Varchenko, *Hypergeometric solutions of Knizhnik-Zamolodchikov equations*, Lett. Math. Phys. **20**, (1990), 279–283.
[Sch-Va2]   V.V. Schechtman and A.N. Varchenko, *Arrangements of hyperplanes and Lie algebras homology*, Invent. Math. **106**, (1991), 139–194.
[Sel]   A. Selberg, *Bemerkninger om et multipelt integral*, Norsk Mat. Tidsskr. **26**, (1944), 71–78.
[Ser]   J.P. Serre, *Faisceaux algébriques cohérents*, Ann. Math. **61**, (1955), 197–278.

[Si]   C.L. Siegel, *Topics in Complex Function Theory I,II,III*, Wiley, 1973.

[Sh]   Y. Shibuya, *Linear Differential Equations in the Complex Domain: Problems of Analytic Continuation*, Transl. Math. Monogr. **82**, Amer. Math. Soc., Providence, 1990.

[Sl]   L.J. Slater, *Generalized Hypergeometric Functions*, Cambridge, 1960.

[Sma]   S. Smale, *Stable manifolds for differential equations and diffeomorphisms*, Ann. Scuola Norm. Pisa **17**, (1963), 97–115.

[Sm]   V.I. Smirnov, *A Course of Higher Mathematics, Vol. III, Part two*, translated by D. E. Brown, Pergamon Press, London, 1964.

[Sp]   E. H. Spanier, *Algebraic Topology*, Springer-Verlag, 1966.

[Stee]   N. Steenrod, *Topology of Fibre Bundles*, Princeton, 1957.

[Ster]   S. Sternberg, *Lectures on Differential Geometry*, Prentice Hall, 1964.

[Tan]   S. Tanisaki, *Hypergeometric systems and Radon transforms for Hermitian symmetric spaces*, Adv. Study Pure Math. **26**, (2000), 253–263.

[Ter1]   T. Terasoma, *Exponential Kummer coverings and determinants of hypergeometric functions*, Tokyo J. Math. **16**, (1993), 497–508.

[Ter2]   T. Terasoma, *Fundamental groups of moduli spaces of hyperplane configurations*, preprint, 1994.

[Ter3]   T. Terasoma, *On the determinant of Gauss-Manin connections and hypergeometric functions of hypersurfaces*, Invent. Math. **110**, (1992), 441–471.

[Tsu]   A. Tsuchiya, *Introduction to conformal field theory (in Japanese)*, notes taken by K. Nagatomo, Lect. Notes **1**, Osaka Univ.

[Tsu-Ka1]   A. Tsuchiya and Y. Kanie, *Fock space representations of the Virasoro algebra – Intertwining operators*, Publ. Res. Inst. Math. Sci. **22**, (1986), 259–327.

[Tsu-Ka2]   A. Tsuchiya and Y. Kanie, *Vertex operators in conformal field theory on $\mathbb{P}^1$ and monodromy representations of braid groups*, Adv. Stud. Pure Math. **16**, (1988), 297–372.

[V1]   A.N. Varchenko, *The Euler beta-function, the Vandermonde determinant, Legendre's equation, and critical values of linear functions on a configuration of hyperplanes I*, Math. USSR Izvestiya **35**, (1990), 543–571.

[V2]   A.N. Varchenko, *The Euler beta-function, the Vandermonde determinant, Legendre's equation, and critical values of linear functions on a configuration of hyperplanes II*, Math. USSR Izvestiya **36**, (1991), 155–167.

[V3]   A.N. Varchenko, *Multidimensional hypergeometric functions, the representation theory of Lie algebras and quantum groups*, World Sci., 1995.

[V4]   A. Varchenko, *Critical points of the product of powers of linear functions and families of bases of singular vectors*, Compositio Math. **97**, (1995), 385–401.

[Va]   V.A. Vassiliev, *Complements of discriminants of smooth maps: Topology and applications*, Translations of Mathematical Monographs **98**, Amer. Math. Soc. 1991.

[Wa]   T. Watanabe, *On a difference system of integrals of Pochhammer*, Proc. Japan Acad. **56**, (1980), 433–437.

[Wat]   G.N. Watson, *Theory of Bessel Functions*, Cambridge Univ. Press, 1922.

[W-W]   E.T. Whittaker and G.N. Watson, *A Course of Modern Analysis*, Cambridge Univ. Press, 1963.

[Wi]   E. Witten, *Quantum field theory and the Jones polynomial*, Comm. Math. Phys. **121**, (1989), 351–399.

[Yos1]   M. Yoshida, *Euler integral transformations of hypergeometric functions of two variables*, Hiroshima Math. J. **10**, (1980), 329–335.

[Yos2]   M. Yoshida, *Fuchsian Differential Equations*, Vieweg, Wiesbaden, 1987.

[Yos3]   M. Yoshida, *Hypergeometric functions, my love. Modular representations of configuration spaces*, Vieweg, 1997.

[Yos4]   M. Yoshida, *Intersection theory for twisted cycles, III, Determinant formulae*, Math. Nachr. **214**, (2000), 173–185.

Index

Morse index, 231
multiple zeta function, 303

n-point function, 292
type $(n+1, m+1)$
    − hypergeometric integral, 128
    − hypergeometric series, 105
    − hypergeometric function, 123
normal crossing, 229
normal order, 290

OPE, 292
operator product expansion, 292

partial fraction expansion, 89, 174
pentagonal relation, 296
period matrix, 250
Poincaré duality, 34, 35, 40
principal affine space, 189
proper morphism, 229
pure braid group, 285, 288

quadratic relations of hypergeometric
    functions, 151
quantum group, 298
quasi-Hopf algebra, 298
quasi-triangular quasi-Hopf algebra, 299

random matrix, 279
rank, 153
reduction of poles, 93
regular direction, 204
regular sequence, 63
regularization, 28, 116
representation of logarithmic differential
    forms, 71, 89
Riemann−Roch theorem, 44

saddle point method, 217
Selberg integral, 279

spectral sequence, 57
steepest descent method, 217
Stirling's formula, 2
Stokes' theorem, 23, 28
Sugawara form, 290
symmetric group, 280
symplectic form, 233

toric multinomial theorem, 264
toric variety, 263
transversal, 233
twist, 22
twisted
    − chain, 38
    − cocyle: intersection number, 147
    − cohomology: basis, 142
    − Riemann's period relations, 150
    − Čech−de Rham complex, 56
    − chain group, 26
    − cycle, 27
    − cycle: construction, 50, 112
    − homology group, 27

vanishing
    − twisted de Rham cohomology, 84
    − cohomology, 78
    − logarithmic de Rham cohomology, 83
    − twisted cohomology, 87
vanishing theorem (ii), 88
Varchenko's formula, 145, 146
variational formula, 173
Vassiliev invariants, 300
vertex operator, 291
Virasoro algebra, 290

Wronskian, 138–141, 144, 147